Multi-Resilience - Development - Sustainability

Karim Fathi

Multi-Resilience - Development - Sustainability

Requirements for Securing the Future of Societies in the 21st Century

Karim Fathi
Berlin, Germany

ISBN 978-3-658-37891-2   ISBN 978-3-658-37892-9 (eBook)
https://doi.org/10.1007/978-3-658-37892-9

© The Editor(s) (if applicable) and The Author(s), under exclusive licence to Springer Fachmedien Wiesbaden GmbH, part of Springer Nature 2022
This work is subject to copyright. All rights are solely and exclusively licensed by the Publisher, whether the whole or part of the material is concerned, specifically the rights of translation, reprinting, reuse of illustrations, recitation, broadcasting, reproduction on microfilms or in any other physical way, and transmission or information storage and retrieval, electronic adaptation, computer software, or by similar or dissimilar methodology now known or hereafter developed.
The use of general descriptive names, registered names, trademarks, service marks, etc. in this publication does not imply, even in the absence of a specific statement, that such names are exempt from the relevant protective laws and regulations and therefore free for general use.
The publisher, the authors, and the editors are safe to assume that the advice and information in this book are believed to be true and accurate at the date of publication. Neither the publisher nor the authors or the editors give a warranty, expressed or implied, with respect to the material contained herein or for any errors or omissions that may have been made. The publisher remains neutral with regard to jurisdictional claims in published maps and institutional affiliations.

This Springer imprint is published by the registered company Springer Fachmedien Wiesbaden GmbH, part of Springer Nature.
The registered company address is: Abraham-Lincoln-Str. 46, 65189 Wiesbaden, Germany

For my children
Ophelia
Rebeca

Preface

This book analyses and discusses the three most important guiding concepts driving societies in the twenty-first century: (multi-)resilience, sustainability and development. All three concepts essentially contribute to building societal future preparedness in highly complex and unpredictable times. They are likely to prove indispensable in the face of today's global challenges which manifest as so-called bundles of crises (I call it 'crisis bundles'), multi-dimensional crises (I call it 'bundled crises') and man-made rebound effects.

This book was firstly published in Germany in October 2019 under the title 'Resilienz im Spannungsfeld zwischen Entwicklung und Nachhaltigkeit: Anforderungen an gesellschaftliche Zukunftssicherung im 21. Jahrhundert' (literally translated: 'Resilience in the Area of Tension Between Development and Sustainability: Requirements for Securing Society's Future in the 21st Century'). It focuses on the concept of the 'resilient society' and contrasts it critically and integratively with the other concepts of 'the sustainable society' and 'the developed society'. For the first time, this book introduced the concept of multi-resilience which has received increasing attention in the last years.

In this translated version, some sources have been updated and two sub-chapters have been added:

1.3. The Coronavirus Crisis: Crisis Bundle and Bundled Crisis
9.2.2 Five Milestones for the Further Political Practice

The currently omnipresent Corona crisis and the Russia-Ukraine war are partially considered. A more in-depth analysis on the Corona crisis can be found in another publication, *The Coronavirus Crisis and Its Teachings: Steps towards Multi-Resilience*, which has been published in December 2021.

Berlin, Germany
01 November 2022

Karim Fathi

Acknowledgements

I would like to thank all those people who have contributed directly or indirectly to this book. My special thanks go to Gitta Peyn – she took over the scientific editing of the German version of this book in a very thoughtful and competent way. Many thanks!

I would also like to thank my many teachers and emotional and intellectual sparring partners who support, inspire and encourage me in my work. Daniel Dahm and Roland Benedikter deserve special mention. Daniel Dahm – in my experience one of the best-connected sustainability experts of his generation – was the first to introduce me to the concept of the resilient society many years ago. Without him, my resilience career would have taken a different course. Special mention should also be made of Roland Benedikter – he is one of the last polymaths of our time and the most prolific intellectual I know (his impressive CV and 60+ page publication list speak for themselves). He introduced me to resilience in the context of multi-disciplinary policy analysis, and I greatly appreciate working with him.

The best thing at the end: The people who are especially close to us are not only our greatest sources of strength but often also our greatest teachers in terms of conflict skills and resilient relationships. In this sense, I thank the dearly beloved women in my life, my wife Oriane and our daughters Ophelia and Rebeca. You are the most precious people in my life and remind me every moment what makes life worth living.

30 November, 2018　　　　　　　　　　　　　　　　　　　　　　　　　　Karim Fathi
Berlin, Germany

About This Book

Multi-Resilience – Development – Sustainability

Requirements for Building Societal Future Preparedness in the Twenty-First Century

Be it the refugee crisis, climate change, social and international conflicts, economic crises, (cyber) terrorism, threatening resource scarcity, or disruptive technological innovations – in the increasingly interconnected world of the twenty-first century, societies face an unprecedented variety of challenges. The inherent volatility, uncertainty, complexity and ambiguity of the phenomena (VUCA) does not make crises any easier to deal with, resulting in new demands on the future viability of societies. In recent years, the relatively new concept of the 'resilient society' has been adopted by more and more disciplines and has increasingly been allowed to serve as a 'universal answer' to today's diverse and multi-dimensional global challenges.

Since the resilience concept has been increasingly applied to the societal contexts at the turn of the millennium, it marked a fundamental change in the way we look at the crises of our time. Until then, researchers, politicians and civil society activists had focused on 'risk factors' and the 'vulnerability' of societies resulting from global interdependence. With the increasing focus on societal resilience, the so-called 'protective factors' enabling a society to deal with complex and uncertain environments and to flexibly withstand crises are coming to the fore.

Using the concept of resilience in a versatile way is not unproblematic. Can a concept being originally understood in terms of 'psychological resilience' be plausibly applied to societies? To what extent can a society having developed many protective factors against natural disasters (e.g. earthquake-resistant architectures, decentralised urban planning and robust critical infrastructures), but is hardly immune to economic crises, be described as 'resilient'? Although the concept of resilience has been adopted by an increasing number of disciplines, there is a noticeable lack of cross-disciplinary, particularly transdisciplinary approaches. Given that today's diversity and multidimensionality of interconnected crises cannot be adequately captured and addressed by just one discipline, cross-disciplinary conceptions in terms of a 'multi-crisis-resilient' society appear to be of increasing importance. What characterises such a society?

The current discourse on building future preparedness of societies in the twenty-first century is by no means limited to resilience. Not to be neglected are the comparatively older political guiding concepts of the 'sustainable society' and the 'developed society'. To date, it has hardly been explored how all three concepts contradict, but in a certain way also complement each other. What distinguishes 'the developed society' from 'the sustainable society' and the 'resilient society' and what do they have in common? How can they essentially contribute to building societal future preparedness in the twenty-first century? Which core principles for societal future preparedness can be derived by integrating all three concepts?

This book is devoted to all the questions outlined above. Chapter 1 provides an overview of the multidimensionality and diversity of current crises. Chapter 2 discusses the multifaceted transferability of the resilience concept and critically examines its claim as a 'one-word-answer' to the diversity of crises. Chapter 3 illustrates methodological approaches from transdisciplinary research and discusses cross-disciplinary and cross-crisis concepts for understanding universally applicable resilience. Building on this, Chap. 4 outlines different 'principles' of systemic societal resilience, say, multi-resilience, which can be applied to different crises. From a broader perspective, the concept of societal (multi-)resilience is compared with the concepts of developed society and sustainable society in Chaps. 5, 6 and 7 and discussed in terms of mutual contradictions, commonalities and points of complementarity. Following on from this, Chap. 8 outlines a number of principles which should be taken into account in building societal future preparedness in the twenty-first century. Finally, in Chap. 9, five leverage points for initiating necessary social changes are derived, the main conclusions of this book are summarised, and five milestones for the further political practice are drafted as an outlook.

Contents

Part I The Multi-resilient Society   1

1 Framework: Most Likely and Threatening Challenges in the Twenty-First Century.. 3
2 Resilience: A Universal Answer to the Crises of Our Time? 29
3 Excursus: Two Approaches to a Cross-Disciplinary Resilience Model .. 59
4 Five Principles of a Multi-resilient Society...................... 81

Part II Future Preparedness in the Twenty-First Century: Development, Sustainability, Resilience   125

5 The Developed Society....................................... 127
6 The Sustainable Society 169
7 Development Versus Sustainability Versus Resilience: Commonalities, Intersections and Contradictions 201

Part III What Does Societal Future Preparedness Mean in the Twenty-First Century? An Outlook 219

8 Future Preparedness in the Twenty-First Century: Contours of an Integrative Concept....................................221

9 Outlook: Five Leverage Points to Trigger Social Change..........267

References...309

Abbreviations

AI	Artificial Intelligence
approx.	approximate
BBK	Bundesamt für Bevölkerungsschutz und Katastrophenhilfe (German Federal Office of Civil Protection and Disaster Assistance
BEH	Bündnis Entwicklung Hilft (German Alliance Development Helps)
cf.	compare
CPI	Corruption Perception Index
e.g.	for example
etc.	et cetera
ff.	following off
FAO	Food and Agriculture Organization
Fig.	Figure
FMER	German Federal Ministry of Education and Research
HDI	Human Development Index
HRO	High Reliability Organisation
GDP	Gross Domestic Product
i.e.	that is
ibid.	in the same place
ICRC	International Committee of the Red Cross
ICT	Information and Communication Technologies
IDGs	Inner Development Goals
incl.	including
lat.	latin
MDGs	Millennium Development Goals
MENA	Middle East and North Africa

MIPEX	Migrant Integration Policy Index
MPI	Multidimensional Poverty Index
n.d.	no date or year
OBOR	One Belt One Road
OECD	Organisation for Economic Co-Operation and Development
OPHI	Oxford Poverty & Human Development Initiative
SD	Spiral Dynamics
SDGs	Sustainable Development Goals
SIPRI	Stockholm International Peace Research Institute
Tab.	Table
TAI	Technology Achievement Index
UK	United Kingdom
UNCTAD	United Nations Conference on Trade and Development
UNDP	United Nations Development Programme
UNISDR	United Nations Office of Disaster Risk Reduction
US	United States
vs.	Versus
WEF	World Economic Forum
WIPO	World Intellectual Property Organization

Part I
The Multi-resilient Society

Framework: Most Likely and Threatening Challenges in the Twenty-First Century

Let me say it right from the start: The situation in the outgoing twenty-first century is not as bad as it appears in the daily news coverage of wars, environmental disasters and famines. Are the people of the world getting worse? Not necessarily. In January 2019, US journalist Nicholas Kristof argued in a New York Times article that 2018 was actually the best year in human history (Kristof, 2019). Recalling data from the platform 'Our World in Data', he cites, among other things:

- Every day in 2018, around 295,000 people around the world gained access to electricity for the first time, around 305,000 people gained access to clean drinking water for the first time, and around 620,000 people gained access to the internet for the first time. Only about 4.5% of children worldwide still die before the age of five. In 1960, the figure was 19%, in 2003 it was still 7%.
- Until 1950, a large part of humanity lived in extreme poverty, i.e. on less than EUR 1.66 per day. In the 1980s, it was still 44% of the world's population. According to the World Bank, the share of the extreme poor fell for the first time in 2015 to 9.6% (Kristof, 2019).

Matthias Horx, an optimistic future scientist, came to similar conclusions. At the turn of the millennium, he saw positive global trends not only in terms of human prosperity and health, but also in ecological development:

- 'Not only in Central Europe, no significant environmental parameter has worsened for twenty years, be it air pollution, water pollution or contamination with poisons and heavy metals'. The important levels of pollutants in the blood – lead, dioxin, DDT, PCBs, cadmium – have been falling for many years. Rivers

and other bodies of water are, with very few exceptions, becoming cleaner and richer in species.
- In Europe and Eurasia there is a net forest gain of considerable proportions. Rainforest clearing has largely stopped or is moving towards ever lower levels.
- The halt to whaling has allowed this species to recover to a large extent, so that apart from a few subspecies it no longer belongs to the endangered species today. The great extinction of species, which reached its peak in the seventeenth century, has been stopped.
- The population explosion of the human species is not taking place. Humanity will reach its numerical zenith between 2050 and 2060 at about 8.5 billion people and then shrink again. Homo sapiens will only be represented by five billion specimens on this planet in 2150.
- (…) In absolute terms, the number of poor people is falling; today there are still 800 million (1.2 billion in 1990), and every year there are between 5 and 20 million fewer.
- In the medium term, the global gap between rich and poor will weaken because the countries with the strongest growth rates are developing countries – China, India, and even countries like Botswana, have been beating the growth rates of industrialised countries for decades.
- Three-fifths of the world's population now lives in democracies – 10 years ago it was less than one-third. In 1955, there were only 22 democracies (14.3%). In 2000, there were 120 (62.5%). Because democracy is the key to prosperity, the likelihood of economic recovery has tended to increase.
- Across the board, the global quality of life is rising, and infant mortality is falling even in the poorest countries. In 84 of the planet's now 193 countries, people reach an average age of more than 77, up from only 55 in 1990. 76% of the world's adults can now read, up from just 64 in 1990 and 42% in the 1960s. More and more children worldwide are going to school, especially girls: their enrolment rate at secondary level has risen from 36% to 61% (Horx, 2003, pp. 250–270).

He confirms a continuation of these global development trends in his later published book 'Megatrends' (Horx, 2014).

If it can be assumed that the situation of humanity in many places is characterised by the development trend towards greater prosperity, progress, access to education and knowledge – why shall this book address 'future preparedness' or even 'resilience'? In answer to this question, four theses can be put forward, which are also the starting point for this book:

1. The world of today has become more developed, but at the same time much more complex and unpredictable.
2. Every increase in development and complexity also brings with it correspondingly more complex challenges being increasingly difficult to manage.
3. What makes them increasingly difficult to manage is their interconnectedness ('crisis bundles') and multi-dimensionality ('bundled crises').
4. Complex challenges are often man-made and lead to unprecedented rebound effects. They can also prove to be more dangerous than ever. In view of today's technical possibilities, they can also turn into so-called 'existential risks'.

In other words, global development is moving forward, with many clearly positive phenomena and trends. At the same time, the world is more interconnected, fast-moving, unmanageable and unpredictable than ever before, bringing with it an unprecedented diversity of known and as yet unknown challenges. In the twenty-first century, today's societies are more than ever required to take a step back and systematically address how they will secure their future. In the context of this book, building societal future preparedness means to strengthen the capacities of societies to cope with current and future increasingly complex global challenges while continuing to develop. In this context, the focus is primarily on dealing with 'risks' and 'crises'. The following subchapters provide an overview of the most probable and dangerous crisis potentials of our time.

1.1 Most Likely and Most Dangerous Risks and Crisis Potentials

The term 'risk' has an unclear origin and refers to an event with a possible negative (danger) or positive effect (opportunity) across all disciplines (Deutscher Alpenverein, 2004). Usually, not all factors influencing a risk are known or dependent on chance. Therefore, risk is often associated with a venture, i.e., 'getting involved in a risky situation' (Warwitz, 2016, p. 16). 'Crisis' comes from the Greek 'krínein' (= to separate, [to] divide) and refers to a problematic decision-making situation associated with a turning point. If the development takes a permanently negative course, it is often associated with the notion of a 'catastrophe' or 'decline'. However, crises also hold the chance of a positive outcome, if we manage to develop new responses and solutions to the problematic situation (Gredler, 1994). Risks and crises have in common that they are accompanied by enormous stress and potential danger threatening the affected system. However, they also hold the opportunity for a positive development by leading to new problem-solving approaches that make similar situations more manageable in the future. However, both involve the challenge of incomplete information. A key distinction between the two terms is that the risk precedes the crisis.

In recent years, the number of available analyses of the global risks of our time has increased – and they seem to come to largely consistent assessments. Currently, the Global Risk Report of the annual World Economic Forum in Switzerland might be considered among the world's most influential analyses assessing global risks in the present and near future. The latest report before the Coronavirus crisis, released in January 2019, already highlighted that the world has reached its breaking point in several places. Almost 60% of the nearly 1000 risk experts surveyed worldwide expected global risks to increase in 2019, with only 7% expecting the situation to ease. Acceleration and networking 'in almost every field of human activity' is cited as a key factor. Figure 1.1 gives an initial impression of the respondents' assessment of the most likely and greatest global risks (WEF, 2019a). Despite the dominance of the Coronavirus crisis since the beginning of 2020, the assessment in the subsequent reports of 2020–2022 has not significantly changed (WEF, 2020, 2021), apart from the increased danger of a regional or maybe global hybrid war with Russia since the beginning of 2022 (WEF, 2022).

1.1.1 Most Likely Risks

According to the WEF's Global Risk Report 2021, the threats that are most likely to occur include the (1) extreme weather events, (2) climate action failure, (3) human environmental damage, (4) infectious diseases, (5) biodiversity loss, (6) digital power concentration, (7) digital inequality (WEF, 2021). The assessments for the first two positions haven't changed over the last three years (WEF, 2019a, b, 2020).

The overall assessment is more or less in line with other risk studies, such as Allianz's new Risk Barometer 2018, updated in 2022 – an annual survey of the most important corporate risks among almost 2000 risk experts from 80 countries. There, too, environmental risks (including natural hazards) are mentioned among the most significant. However, the Allianz Risk Barometer 2018 and 2022, stress a higher significance of cyber-attacks. They consider both environmental risks and cyber-attacks not only as the most likely, but also as the most important causes of business interruptions (Allianz, 2018, 2022). While cyber crises have only been

1.1 Most Likely and Most Dangerous Risks and Crisis Potentials

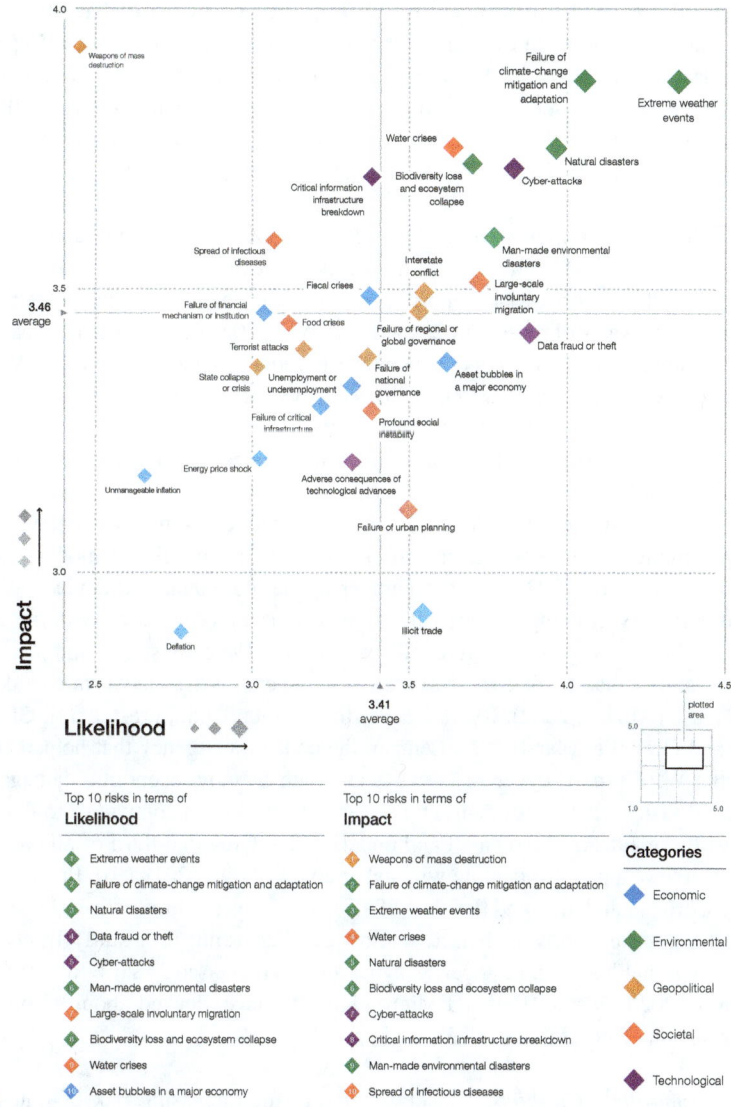

Fig. 1.1 Most likely (likelihood) and greatest hazards (impact) in context. (WEF, 2019a, p. 5)

considered among the most likely in 2019, environmental crises have been regarded at the top of the most significant crises in all previous Global Risk Reports of the last years (WEF, 2021). Due to their particular importance, the next two subchapters will deal separately with the various environmental risks. The following is a brief list of the most important global risks expected by observers in the short and medium term, according to various trend studies.

Cyber Hurricanes/Cyber Attacks The risk of cyber crises has increased in recent years and was ranked among the top five of the most significant crises for the first time in the 2018 Global Risk Report. The cyber crisis has now replaced the refugee crisis that was still present in 2015–2017 (WEF, 2021). Along with increasing urbanisation and technological dependencies of modern societies (WEF, 2019a, b; partly also UK Ministry of Defence, 2014), cyber crises will become more relevant and frequent in the future. To illustrate the scope of cyber crises, the Allianz Risk Barometer explicitly mentions 'cyber hurricanes', which included events like the Mirai botnet and the WannaCry and Petya attacks in 2017. Moreover, the recently identified security vulnerabilities in computer chips in almost every modern communication device highlighted the digital vulnerabilities of modern societies (Allianz, 2018, 2022). As the number of internet-connected devices in the world is rapidly growing, digital vulnerability appears to increase. With the current escalation of geopolitical conflicts between Western states and Russia, Belarus and China, the risk of cyber attacks as a means of hybrid warfare is likely to increase (Allianz, 2022). Overall, according to John Drzik, President of Global Risk and Digital at Marsh & McLennan, the consulting agency that helped conduct the WEF report, cyber risks cause as much or more economic damage as natural disasters. Yet the infrastructure of governments and companies to defend against these attacks is too small and underfunded. Only one-third of all companies even have a plan for dealing with cyber attacks (WEF, 2019a, b). In the wake of the Corona pandemic and the shift of all workplaces to home offices, cyber attacks have significantly increased worldwide. According to a study by Trend Micro, a global leader in IT security, cyber attacks increased by 20% to over 62.6 billion in 2020. For 2021, Trend Micro claims to have blocked about 94 billion cyber threats (Trend Micro, 2022).

Wars/Geopolitical Conflicts Closely related to the increasing risk of state-led cyber attacks, various analyses of the last years forecasted an intensification of the geopolitical environment. According to the latest report of the Stockholm International Peace Research Institute (SIPRI), the number of armed conflicts involving at least one state doubled in the 2010s, as did the number of deaths in conflicts (SIPRI, 2022). After years of decline, the number of operational nuclear warheads rose again in 2020. Last year, global military spending reached a peak of more than two trillion US dollars. The US CIA and the UK Ministry of Defence counted the intensification of conflicts (particularly inter-state) as one of the most significant risk trends of our time (CIA, 2018; UK Ministry of Defence, 2014). 93% of risk experts surveyed by the WEF expected political or economic disputes between major powers to intensify, and nearly 80% anticipated a further escalation and increase in associated risks in the form of war involving major powers in the coming years (WEF, 2019a, b). Observers stated diminishing influences of multilateral rules, making it increasingly attractive for many countries to re-establish national egoism as the focal point of power and legitimacy. At present, in the absence of a global governance, there appear to be no binding norms or adequately potent institutions that could unite the major powers. This leads to new risks and uncertainties: growing military tensions, economic and trade disruptions, but also new cyber options for hard and soft power, proxy conflicts (such as in the MENA region) (CIA, 2018; WEF, 2019a, b). The first half of 2022 is still characterised by an ongoing rivalry between the old superpower USA and the emerging new superpower China and an escalating conflict between the EU and the USA against Russia and Belarus in the wake of the Russian invasion of Ukraine since February 2022.

Future geopolitical conflicts, as stated by the two science journalists Christian Schwägerl and Andreas Rinke in their analysis '11 Looming Wars', could be largely driven by environmental factors. War scenarios over scarce resources such as water, raw materials (e.g. conflicts about energy resources in the Arctic region or about manganese nodules in the sea), food (e.g. so-called 'protein wars' as a result of overfished seas) and even about the radius of action in outer space would be conceivable. In many of their scenarios, Rinke/Schwägerl also address the political temptation to use superior military technology as an opportunity to eliminate a competitor (e.g. the USA versus the EU in the fight for the last fish reserves in the sea [protein war]). Moreover, in almost all their scenarios they see the economic and military rise of China as a key variable in the intensification of conflicts (e.g. in the scenario of a cyber war between the USA and China) (Rinke & Schwägerl, 2015).

Economic and Financial Crises In the context of economic and financial crises, many observers stress long-standing structural challenges, such as increasing global debts and a fragile global finance system. Many publications tend to focus on economics and financial markets and assume a high probability of occurrence of an economic crisis with global impact. Examples of these include analyses by Ernst Wolff (2017), Markus Krall (2017), Florian Homm (2016), James Rickards (2012) or Ulrike Herrmann (2015). They all emphasise that the emergency measures that prevented the collapse of the global financial system in 2008 did not eliminate the risk of an impending crisis. On the contrary, the immanent systemic flaw of the global economic and financial system will – should there be no profound structural changes – inevitably lead to a collapse with serious consequences for the social order of all societies concerned. The authors describe the systemic failure as a mismatch between the stagnating real economy (creating value through the production of goods) and the speculative financial economy. The latter is built on credit, does not create value, but is dependent on uninterrupted growth. As the financial economy is not able to generate the money to serve the constantly rising number of credits, the central banks steps in, 'generates' money without equivalent value in the real economy 'out of nothing' and makes it available to the banks (quantitative easing). At the same time, central banks lower the key interest rate at which they lend the money to banks (which has happened a total of 660 times between 2007 and 2016) (Wolff, 2017). But since this money mostly flows directly into financial speculation, the lowering of interest rates results into new speculative bubbles. The dilemma here is that a reversal of the destructive financial policy – through an increase of the key interest rate or/and a freezing of the money supply – will not solve the problem and is also likely to lead to new crises.

Pandemics and Technological Side Effects Other crisis contexts mentioned in many studies address, on the one hand, the risk of side effects and misuse of new technologies and, on the other hand, the scenario of a global pandemic (Rinke & Schwägerl, 2015; UK Ministry of Defence, 2014; CIA, 2018; WEF, 2021; Orrell, 2007). A global pandemic with the infection rate of Covid-19, but much deadlier, would count as a potential existential crisis. Long before Covid-19, the global pandemic has been considered to have a relatively high probability of occurrence (e.g. Orrell, 2007).

Multiple Social Risk Trends In addition to the risk potentials listed above, there are several multiple risk trends centred around the social dimension. Here are some examples:

1.1 Most Likely and Most Dangerous Risks and Crisis Potentials

- *Employment, social equality:* Employment rates and social equality have always been among the most important internal indicators of a societal crisis (see Chap. 8 for more details). In terms of employment, it is apparent that in the relatively wealthy societies of the First and Second Worlds, the working-age population is shrinking, while in the poorer Third World societies, which tend to experience higher unemployment, it is still growing (CIA, 2018; WEF, 2019a; Orrell, 2007). Moreover, in the coming years, as digitisation and automation of processes advance, jobs are expected to disappear, for which crisis-preventing labour market policies should prepare in time (Brynjolfsson & McAfee, 2014). As far as social equality is concerned, it can currently be observed that while prosperity is increasing worldwide, social inequality is on the rise in most societies, especially in the populous regions of the developing world (CIA, 2018; UK Ministry of Defence, 2014; WEF, 2019a). Employment and social equality have been key themes of social protests since 2011 both in wealthy First World societies (Occupy movements) and especially in the MENA region (Arab protests) (discussed in more detail in Chap. 8). More than ever, they are regarded as central challenges for a sustainable welfare policy. Today, as a result of these challenges, many risk experts are observing a trend towards increasing populism, social polarisation, nationalism and religious fanaticism, as a potential social crisis factor (CIA, 2018; WEF, 2019a; partly also UK Ministry of Defence, 2014).
- *Integration policy:* Closely related to the challenges of welfare systems outlined above is the need to pursue crisis-proof integration policies. In this context, the wealthier immigration societies in particular will be affected by continuing 'involuntary migration movements'. Although for the first time since 2018 the refugee crisis is no longer counted among the top five most important crises, its explosiveness has by no means diminished. On the contrary, experts expect an increase in involuntary migration in the future. The main reason for this is that the refugee crisis is closely linked to other global risks that are currently receiving greater attention, in particular geopolitical crises (war refugees) and environmental crises (e.g. climate refuge[1]) (Rinke & Schwägerl, 2015; CIA, 2018; WEF, 2019a).

---

[1] The World Bank estimates that more people are already fleeing from the effects of climate change than from war. In Sub-Saharan Africa, Latin America and South Asia alone, the World Bank estimates that there will be more than 140 million refugees by 2050 due to droughts, crop failures, storm surges and rising sea levels. Above all, the World Bank expects so-called 'internal migrations', i.e. movements within countries from rural to urban regions (World Bank, 2018).

- *Health:* Mental stress disorders (especially occupational burnout and depression) have significantly increased over the last two decades – and the trend is rising. In conjunction with increasing pressure to meet deadlines, to perform, to consume and to compete, as well as growing demands for mobility, speed and flexibility, it can be observed that chronic mental stress disorders, in particular occupational burnout and depression, are worldwide on the rise, especially in the wealthy societies (ILO, 2016; Knieps & Pfaff, 2016; Lohmann-Haislah, 2012; Galuska et al., 2010). Mental stress disorders in professional and private life are also closely associated with a suicide rate of – according to the latest study by the World Health Organisation (WHO) – up to 800,000 people per year.[2] This rate even exceeds the annual global homicide rate (ILO, 2016). Mental stress disorders increase the risk of heart disease by 50% among people affected by stress at work, according to the authors of the ILO Stress Report. Heart disease is the leading cause of death worldwide, accounting for nearly 17.5 million deaths annually compared to all chronic diseases (ILO, 2016).

These examples of multiple risk trends illustrate the interrelation of current national and global risks posing enormous challenges to a complexity-adequate sustainability and resilience policy of societies.

1.1.2 Most Impactful Risks and Existential Risks

Among the five most impactful threats, analysts of the Global Risk Report 2021 listed: (1) infectious diseases, (2) climate action failure, (3) weapons of mass destruction, (4) biodiversity loss, (5) natural resource crises, (6) human environmental damage, (7) livelihood crises (WEF, 2021).

In addition to this assessment, the current debate also includes a large number of other particularly dangerous risks, which are referred to as 'existential risks'. An existential risk can be understood as an event that is capable of wiping out all intelligent life on Earth or drastically and permanently restricting its desirable development (Bostrom, 2002, 2013). While individual threats (such as those listed above) have been intensively studied, systematic analysis of existential risks has only begun in the early 2000s, notwithstanding their enormous importance (Bostrom, 2002). This may also be due to the fact that re-

---

[2] 75% of those affected belong to low- and middle-income societies (ILO, 2016).

1.1 Most Likely and Most Dangerous Risks and Crisis Potentials

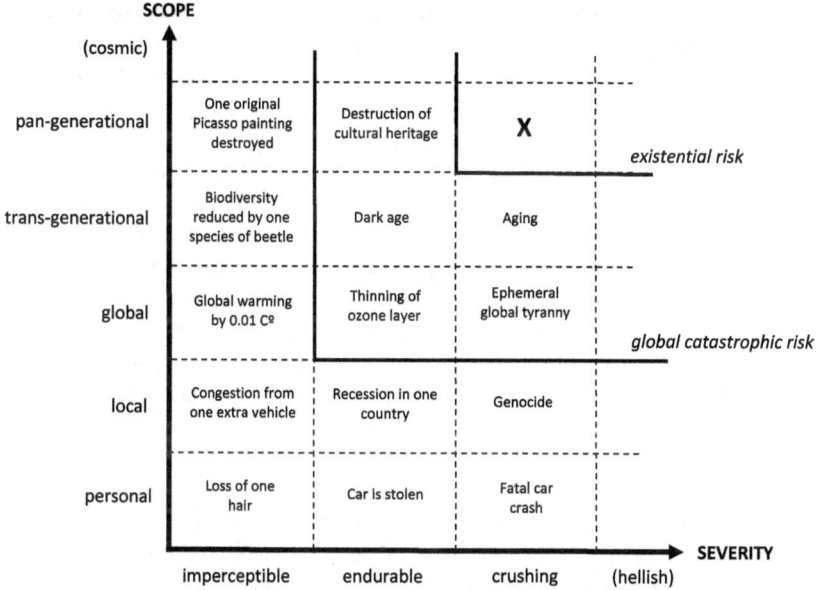

Fig. 1.2 Classification of existential risks. (Bostrom, 2013, p. 17)

search into existential risks would require a pooling of highly cross-disciplinary knowledge.

In his analysis, Nick Bostrom, a philosophy professor who is considered a pioneer in this field, distinguishes the concept of 'existential risk' from 'global catastrophic risk'. Compared to all other types of risk, existential risks are characterised by the greatest spatial and temporal scope and by its lethality (Fig. 1.2).

The following is a brief description of the deadliest risks that are currently most frequently discussed:

Weapons of Mass Destruction (Including Nuclear Holocaust) Weapons of mass destruction include nuclear, biological or chemical weapons, but also those based on completely new technological principles, such as nano weapons or gene bombs (Rinke & Schwägerl, 2015). Nuclear weapons are currently considered to be among the most threatening weapons in the current debate. Thus, the use of even a fraction of the world's arsenals could kill a large part of humanity in its direct effect, while the consequential effects of the explosions would threaten humanity existentially for generations. These would include poisoned water, germline mutations, rising cancer rates, a collapsing ecosystem, the collapse of social institutions, and a nuclear winter (Human Wrongs Watch, 2012). Moreover, developments in the fields of genomics, genetics, and biotechnology are believed to have lasting implications for global security, and the means of hyper-disease as a weapon (e.g. in the hands of bioterrorists) poses a global threat (van Aken & Hammond, 2003).

Environmental Disasters While the use of weapons of mass destruction assumes a certain intention, unintended man-made existential risks, such as environmental disasters, are also conceivable and not unlikely. An environmental catastrophe is usually triggered by an industrial accident – examples include the nuclear catastrophe at Chernobyl in 1986, or the rather hypothetical doomsday scenario 'Gray Goo'. However, environmental catastrophes also include creeping environmental pollution – such as the greenhouse effect, which in turn could lead to escalating global warming or even a new ice age. These last three existential risks are explained in more detail below.

- The *'Gray Goo' scenario* was first introduced in 1986 in the book 'Engines of Creation' by nanotechnology pioneer Eric Drexler, and describes the existential risk of molecular nanotechnology getting out of control (Drexler, 1986). Here, self-replicating assemblers consume all the matter on Earth to create more and more copies of themselves (Institute of Physics, 2004). This scenario is also called ecophagy ('eating up the environment') (Freitas, 2000).
- *Global warming* is the increase in the average temperature of the Earth's atmosphere and oceans. According to current research, during the transition from the ice age to an interglacial period, the Earth warmed by about 4–5 °C within approximately 10,000 years. However, with man-made global warming, the temperature is expected to increase by 4–5 °C within 100 years. 2016 was the warmest year since systematic measurements began in 1880

(Washington & Cook, 2011; Graßl 2007). In the worst-case scenario of escalating global warming, the release of greenhouse gases into the atmosphere could prove to be a strongly self-reinforcing feedback process that, beyond a certain point of no return, can no longer be averted by currently known interventions (such as geoengineering). The Earth's atmosphere would then become similar to that of Venus, and life would be impossible (Umweltbundesamt, 2017).

- An *ice age* is a period of the Earth's history in which, depending on the definition, at least one pole of the Earth is glaciated (broad definition [Murawski & Meyer, 2004]) or when glaciations occur in both poles (narrow definition [Imbrie & Imbrie, 1979]). Depending on the definition, there have been about four to seven ice ages in the history of the Earth, with warm periods of varying length in between. Should the current warm period abruptly end, the Earth could become a single 'snowball'. The temperature would drop and its impact on the food supply would endanger the continued existence of humankind (for the 'snowball earth' hypothesis, see Ward & Kirschvink, 2018).

Similar to or even accompanying the Gray Goo scenario, an extinction through the influence of *superior artificial intelligence* ('superintelligence') would also be conceivable. The assumption is that the artificial intelligence would either deliberately extinct all life on earth or accidentally destruct the basis of life in the course of fulfilling its tasks. In this context, Nick Bostrom's thought experiment is one of the best known, according to which a superintelligence transforms the entire planet and possibly also the solar system into computronium[3] without having any direct malicious intentions. In his widely acclaimed book 'Superintelligence' (2016), Bostrom argues that and how human

---

[3] In conjunction with this term, which was originally coined by Norman Margolus and Tommaso Toffoli of the Massachusetts Institute of Technology, Bostrom describes the following scenario: A machine programmed to produce as many paper clips as possible becomes incredibly intelligent. Given its programmed goals, it might decide to develop newer and more efficient paperclip-making techniques, until eventually it would have processed the entire world into paperclips. Even a programmed goal to stop after exactly one million paperclips would not be enough to stop the artificial intelligence. Indeed, the super intelligent machine might decide to review its work, and to be really sure, it would evolve the secondary goal (instrumental goal) of becoming smarter. Subsequently, the super intelligent machine would develop a new type of computer component – 'computronium' – that would check all cases of doubt. As new doubts emerged, the whole world would be transformed into computronium in the course of the process – down to a million precisely counted paperclips (Bostrom, 2017).

intelligence could not compete with an emerging superintelligence if it approaches human capabilities to the extent that it could begin to develop machines that improve themselves (Bostrom, 2017). Current projects simulating brain processes (e.g. the Blue Brain project),[4] modelling cognitive architectures (e.g. ACT-R and Soar),[5] or developing knowledge bases (e.g. Cyc),[6] search en-

---

[4] The Blue Brain project regards itself as a pioneering project to virtually simulate the human brain through large-scale computer models. The goal is to understand and replicate how the human brain works. It was launched by a collaboration between Henry Markram's Brain and Mind Institute at Ecole Polytechnique (Switzerland) and IBM (USA) in May 2005 and has been called the Human Brain Project since receiving EUR 1 billion EU funding (Markram, 2006). In October 2015, it was already possible to simulate the activity of around 31,000 neurons from the somatosensory cortex of a rat brain – i.e. the part of the cerebral cortex that serves the central processing of haptic perception (Markram, 2015). According to Bostrom, progress in this research direction could pave the way to 'brain emulation' ('uploading') and thus to a type of artificial intelligence that appears relatively human (Bostrom, 2017).

[5] These are computer-based models for modelling and simulating complex human cognitive processes (memory, language, perception, problem solving, etc.). The project *'Adaptive Control of Thought-Rational' (ACT-R)* which was significantly developed by the cognitive psychologist John Anderson (Anderson et al., 2004), is well known. By contrast, *Soar* is a cognitive architecture dating back to Allen Newell, John Laird, and Paul Rosenbloom that defines and attempts to simulate all primitive principles underlying human cognition. Primitive principles remain constant over long periods of time and across different application domains. An example of such a principle is: 'New goals are generated only when impasses occur.' It is only on the basis of such an architecture that more complex human capabilities can be modelled, such as mental arithmetic, language processing, learning processes, etc.). The assumption is: if these models are mature and complete, it might be possible to create an artificial intelligence that exhibits all human behaviours (Ritter et al., 2001).

[6] *Cyc* (from the English *encyclopaedia*) is a machine-interpretable knowledge database of everyday knowledge. It has been under development since 1984 and contains a logic of reasoning. For this purpose, all objects in this world are described by unique objects and then the relationships between these objects are specified by rules, e.g. 'water is wet'. Thus, Cyc might infer that the person in question is wet from statements that Peter is swimming in the sea and that the sea is mostly water. Since 1995, Cyc has been published by Cycorp Inc (http://www.cyc.com/).

gines (e.g. Watson),[7] or chat programs (e.g. A.L.I.C.E.)[8] could pave the way for the emergence of artificial general intelligence and possibly superintelligence before the end of this century (Bostrom, 2017). All these and other projects have shown surprising and enormous progress in recent years, but at the same time the need to put self-learning artificial intelligence under human control. A recent and popular example is the chat programme 'Tay' developed by Microsoft, which went online on Twitter in 2016 to casually converse with and learn from chat users. The result was that within 24 h the chat program adopted a variety of racist and sexist attitudes, posted corresponding comments and insulted individual users. The experiment therefore had to be stopped (Horton, 2016).

In addition to man-made scenarios, existential risks can also be caused by purely natural processes that are not influenced by humans.

This would be the case, for example, with the eruption of *supervolcanoes* which are considered the largest known volcanoes. In contrast to 'normal' volcanoes, they do not build up volcanic cones during eruptions due to the size of their magma chamber, but leave behind huge calderas (collapse cauldrons) in the ground.

---

[7] *Watson* was developed by IBM, named after Thomas Watson, one of its first presidents. A powerful search engine, it is designed to provide answers to questions entered in digital form in natural language. The software could assist with complex decisions in many fields, such as medical diagnostics, especially under time pressure. Japanese insurance company Fukoku Mutual Life Insurance has replaced 34 employees with IBM's Watson Explorer on January 1, 2017. In the future, the AI will collect hospital records and other documents, among other things, in order to calculate possible pay-outs (Haase, 2017). It gained notoriety when it successfully competed against two of the world's top players in three episodes of the quiz show Jeopardy! broadcasted from February 14 to 16, 2011, and won by a wide margin (IBM). The game is now compared to the duel of the world chess champion Garri Kasparov against the IBM computer Deep Blue (Bostrom, 2017) or most recently the duel of the Go world champion Lee Sedol against the Google computer AlphaGo. It is interesting to note that AlphaGo is said to have played at a much worse level five months before the duel and to have significantly improved within this short time period via deep learning. Rapid improvements could particularly be observed during the matches with Lee Sedol (Chouard, 2016).

[8] *A.L.I.C.E.* (acronym for *Artificial Linguistic Internet Computer Entity*) is a natural language chatbot. It is the brainchild of programmer Richard Wallace and was inspired by Joseph Weizenbaum's ELIZA psychotherapy program developed in the 1960s. This was based on a structured dictionary (thesaurus) that scans the input sentence for words it knows, matches them with synonyms or generic terms, and integrates them into a present question or prompts (e.g. user: 'I have a problem with my father.' ELIZA: 'Tell me more about your family!') (Weizenbaum, 1966). First presented in November 1995, the A.L.I.C.E. program is now considered one of the most powerful programmes of its kind and has won the Loebner Prize three times (2000, 2001, 2004). However, A.L.I.C.E. has so far not been able to pass the Turing test – even an occasional user quickly recognises that the interlocutor is a machine (Wallace, 2009).

Eruptions with a Volcanic Explosivity Index value of 8 (VEI-8) are referred to as super eruptions, although VEI-7 eruptions are occasionally included. The last eruption of a supervolcano occurred in the Lake Taupo area of New Zealand about 26,500 years ago (Gualda et al., 2012). The eruption of an entire volcanic complex could cause effects comparable to a nuclear winter and/or catastrophic climate change, thus endangering the continued existence of humankind.

The same applies to the scenario of the impact of a meteorite. *Impact* (from Latin impactus = to strike) is when two celestial bodies collide at very high speed. Objects with a diameter of more than five hundred metres are considered globally dangerous. Scientists in New Mexico (USA) counted more than 1100 asteroids with a diameter of more than one kilometre that are in an orbit that could bring them dangerously close to the earth. In the event of an impact, the probability of such a meteorite striking the ocean would be relatively high, especially since 71% of the Earth's surface is covered by water (Garshnek et al., 2000; Morrison, 2006). The result would be a mega tsunami with wave heights in shallow water areas of one hundred meters and above, which would flood entire coastal landscapes and their hinterlands over a wide area (Yabushita & Hatta, 1994). An impact on land would potentially cause widespread and prolonged darkening of the sky as the churned-up earth would lead to species extinction (Global Challenges Foundation, 2017).

The probability of the first two scenarios occurring is likely to be the highest compared to the others. The latter two existential risks are relatively unlikely, even though there relatively accurate estimates of how likely they are to occur. For example, astronomer James Gavarick Matheny estimates the probability of an existentially dangerous asteroid impact at 0.000001% (one in a million) in the next 100 years (Matheny, 2007). The Global Challenges Foundation estimates the probability of such an impact at a cycle of once every 120,000 years (Global Challenges Foundation, 2017). Comparable to this are supervolcano eruptions, the frequency of which Rampino/Ambrose estimate at about once every 50,000 years (Rampino & Ambrose, 2002). The probabilities of other threats are likely to be more ambiguous to estimate. Additionally, threats that are not yet known today and therefore cannot be predicted a fortiori must always be taken into account (Bostrom, 2002). A distinctive feature of existential risks is that, unlike most other crisis events, their absence in the past is not evidence that the probability of existential risks occurring in the future is low. This is because if an existential risk had occurred in our past, there would be no people left to observe it (Bostrom 2013).

1.2 Environmental Hazards at the Centre of Attention

Out of 30 identified global risks, environmental hazards (natural and man-made) appear to be the most worrying in the Global Risk Reports, as they are both very likely and impactful. This assessment is also broadly in line with the analyses of Rinke and Schwägerl (2015), UK Ministry of Defence (2014), CIA (2018), and the US Department of Defense (Department of Defense, 2015).

In the course of this emphasis on ecological risks, the German World Risk Report (Weltrisikobericht) came into being, which has been very influential in the German-speaking world. It presents an index for 181 countries that shows how vulnerable societies are to floods, droughts, storms or earthquakes. This report has been published annually since 2012 by an alliance of several large aid organisations, the 'Bündnis Entwicklung Hilft' (BEH, in English: Alliance Development Helps).

The authors of the BEH report do not base their projections solely on the likelihood of natural forces hitting a particular region, as other organisations do. They link this category, called 'exposure', to the 'vulnerability' of society. Thus, on the one hand, the authors asked how vulnerable a country is (e.g. how many people are at risk and how intensely?), and on the other hand, the index covers a country's ability to adapt to the consequences of climate change and to cope with a disaster (e.g. are there early warning systems and other preventive measures such as agricultural extension, building codes, dikes, food and medicine storage, education and structures at the local level?) (BEH, 2021). Figure 1.3 gives an impression of the composition of the World Risk Index:

Accordingly, island states such as the Philippines, Tonga and Vanuatu are considered the most vulnerable countries (BEH, 2021), also in prior reports (such as BEH, 2017). They are particularly vulnerable to cyclones and sea-level rise. Developing countries such as Bangladesh, Guinea-Bissau and many poor nations in Africa, which lack money and efficient government structures to prevent risks, have similarly high-risk scores. Highly developed countries like Japan are the exception. Japan, in 46th place in 2021 (BEH, 2021) and in 17th place in 2017 (BEH, 2017), has all the means of disaster prevention, but is surrounded by sea and is threatened by typhoons and tsunamis as well as earthquakes and volcanic eruptions (Fig. 1.4).

Similar to prior reports (such as BEH 2017), island states are at the top of the global risk ranking, as many of these countries are not only highly exposed to natural hazards such as earthquakes, cyclones, floods, and droughts, but are also increasingly threatened by rising sea levels due to climate change. The 2021 report

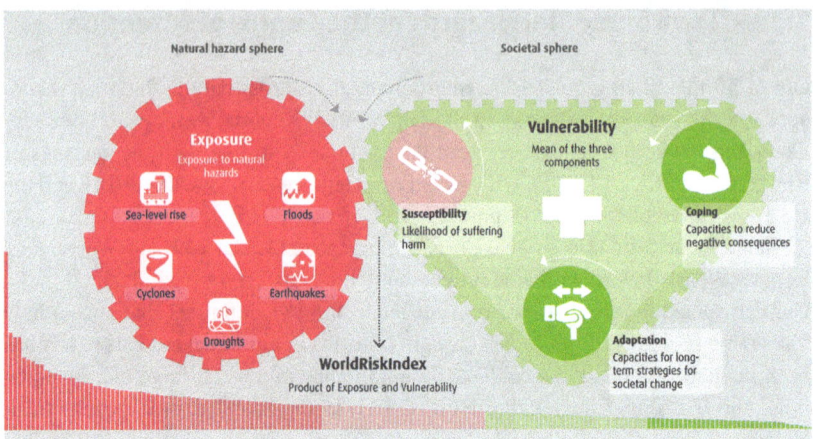

Fig. 1.3 Components of the World Risk Index. (BEH, 2021, p. 15)

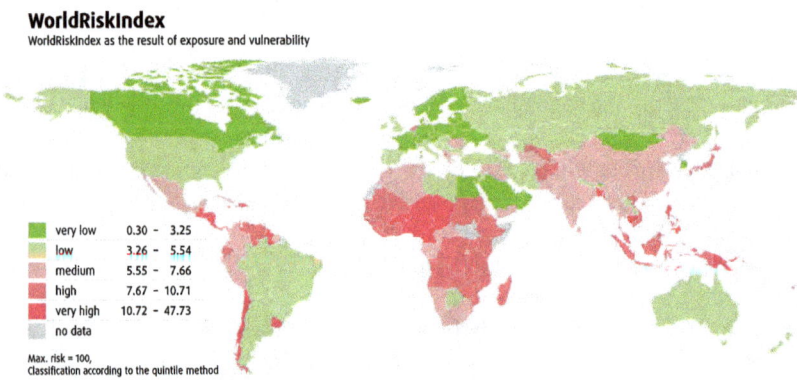

Fig. 1.4 Global distribution of risk potential. (BEH, 2021, p. 69)

shows a great heterogeneity of global disaster risks, but also highlights a strong relationship of geographic location, disaster risk and social aspects including inequality and poverty. With Vanuatu, the Solomon Islands, Tonga, Dominica, Antigua and Barbuda, Brunei Darussalam, the Philippines, Papua New Guinea, Cape Verde, and Fiji, 10 island states are among the 15 countries with the highest risk. Comparing the ranking of continental medians shows that Oceania carries the highest risk, followed by Africa, the Americas, Asia, and Europe (BEH, 2021).

The authors conclude by emphasising that the general vulnerability in many societies has changed slightly over the years. Moreover, they stress that disaster risk and vulnerability can be disrupted by targeted measures to strengthen social capacities. This is shown by the examples of the Netherlands, Japan, Mauritius or Trinidad and Tobago (BEH, 2021). The authors also conclude that 'against the background that the risk profiles of countries are becoming more complex due to climate change and many regions are facing new hazards, this is not only a challenge for the international community for the present, but a matter of great importance for the future (BEH, 2021, p. 49).'

Given the increase of natural hazards, it is not without reason that ecological crises are the focus of current risk reports like the World Risk Reports or the Global Risk Reports, since they are currently classified as the most likely and impactful crises and in the near future. Thus, it is not surprising that climate change and natural hazards have traditionally been a particular focus in the discourse on the resilient society. At first glance, this seems plausible, but on closer inspection, this approach proves to be short-sighted, as today's societies are facing many challenges at the same time, which are also interconnected and multidimensional.

1.3 The Coronavirus Crisis: Crisis Bundle and Bundled Crisis

The complexity of the contemporary and future landscape of risks and crises is shaped by various complex risks and bundles of risks. The authors of the latest SIPRI report call it a 'new area of risk' describing it as follows:

> Unrelated events in two different parts of the world combine to create an unforeseen risk in a third location. A climate change-fuelled weather disaster creates shocks throughout a global supply chain. Violence and crop disaster jointly make life unliveable for many thousands of farmers, contributing to migration. All over the world, predominantly in places already hampered by poverty and poor governance, the combination of rising insecurity and environmental degradation is creating a new era of risk (SIPRI, 2022, p. 28).

Hereby, they stress that

> In today's interconnected world, stresses, impacts and risk factors can combine in a number of ways (SIPRI, 2022, p. 32).

The authors of the SIPRI report describe five ways how risk factors can combine and bundle into full-blown complex crises. They include:

- A *compound risk* is a result from the interaction of two or more factors in a given region to generate a new type of risk that would not otherwise exist. The authors cited the 2015 Chennai floods in India, which killed nearly 200 people and displaced 200,000 from their homes, a typical compound risk.
- A *cascading risk* emerges like 'a stream coursing down a mountainside [...]. A storm can lead to a power cut, which closes a hospital, which is then unable to stem a disease outbreak, which ... and so on, through a sequence of events unfolding across time and space, including across national borders.' (SIPRI, 2022, p. 33)
- An *emergent risk* results from the combination of two or more seemingly independent factors in different parts of the world. As these factors are unrelated makes such risks typically hard to predict. The Arab Spring in 2011 can be regarded as a typical example resulting from a combination of globally soaring food prices and a dissatisfaction with governments that many citizens saw as oppressive and corrupt (ibd.).
- A *systemic risk* occurs when multiple shocks interact with sufficient severity and on a sufficient scale (SIPRI, 2022, p. 35), putting the very stability of a system such as a community or even a country at risk. As a typical example, the Corona crisis has shown how a new disease can affect and every aspect of society and every part of the world (ibd.; Benedikter & Fathi, 2021).
- In an existential risk, 'a mixture of drivers creates conditions so severe that they threaten the end of something profoundly important such as country or a culture or they may lead to a large number of deaths.' (SIPRI, 2022, p. 36)

In this book, these different combinations will be called 'risk bundles' or 'crisis bundles' (Fig. 1.5).

Besides 'crisis bundles', there are also 'bundled crises'. While the first term describes the interrelation between different risks and crises, the second describes the multidimensional nature of individual but complex risks and crises. Both, crisis bundles and bundled crises, can be illustrated using the example of the Coronavirus crisis.

In the context of crisis bundles, the Coronavirus pandemic deeply related to the trend of social risks in terms of political radicalisation and the increasing spread of fake news and conspiracy beliefs ('infodemic'). Among the conspiracy beliefs, QAnon or Q might be considered as one of the most influential ones worldwide. QAnon has been active since 2017 and experienced a considerable boost during the

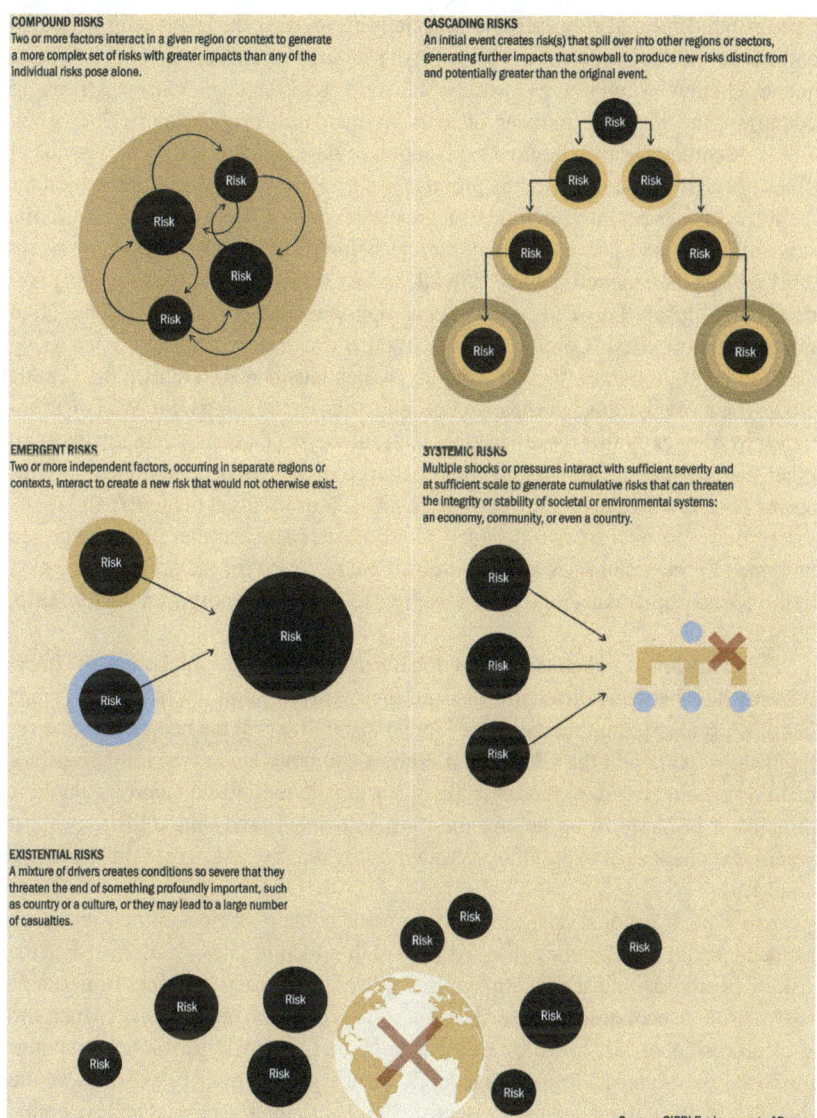

Fig. 1.5 Different types of risk bundles. (SIPRI, 2022, p. 34)

Corona pandemic, also in Europe, especially in parts of the right-wing extremist and lateral thinking scene in Germany. In the summer of 2020, three of the 10 most active QAnon communities worldwide were registered in German-speaking countries. In 2021, the number of activists increased even further (Huesmann, 2021). A common psychological explanation of the spread of conspiracy myths in times of crisis is that these myths promise a clear enemy image and orientation for many people, especially in times of great uncertainty. Here, internet platforms serve all the more as accelerators of radicalisation (Bundesstelle für Sektenfragen, 2021). More than the refugee crisis in 2015, the Corona crisis has led to a polarisation of the debate between 'mainstream' opinions and 'sceptic' opinions, often falsely categorised as 'Corona deniers'. Another driver of radicalisation and social division is proving to be social inequality, which has increased during the Corona crisis. In an international comparison of industrialised countries, the WSI distribution report suggests that wealth is more concentrated in Germany than in almost all other European countries: While about a third of households in Germany have reserves for a few weeks or a few months at most, households at the upper end of the scale could last at least two decades. Other studies suggest that people who are burdened by income losses are significantly more critical of the political situation in the country and on average more receptive to pandemic conspiracy myths (Otto, 2021).

The chart below, taken from the Global Risk Report 2019, illustrates the interrelationships between different risks and their condensation into so-called 'crisis bundles'. It is striking that the risk of social instability is at the heart of all interrelationships. Although the Global Risk Report and other studies currently rank social crises neither among the most likely nor among the most dangerous, this risk dimension is likely to be among the most immanent, and thus ultimately most significant challenges to building societal future preparedness in present and future (Fig. 1.6).

As a typical bundled crisis, the Corona crisis contains various dimensions such as medical and health policy risks, expressed in infection rates, death rates and the risk of overburdening the health system, as well as impacts on other dimensions such as social, economic and psychological. This does not make it easy to measure societal resilience, as shown by the discussion on the 'Covid Resilience Ranking' proposed by the business news agency Bloomberg. This model measures 'where the virus is being managed most effectively and with the least social and economic disruption' (t-online dpa, 2021), taking into account not only infection and death rates, vaccination rates or quality of health care, but also how many people are going shopping or traveling to work, what restrictions there are on social and economic life and how many flights take off. According to the report, Switzerland and

1.3 The Coronavirus Crisis: Crisis Bundle and Bundled Crisis

Risk Interconnection Map

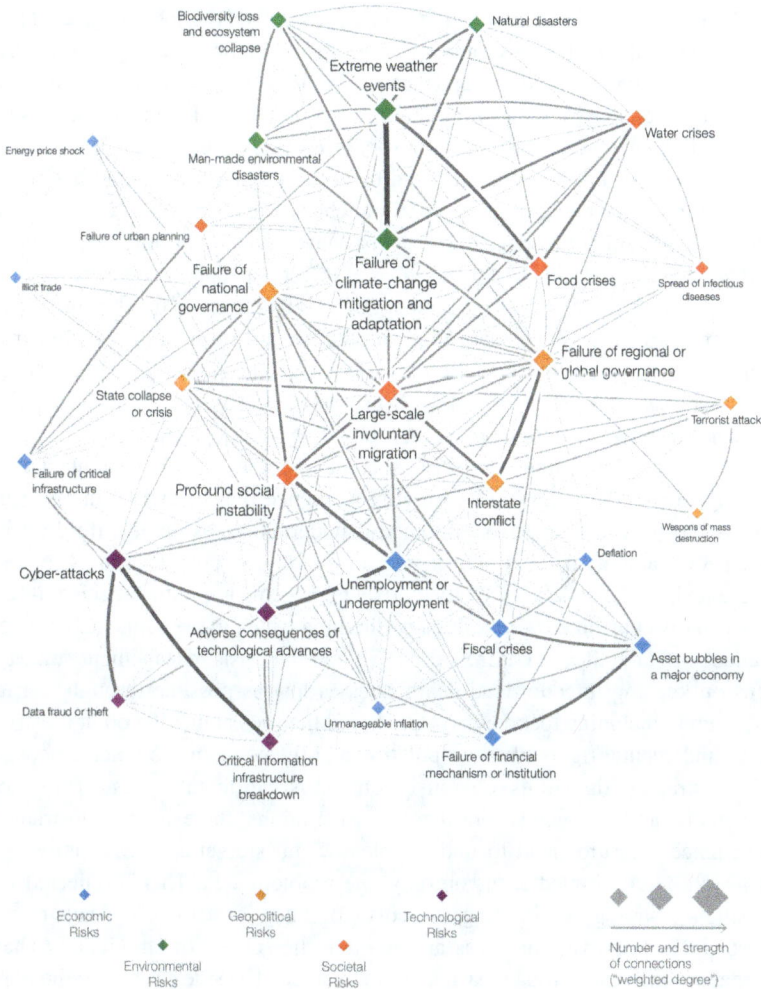

Fig. 1.6 Interrelationships of global risks. (WEF, 2019a, p. 7)

Austria rank second and sixth respectively in July 2021, and Germany only 12th out of a total of 53 countries covered. Norway is in first place, New Zealand, France and the USA in third, fourth and fifth place respectively (Bloomberg, 2021). Here it is noteworthy that Switzerland was considered a pioneer of easing measures in Europe, particularly in March and April 2021. Measures included a relatively short closure of schools (only in spring 2020), while hotels could remain open, as could ski resorts. There was never a compulsory test for shopping or restaurant visits, nor was there a curfew. However, it could be criticised that Switzerland has accepted higher infection rates, more Corona patients in intensive care units and more deaths. The number of deaths in the first wave of the pandemic was about 60% higher than in Germany and the number of infected people was practically always higher than in Germany – in November 2020 even up to three times higher – so that in some cases it was necessary to decide which patient got an intensive care bed and who did not (t-online dpa, 2021). In Germany on the other hand, which according to the Covid Resilience Ranking is in a much worse position, the health system was never overloaded to the point that such triage had to occur.

This example illustrates two typical challenges that arise in the context of measuring resilient societies and designing resilience policy: Firstly, the measuring proves to be a difficult, even ambiguous undertaking, as the manifold crisis dimensions and associated objectives for resilience policy cannot be clearly weighted. In the context of Corona crisis management, political decision-makers are usually confronted with a strategy dilemma which in the public debate is often reduced to an either-or confrontation: Should the crisis management strategy focus on keeping infection and death rates as low as possible through restrictive measures (including lockdowns)? Or should the focus rather be on not restricting the fundamental rights of the population and the economy as much as possible? Both strategy directions obviously emphasise different types of resilience. Secondly, and following from these considerations, the example illustrates that resilience measures lead to undesirable societal side-effects, especially if they have an overly limited focus on only one problem area. This is reflected in the resilience strategy of keeping the curve of infection case and death rates flat through lockdowns which was and is often the subject of emotionally charged debates about the Corona measures implemented. The side effects for the population included economic problems and existential fears, increasing social isolation and loneliness as well as an increase in frustration, boredom, domestic violence and depressive disorders (Friebe et al., 2020). 'Is the cure worse than the disease?' is a question that is indeed raised in many debates about the 'right' or 'resilient' Corona policy (Dorling, 2020).

Given the described multi-contextual challenges, which are characteristic for the present and the imminent future of global crises, we assume that contemporary societies require a 'broader' concept of resilience which is trans-disciplinarily applicable to multiple contexts. I call it 'multi-resilience', referring to a competency inherent in the resilience of a given societal system itself as well as to its interconnections to other systems, which can be applied to simultaneous crisis contexts. As we will see in later chapters, there are difficulties in the current debate on the resilient society to interpret and implement the concept of resilience in an integrative way, which will, however, be crucial to respond to the complexity of interrelated, unpredictable and multidimensional global risks and threats.

Resilience: A Universal Answer to the Crises of Our Time? 2

In the face of manifold crisis bundles and bundled crises, the question of preventive 'protective factors' that enables societies to cope with multiple and diverse uncertainties and adversities has increasingly arisen in recent years. The concept of resilience, which is used in many ways, is currently being discussed as a counter-concept to the concept of vulnerability that prevailed for decades. Where does the concept of resilience come from and how is it used? Which contributions does the resilience concept offer as a 'one-word answer' to different types of problems (psychological, political, economic, ecological, social and other crises) and different system levels (individuals, organisations, societies)?

2.1 Different Definitions and Uses of the Term *Resilience*

Although the concept of resilience is increasingly finding its way into the sustainability discourse, also being treated as a concept of 'new sustainability', both concepts represent different ideas. While sustainability aims at shaping economic and political behaviour in a way that any ecological, social and economic crises in the future do not occur and the needs of future generations are not compromised, resilience is fully expecting a future with multiple problems. Hence, resilience is focused on the question how any system, such as persons, organisations or societies, can prepare for inevitable and unpredictable problems. It is thus assumed that sustainable policies alone can no longer sufficiently prevent crises and that societies should develop adaptive and robust structures to cope with unforeseen events. In view of current global challenges, the concept of resilience is becoming increasingly important.

The term 'resilience' comes from the Latin 'resilire' and means, directly translated, 'to bounce back'. Its origins can be traced back to the natural sciences of the 1950s, but other observers confirm that the resilience concept has been developed in parallel in both the natural sciences and the humanities since the beginning of the twentieth century (Tusaie & Dyer, 2004). What is certain is that the resilience discourse has gained widespread public attention in education and psychology since the 1970s and – what is less well known – that resilience research has developed in four phases to date:

1. In the first phase, the aim was to identify which factors constitute a resilient person in the first place. The famous study by Emmy Werner made a significant contribution to this. Based on her observations, she assumed that there are universal resilience factors that characterise resilient individuals (Cutuli et al., 2008). According to Diane Coutu, a similar relationship exists in the organisational context. She assumes that the great common denominator of all studies in resilient organisations are the following four criteria: Acceptance, meaning in life, value orientation, and improvisational capacity (Coutu, 2002).
2. The second phase (from around the 1980s) was dominated by the research question of how resilience works. Researchers came to the conclusion that it would be difficult to generalise resilience factors, since resilience can manifest and develop differently depending on the respective context and situation. Also, protective factors such as 'acceptance' or 'optimism' can also turn out to be risk factors in certain contexts, if they are exaggerated. Compared to phase 1, this meant a change in thinking: resilience is not a fixed trait, but rather a dynamic, context-dependent process (O'Dougherty et al., 2013).
3. The third phase of resilience studies (from around the 1990s onwards) is characterised by the development of resilience-promoting interventions and their scientific evaluation.
4. The fourth phase (since about the 2000s) is characterised by interdisciplinary research at several system levels. How do collective and individual resilience interact? This gives rise to further opportunities for cross-disciplinary research of resilience and resilience promotion.

The last two phases are still in their infancy and are thus little explored to date (O'Dougherty et al., 2013). This book ties in with the fourth wave of resilience studies and examines societal future viability from a transdisciplinary scope – which is also currently little explored.

In the current practical and academic discourse, the resilience principle is thus not only applied to the individual, but to all system levels – individual, organisa-

2.1 Different Definitions and Uses of the Term Resilience

tion, city, society. Initially, the term described the ability of elastic materials or ecosystems to withstand external pressure without damage. Accordingly, resilience can be defined in the words of resilience scientist Ann Masten as 'the capacity of a dynamic system to withstand or recover from significant challenges that threaten its stability, viability, or development' (Masten, 2011). Similarly, Zolli/Healy refer to resilience as 'the capacity of a system, enterprise, or a person to maintain its core purpose and integrity in the face of dramatically changed circumstances' (Zolli & Healy, 2013, p. 7). The 'resilience term' often comes up in the context of 'crisis.' As stated above, this term translates from the Greek 'krísis', meaning 'decision' or 'turning point'. A crisis refers to a situation that is not only perceived as threatening by those affected, but has become so acute that they are unable to resolve it satisfactorily with their previous coping strategies. They therefore find themselves at a turning point, at a point of 'decision'. In this situation, resilient systems have the opportunity to perceive their situation differently, to discover opportunities if necessary, and to discover and develop new ways of dealing with the crisis.

An overarching model that helps to categorise different aspects of resilience (here e.g. Birkmann et al., 2011; CSS, 2009) is a phase model:

- Before the crisis: To anticipate and observe possible risks, and to prepare accordingly (preparation). If possible, identify dangers at an early stage and effectively prevent crises by reducing risk factors or strengthening corresponding protective factors (prevention).
- During the crisis: To be protected from the negative effects of an adversity, to remain fully functional (protection), and to be able to respond quickly and effectively to challenges posed by a crisis (reaction).
- After the crisis: To recover quickly from a crisis and be able to learn from past events (recovery).

Another model classifies different notions of resilience along a spectrum between 'persistence' and 'change'. Accordingly, on one side of the spectrum, a system can be defined as resilient if, despite unforeseen influencing factors, it retains its ability to function and its essential properties while 'somehow remaining as it is' – this understanding of resilience emphasises the value of stability and is frequently associated with the image of being 'firm as a rock' (e.g. in Wellensiek, 2011). In the middle of the spectrum, definitions of resilience stress the aspect of flexibility. The image often associated with it in popular literature is that of a skipjack – this principle is also implicitly related to the above-mentioned definition of Ann Masten. This notion emphasises the ability of a system to come back and to find a way back to its 'normal initial state' after it has been knocked down by the crisis. Another definition, at the middle end of the spectrum, defines resilience as the capacity of a system to gradually adapt to the new conditions associated with the crisis (Bené et al., 2012). At the end of the spectrum, resilience is described as the capacity of a system to transform completely in the face of crisis and, in the broadest sense, to emerge stronger (Hodgson, 2009; Bené et al., 2012), which is also called the capacity to 'bounce forward' to a 'new normal' (Gilan et al., 2021). Figure 2.1 shows two current examples of a spectrum of different resilience concepts between persistence and change.

In recent years, the focus of the debate has increasingly shifted towards the latter concept of resilience, mentioning 'Transformative resilience' (Gotham & Campanella, 2010), '3D resilience' (Bené et al., 2012) or 'Resilience 2.0 (Hodgson, 2009)'. Overall, this description already suggests that resilience is very context-dependent – in other words, a phenomenon that depends on very different concrete variables implying both, the resilient system itself and its environment.

What is certain is that the research of US American developmental psychologist Emmy E. Werner laid the foundation for the currently widespread understanding of resilience. In her long-term study, she observed the development of 698 children born in 1955 on the island of Kauai (Hawaii) until they were 40 years old. She was particularly interested in the 201 children from socially disadvantaged families. Werner found out that, despite a difficult childhood, one third of these children succeeded in leading an orderly life in adulthood (Werner, 1977). Since then, psychologists have been using the concept of resilience to research the factors that must coincide in order for people to develop inner strength and to not break down in crises or traumatic situations. In addition to the findings of Emmy Werner, who observed several 'inner strength' factors of resilient people, in the late 1970s Israeli

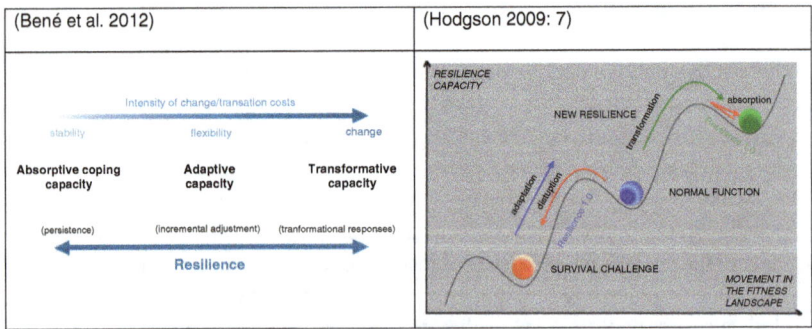

Fig. 2.1 Resilience in the spectrum between persistence and change – two examples. (Bené et al., 2012; Hodgson, 2009, p. 7)

2.1 Different Definitions and Uses of the Term Resilience

medicine sociologist Aaron developed the concept of *salutogenesis*. He observed three psychological factors or attitudes that are most important to cope with any stressful event. The term 'salutogenesis', directly translated from Latin 'emergence of health', initiated a change of perspective, away from the usual focus on the analysis and elimination of diseases (pathogenesis) towards the development of overall health. Among the three most important attitudes reducing stress, Antonovski mentions the feeling that one's life can be understood, the feeling that it can be managed, and thirdly, that it has meaning. Taken together, these three feelings create what is known as the 'sense of coherence' or 'SOC' (Antonovski, 1997). In accordance with Viktor Frankl's Logotherapy (2004), the key to salutogenesis and, closely related to this, to the resilience of individuals therefore lies in the development of one's own ways of thinking, perceiving and behaving as an answer to the questions of life. In total, the context of psychological resilience is relatively well researched. There is a broad consensus here that personal resilience is characterised by about 7–10 so-called 'resilience factors', such as optimism, composure, the ability to see meaning in crises, leaving the role of the victim, taking responsibility for one's own situation, network orientation, future orientation, solution orientation, improvisation ability, etc. (Gilan et al., 2021; Werner, 1977; Zander, 2011). With the trend of increasing mental stress disorders (Lohmann-Haislah, 2012; pronova BKK, 2018), stress management and burnout prevention are increasingly coming into focus. Thus, today, an increasing number of resilience training and counselling programmes can be observed that aim to develop the psychological resilience of managers and employees and thus counteract the trend of increasing stress-related illnesses in the workplace, including occupational burnout (cf. e.g., Wellensiek, 2011).

Since around the end of the 1990s, the concept of resilience has also been applied to the system level of organisations. Since then, the question of which criteria an organisation must fulfil to be robust and flexible in order to withstand and overcome unpredictable crisis situations has been pursued with growing interest in the USA and Europe. Against this background, the term 'black swan', coined by the mathematician Nicholas Taleb, has received increasing attention. The term describes phenomena that are statistically highly improbable and therefore hardly predictable, but at the same time have a significant impact (Taleb, 2008). As companies are becoming more vulnerable to unexpected external shocks, such as financial crises, market trends or disruptive technological developments, observers note that the lifespan of companies has shortened significantly over the last 100 years (Probst & Raisch, 2004). As a response to the challenge of coping with unpredictable events, various concepts of organisational resilience have emerged. Here, too, a discussion about resilience factors can be observed, which has come to increasingly concrete results in recent years. Mainly about 5–10 factors are assumed to make organisations as social systems more resilient to various unexpected crises. These include: decentralised decision-making, time-efficient decision-making methods, a shared vision, error-friendly corporate culture, complexity-appropriate error analysis, and adaptive and resilience-promoting leadership. The best-known concepts behind these resilience factors include Peter Senge's 'Five Disciplines' (1990), popularised in the early 1990s; the concept of 'High Reliability Organisations' (HROs) popularised by Karl Weick and Kathleen Sutcliffe in the early 2000s (Weick & Sutcliffe, 2001); and the ISO standard 22,316 on corporate resilience published in 2017 (ISO:22316, 2017).

The field of 'resilient societies' or 'resilient cities' (quasi as a sub-form of resilient societies) is relatively unexplored, however, this system level is more complex for at least two reasons. Firstly, the society level encompasses and integrates the other system levels; secondly, it represents a new contextual phenomenon, which accordingly involves new factors, framework conditions and challenges for resilience.

It is noteworthy that the above-mentioned 7–10 factors of psychological or organisational resilience are more or less applicable to a broad range of multiple, maybe also simultaneously, emerging problems. Is there a multidimensional, universally applicable set of resilience factors for the society level in response to a broad range of crisis bundles and bundled crises? This will be discussed in the further sub-chapters and chapters.

2.2 Hard and Soft Dimensions of the Resilient Society

In the discourse on societal resilience, the term is mostly considered as a counter-term to 'vulnerability'. At the 2013 World Economic Forum, which was held under the motto 'Resilient Dynamism', the resilience concept was even referred to as the 'Twenty-first Century Imperative' (Huffington, 2013).

Even before the term 'globalisation' was on everyone's lips, the research discipline of international relations kept its eyes on 'risk factors' resulting from increasing global interdependence. Since the 1970s, interdependency theory has emphasised that increasing globalisation of the economy, technology and transport systems, as well as communication would also globalise undesirable developments. Thus, problems in seemingly distant regions would have regional and global spill-over effects (Nuscheler et al., 2007). Today, the focus has slightly shifted from the Third World (Bohle et al., 1994) to the question of how all countries in the world, particularly the modern, technology-dependent societies of the First World, can prepare for unpredictable events in an increasingly complex, interconnected world (Edwards, 2010). At the same time it can be observed that more and more disciplines are involved in analysing societal resilience, increasingly addressing not only hard factors like 'economy', 'critical infrastructures', but also intangible soft factors like 'culture', 'spirituality', 'solidarity' or 'collective learning'. What dimensions and factors are most important for a resilient society in response to contemporary challenges?

2.2.1 Hard Factors: The Prevailing Discourse on the Resilient Society

In response to this question, contributions from social science spatial research, especially planning sciences and geography, dominate. The thematic focus is primarily on the development of orientation knowledge and application-related questions for policy and planning in dealing with environmental changes (e.g. climate change) and natural disasters (Birkmann, 2008). In addition, man-made security risks also play a role – such as terrorism (Coaffee & Wood, 2006). Furthermore, social science vulnerability and resilience studies examine the concept as an outcome of social action. The focus here is on how vulnerable and resilient individuals are in the face of socio-psychological risk factors ('individuals at risk' [Luthar & Cicchetti, 2000]), such as lack of schooling, child poverty, etc. (Zander, 2011). From a political science and social science perspective, the resilience of authoritarian regimes in the face of civil revolutions has also increasingly become the focus of interest in recent years (Goldstone, 2011). Overall, it is clear that very different disciplines have appropriated the terms 'resilience' and 'vulnerability', however focusing on different aspects. What they all have in common is that they start from 'hard', i.e. objective, threats and corresponding measures.

2.2.2 Soft Factors: The Neglected Component in the Resilience Discourse

To date, the current discourse on the resilient society has largely emphasised 'hard' factors. At this point, Christmann et al. (2011) rightly point out that the entire body of research in the debate on societal resilience is based on an 'essentialist' view that conceives resilience as something factual. According to this view, the vulnerability of a system would not only result from objectively ascertainable factors, but it would also be a construction, i.e. a 'shared assumption of being threatened' (Christmann et al., 2011, p. 5). These 'soft', in other words subjective and intersubjective factors that constitute the collective 'resilience spirit' of a population are only sporadically considered in current resilience studies:

- Contributions from discourses concerned with individual stress resilience relate the resilience and vulnerability of a society to the degree of mental health and stress of a population. The focus of the debate here includes questions such as: How are individual and societal resilience concretely interrelated? For example, could a 'critical mass' of resilient individuals lead to a more resilient society as a whole? In how far do spiritual or meditative practices impact the collective stress resistance and coping capacity of the population as a whole? Against this background, it is important to consider mindfulness and corresponding spiritual practice[1] not only as a key to develop individual resilience, but also as an important factor to foster a resilience-promoting culture and overall collective resilience. In this regard, it can be argued that mindfulness provides meaning and releases attachment to any stressful feelings, thoughts and world views in the face of multiple adversities (Singer, 2016; Kabat-Zinn, 1982), contributing to a more equanimous, healthier and thus resilient society (Dalai Lama, 2013; Wellensiek, 2011). Spiritual stress-releasing practices may include e.g. Mindfulness Based Stress Reduction, a meditation technique which is well-researched (Srivastava, 2020; Kabat-Zinn, 1982), or yoga breathing techniques (Pranayama) (Bönsel, 2012). It is worth noting that spiritual practice and meditation are not panaceas. They cannot replace therapies for dealing with mental illnesses. However, in the context of resilience, meditation and other spiritual

---

[1] The term 'spirituality' (from Latin *spiritus* = spirit, breath, or *spiro* = I breathe) can have different meanings and is therefore difficult to define concisely. Regardless of its specific manifestations in the different religions, it can be stated that spirituality always refers to the principle of a transcendent, non-personal ultimate truth, which cannot be grasped by the senses, but can nevertheless be experienced or imagined. These insights, which take the form of an 'awakening' or 'enlightenment', can be accessed through the practices of various wisdom traditions. These include for example Sufi or Christian Gnostic heart prayers, shamanic trance dances, holotropic breathing practices, Hindu yoga, Kabbalistic number mysticism, Jain asceticism, QiGong/Tai Chi, Buddhist meditation or contemplation. Spirituality is therefore not characterised by following certain unquestionable dogmas of a religion, but rather by overcoming the dual ego-consciousness ('I' versus 'environment') through inner work, as well as the profound insight that the vulnerable ego-self is not the 'real self' (Singer, 2016; Maharshi, 2018).

2.2 Hard and Soft Dimensions of the Resilient Society

practices can contribute to a serene, wise view of one's own self and the world. It should be emphasised that spiritual or meditative practices to promote resilience are not necessarily related to a denomination or dogma, but may express themselves differently in each resilient culture.

- Another strand of discourse is concerned with the contributions of culture and cultural techniques to foster collective resilience. Here, resilience scientists have addressed the question of why some ethnic groups have achieved comparatively higher social status and success despite discrimination from the majority society. Judith R. Kramer and Seymour Leventman, for example, reported that the descendants of Eastern European Jews who had immigrated to the United States not only showed relatively good integration of their own children, despite great poverty, but also committed crimes less frequently than the American majority and were more likely than average to attend university. Nathan Caplan attributed this to cultural factors and strong parental involvement. He came to similar conclusions regarding the children of the so-called 'boat people.' The 'boat people' became known in the 1970s and 1980s as Vietnamese refugees who fled by boat across the South China Sea after the Vietnam War out of fear of the new communist regime. Despite their low education and high poverty, and to the astonishment of scientists, the refugee children performed better on all achievement tests than middle-class children of the American majority society. One of the most striking findings of the study was that children with many siblings proved to be the best performers. The explanation behind this was, and still is, that Vietnamese culture tends to be more collectivist-oriented and that older siblings were expected to help their younger siblings with homework. The children benefited from that enormously in a way that they did not only learn facts from their siblings, but also copied learning strategies and adopted academic values (Caplan et al., 1989, 1992). Questions that remain unanswered in this context include: What role do identity and language and, if applicable, religion play in the resilience of an overall population? What are the characteristics of effective integration policies that increase the resilience of an overall multi-cultural society?
- This gives rise to a largely unexplored discourse field with regard to the question of cultural virtues that promote resilience and which are supported by the society as a whole. Some observers refer to the example of Japan, where the population is characterised by a comparatively high level of discipline and composure in the face of natural disasters, such as the Fukushima incident. 'Ganbaru'

(also: 'Gambaru'), which means 'giving the best until the end', or 'Gaman' (also: 'Gamban') in the sense of 'maintaining dignity and patience even under the most difficult circumstances', are frequently mentioned virtues of Japanese perseverance mentality (Palin, 2011; The Australian, 2011). Another example of resilience-promoting virtues is the common Thai saying 'mai pen rai', which means 'it doesn't matter' (in the sense of 'if something goes wrong, it doesn't matter') to many tourists. A resilient collective mental attitude in times of crisis could be also found in England at the time of the Second World War with the popular motto: 'Keep calm and carry on'. At this point, the strong contextual dependency of resilience becomes apparent: 'Mai pen rai' and 'Ganbaru' are specific expressions of different Asian cultures, which presumably cannot be easily transferred to other cultures, while the case of England is associated with the historical background context of the Second World War. It would be thus interesting for further research to investigate whether and to what extent 'universal' resilience principles can be drawn from a cross-cultural comparison. Do collectivist cultures (such as Japan and Thailand) and individualist cultures (such as continental Europe and the USA) exhibit resilience profiles with characteristic features? Are there systemic commonalities beyond these culture-specific differences that could be valid for all societies?

- A relatively new school of thought in resilience studies emphasises the aspect of culturally constructed, collective perceptions of vulnerability and resilience. According to this view, there are considerable differences in sensitivity to external influences from society to society. Since individuals, but also collectives, can develop differing perceptions of their own vulnerability and resilience, they would 'construct' threats in a certain way. This was one of the main insights of the researchers of the IRS Leibniz Institute for Regional Development and Structural Planning following a three-year study, which was completed in January 2013. Based on these considerations, the project group at IRS developed a relational model based on Bruno Latour's Actor Network Theory (ANT) – a theory that relates the social to the interpersonal, but also to objects (e.g. artefacts, tools, machines and objects developed in the course of cultural history) (Christmann et al., 2011). In terms of vulnerability to hazards, ANT is based on the assumption that there would be 'a relational, socio-technical web including all natural, human, material and immaterial entities which are related to the concrete hazardous situation and the central element (be it a society or a person) would "wave" this web and position itself within it.' This web, according to Oliver Ibert, is not natural, all these connections would be highly artificial and thus be socially constructed 'in a fundamental way' (IRS Leibniz Institute, 2013).

2.2 Hard and Soft Dimensions of the Resilient Society

Beyond these roughly outlined discourses, further topics emerge that could encompass other aspects of the collective resilience spirit, but are not (yet) explicitly considered in resilience studies.

- A widespread (usually populist) debate focuses on 'identity' as the essential core of a society. This usually takes place against the backdrop of the perceived threat of 'alienation' of a society through the cultures of minority groups, in terms of, for example, an 'islamisation' or 'arabisation' of Europe. With advancing globalisation, the societal handling of multiculturalism, social, economic and political integration is becoming an increasingly important political challenge. Domestically, the ethnic social divide is proving to be a particularly explosive one, especially in the context of welfare and labour market policy. The degree of escalation appears to be higher and more emotionally charged than other lines of social conflict, such as 'generation' (old vs. young) or 'gender' (e.g. man vs. woman). In this context, the question arises: What does a resilient identity and culture mean? Is a resilient society characterised more by the fact that a dominant canon of values, perhaps also a dominant 'guiding culture' (in German: 'Leitkultur') of the majority society is cultivated and preserved, which is 'strong' enough to be protected against 'alienation' (if such a phenomenon even exists)? Is identity something that can be lost in an 'either-or'-game and should thus be preserved in a 'we-them'-differentiation?[2] Or, in contrast, is a resilient culture characterised by adaptation and evolutionary change? Is it

---

[2] This question is based on a concept of identity that is defined by demarcation from other identities and that is based on a perception of 'either-or' or 'self-culture vs. foreign culture'. The political consequence is usually a move towards assimilation (e.g. France) or the melting pot (e.g. USA), in which all immigrant cultures are oriented towards an overarching guiding culture. This approach is the exact opposite of multiculturalism, which is rather characterised by a 'coexistence of different cultures' (example: Canada). Externally, the society that identifies itself through demarcation can vary – this would correspond to the 'clash of civilizations' popularised by Samuel Huntington or, in a mitigated form, to 'interculturality' (= constructive understanding between different cultures). What all the approaches presented here have in common is that cultures are always perceived as self-contained, homogeneous and mutually demarcated 'spherical systems" (on this Welsch, 2009).

possible to preserve and maintain multiple fully developed cultural identities in terms of 'both/and'?[3] We will return to this in a later section.
- Another, still largely open discourse area is generally concerned with the topic of the collective coping with traumatic experiences – not only of natural disasters, but also of wars and terrorist attacks. Previous research on 'disaster cultures' has focused almost exclusively on communities in the Third World which are often characterised by being remarkably self-organising and adaptive when confronted with natural disasters.[4] By contrast, the resilience discourse has yet to address the question of how different populations cope with trauma. As one of the few disciplines that at least implicitly addresses this issue, conflict research is providing relevant pioneering work – including contributions of Vamik Volkan (2006), John Paul Lederach (2005) and Johan Galtung (1998). Going through all of these works, it is possible to identify at least three overall forms of collective psychological crisis management that are practiced to varying degrees around the world: First, there is the construction of a collective identity with the help of a delimiting enemy image and a clear assignment of 'victim' and 'perpetrator' roles.[5] Secondly, there is a method of coping with collective

---

[3] This question is therefore not based on the traditional concept of identity, which is characterised by a delimiting and separating 'either/or" view. Rather, the perception of identity is inclusive – the focus is hereby not on what separates, but on commonalities across cultural areas. The comparatively little established approach of 'transculturality' is shaped by this view. This term was introduced by the Cuban anthropologist Fernando Ortiz in the early 1940s, but was not coined in the German-speaking world until the 1990s by Wolfgang Welsch. According to his definition, 'transculturality" means that the encounter of different cultures can subsequently lead to a blurring of boundaries or even to a dissolution of these boundaries. In the course of globalisation, Welsch does not see the emergence of a uniform world culture, but rather individuals and societies that carry transcultural elements within them. Only the combination of different elements of different origins makes each individual transcultural (Welsch, 2000).

[4] Here, the historian Greg Bankoff is considered an important pioneer researcher (e.g., Bankoff & Hilhorst, 2009).

[5] This approach transfigures the collectively experienced traumatic event into an identity-forming narrative. Hereby, both historical traumas and glories can be used for fostering cohesion and identity-building through cultural rituals and symbols (such as folk festivals, folk songs, memorials, etc.) in which, for example, 'fallen heroes', 'helpers' or 'innocent victims' from one's own people are commemorated. The flipside is that in the context of social conflicts, war and terrorism, these strategies can foster a 'we' feeling at the expense of 'the others'. Another disadvantage is that it prevents the possibility of letting go of the past and learning from one's own contributions to the conflict. Thus, this coping method does not prove to be particularly resilient in the sense of crisis transformation. At the same time, it is not particularly sustainable, as it promotes violence through the devaluation of the other – Galtung aptly describes this aspect as 'cultural violence'.

trauma by removing the traumatic event from the collective memory.[6] A third form would be inclusive, i.e. a non-separating but integrating reappraisal of the event. Here it is not a matter of repression or transfer of guilt, but of reconciliation with 'the other' and with the crisis event. This usually involves a psychological (often also spiritual) letting go of the suffering experienced and thus an overcoming of the collective ego.[7]

---

[6] This form of the collective trauma management implies the suppression of any traumatic memory from the historical narration. The traumatic incident is simply no longer spoken about and is, as it were, erased from the collective memory. However, the extent to which this has indirect effects on a collective subconscious, which may also be associated with epigenetic inheritance, has not yet been researched. As an example: How does the relative lack of confrontation with the responsibility of the US founding fathers in the mass death of Native Americans affect the collective psyche of US society today? This trauma coping strategy may have the advantage of being relatively easy to implement. The disadvantage, however, is that it prevents learning and reappraisal processes and thus the chance of 'emerge stronger from the crisis event'.

[7] This form of cultural trauma management involves an inclusive form of identity building; in other words, a crisis management that integrates 'the other" in the course of a reconciliation process. Examples of this are rituals in which victims on both sides (i.e. including the war enemy) are mourned on war memorial days. Peace researcher Johan Galtung considers this kind of trauma management as typical for a 'peace culture'. In contrast, overcoming trauma by transferring all responsibility and guilt to 'the other' and one-sidedly mythicising one's own side would be typical of 'cultural violence', because it only paves the way for further conflicts and thus suffering (Galtung, 1998). That the culture of peace outlined here, which integrates 'perpetrators' and 'victims' into a common identity context, is not purely utopian can thus be observed in the diverse forms of reconciliation work practised in the context of different cultures and religions. Hereby, Galtung's work offers an exemplary overview of different cultural and spiritual traditions of reconciliation work, from the Buddhist karma method to the Christian confession method to the Hawaiian method of Ho'o pono pono. Further evidence of a form of identity building that overcomes the collective ego and also mourns for the opponent can be found in corresponding places in the writings of different wisdom traditions (e.g. Chap. 31 of the Tao Te King). The advantage of the method of collective trauma management presented here is that it enables a thorough coping with and letting go of the traumatic event, while at the same time learning from one's own coresponsibility. The disadvantage might be the relatively difficult and painful implementation. However, the fact that inner reconciliation can make a meaningful contribution to psychological resilience has been proven by the impressive work of the Austrian psychiatrist, neurologist and Holocaust survivor Viktor Frankl. The pioneering method of logotherapy he developed was based, among other things, on his personal experiences with the Holocaust (Frankl, 2004).

It remains to be said that many aspects remain open for further research in the theoretical discourse on resilience, particularly with regard to the 'soft factors' associated with the collective resilience spirit of a population.

2.3 Can the Resilience of a Society Be Measured?

Is Japan more resilient than Switzerland? How can the resilience of a society be measured? Are there qualitative and quantitative indicators that make it possible to compare the resilience of different countries? The reader can imagine that it is not so easy to find answers to these and similar questions. One challenge is that resilience is a context-dependent phenomenon: A society's resilience is always demonstrated in the context of a concrete challenge. This makes it difficult to compare Japan and Switzerland. While Japan has always been exposed to various natural disasters due to its geographical location (and thus has a high World Risk Index) and has developed a high degree of resilience in this context, the same cannot be said of Switzerland, which, however, under the impact of the Second World War and the Cold War, has created a globally unique system of protective bunkers and mountain fortresses (so-called 'réduits'). According to observers, these could be quite effective against nuclear attacks or a military invasion, but not against reactor accidents such as those in Fukushima (Putzier, 2011). Other typical problems associated with indicator-based analyses concern weighting (which is more important: earthquake-proof architectures in Japan or nuclear shelters as in Switzerland?) and the comparability of hard and soft resilience factors (e.g. currency reserves in comparison and the ability to cope with collective trauma).

Despite these strongly limiting considerations, there are several attempts to define the resilience of cities and societies by means of indicators. In the following, I will outline five examples:

- In a study carried out in 2011 by the organisation 'Triple Pundit', study leader Boyd Cohen mainly compared municipal measures to strengthen the regional and local economy and to reduce fossil fuel consumption. As a result, he saw four continental and northern European cities represented in the top 10, namely Copenhagen in the first place, Barcelona in the third place, followed by Stockholm (4) and Paris (6). In the Anglophone sphere, he sees Vancouver in 5th place, the US cities of San Francisco and New York in 7th and 8th place, followed by London in 9th place. Tokyo is ranked in 10th place and the Brazilian city of Curitiba is ranked 2nd (Cohen, 2011). In his ranking, it quickly becomes clear that 'resilience' is confined to a very limited context – mainly to the cli-

2.3 Can the Resilience of a Society Be Measured?

mate issue and natural disasters. Socio-economic or other indicators are rather neglected here.
- Another study from 2010 by the Hanover-based Pestel Institute looked into Germany's inherent regional resilience to crises. The analysis included 18 indicators from the fields of 'social affairs', 'housing', 'energy', 'land use', 'transport' and 'economy' to assess the regions' and cities' capacity to act through flexibility, resource endowment and social capital. The authors of the study concluded that eastern and southern Germany were particularly resistant to crises. They also concluded that each region had its respective strengths and weaknesses. Compared to Triple Pundit, this study comes up with a broader variety of resilience dimensions. However, the open question of how to weight the indicators remains critical, as apparently exactly three indicators are assigned to each of the six dimensions, probably for reasons of symmetry. For example, is the tenant rate (which is considered a positive indicator of flexibility in the face of structural changes) just as relevant for the dimension of 'housing' as the unemployment rate (as a negative indicator of social crisis potential) is for 'social' affairs (Pestel Institute, 2010)?
- A third approach was also developed in 2010 by the ÖAR-Regionalberatung GmbH on behalf of the Federal Chancellery of Austria, which refers to the regional crisis resistance of individual locations in that country. Among other things, this approach distinguishes 26 indicators within the fields of environment (4 indicators), economy (7 indicators) and society (15 indicators). The analysis focuses on the individual assessment of three federal states in Austria for the management of regional resilience. In contrast to the other two studies, the analysis model presented here is less focused on a comparative study of locations, but rather on long-term monitoring and resilience management of the locations. Compared to the other two analysis models, it is therefore striking that the ÖAR study does not amount to a comparative ranking. It is also noticeable that a completely different weighting is applied in the selection of indicators: the focus is primarily on social and socio-economic factors, while urban development, infrastructural and energy-economic factors are not considered in this indicator pool (these are incorporated into the overall analysis in the study in a different way, by means of qualitative-narrative methods) (Lukesch et al., 2010).
- A more recent study was published in 2014 on 'Global Cities – A Grosvenor Research Report', conducted by the 'Grosvenor' organisation. Over 50 cities worldwide were compared; Toronto, Vancouver and Calgary were named as the three most resilient. Compared to the other above-mentioned studies, this one assessed societal resilience based on qualitative criteria such as (1) governance,

(2) institutions, (3) technology and learning, (4) planning systems, and (5) access to funding (Grosvenor, 2014). These criteria are based on a comparatively broader understanding of societal resilience that allows for application to more crisis contexts than environmental disasters.

- One of the most recent and most acknowledged approaches is that of 'Community Resilience', based on a meta-study by Patel et al. (2017). It refers to the community level, and, similar to the above-mentioned Grosvenor Research Report, includes predominantly qualitative criteria, such as local knowledge, community networks, communication, health, governance/leadership, resources, economic investments or mental attitudes (Patel et al., 2017). Similar to the Grosvenor Report, the Community Resilience approach allows for a broader application to more problem contexts than environmental disasters.

All five studies provide exemplary guidelines for the indicator-based assessment of resilient societies. However, none of them could fulfil the claim of a globally recognised analytical framework that can be applied to the whole variety of contemporary crisis bundles and bundled crises. Against the background of the above-mentioned limitations of the studies, but also of their different contributions, it might be worthwhile to look for commonalities and mutual complementarities. The resulting synthesis could contribute to a more broadly applicable concept of resilience. The following (indicator-based) factors could contribute to more multi-responsive resilience:

- *Social:* Both the studies of the Pestel Institute and the ÖAR-Regionalberatung GmbH use indicators such as the unemployment rate, public health indicators (Pestel Institute: supply with family doctors per 100,000 inhabitants; ÖAR-Regionalberatung: life expectancy in good health) and the migration balance of the population, which provide information about the attractiveness of the location. The study of the ÖAR-Regionalberatung also cites criteria such as life satisfaction (housing, work, freedom), income distribution and voluntary work, as well as the crime rate and trust (here: trust in institutions). By contrast, the more qualitative approach of Community Resilience mentions 'community networks' and 'health' as relevant criteria which could be associated to this dimension. Moreover, the study of the Pestel Institute includes the factor of education by taking into account the proportion of school leavers without qualifications. The analysis of young individuals who are exposed to an above-average number of social risk factors ('individuals at risk') is central to the current debate on resilience (Luthar & Cicchetti, 2000). These include not only lack of schooling,

2.3 Can the Resilience of a Society Be Measured?

but also child poverty, ethnic exclusion, shattered families and mental illness. By contrast, integration, education, health and social policy measures aimed at promoting individual resilience could be regarded as protective factors. This discussion primarily focuses on promoting the resilience of children and young people in the relevant educational institutions (Zander, 2011). Another discourse context relates the trend of increasing mental illnesses as a social challenge. Observers like Galuska et al. (2010) relate this to changes in social conditions, such as global competitive pressure and, as a result, job insecurity and performance pressure as well as a decrease of sustainable social relationships (in detail Galuska et al., 2010). In this regard, a preventive healthcare system would significantly contribute to societal resilience.[8]

- *Economy:* The economic dimension is considered in all above-mentioned approaches, but with different foci. The Grosvenor study emphasises access to finance as a key economic resilience factor (Grosvenor, 2014). The studies of the Pestel Institute and the ÖAR-Regionalberatung both take into account the debt ratio (Pestel Institute, 2010; Lukesch et al., 2010). In addition, the ÖAR study mentions indicators related to the innovative capacity of the population, such as the mix of businesses and the share of the creative class in the labour force (Lukesch et al., 2010). It is striking that economic resilience is defined on the basis of criteria that can also be primarily attributed to the concept of the 'developed society'. These include indicators relevant to competition, such as the value added and investment ratio, household income and purchasing power per capita (Lukesch et al., 2010). By contrast, the Pestel Institute's analysis places greater emphasis on the aspect of decentralisation and self-sufficiency. These include, for example, the proportion of local employees who do not commute beyond the regional border and the industrial employment rate (both should be as low as possible). The latter means that while high industrial turnover has a positive impact on growth, heavily industrialised regions are much more affected by times of crisis and fluctuations in the world market (Pestel Institute, 2010). A recent study by Rüdiger Wink, an expert on economic resilience, also mentions the diversity of industries as a key structural characteristic of economic resilience to crises (Wink et al., 2016). A more radical approach, which is not considered in the studies mentioned above, however, also relates economic resilience to a certain independence from the fragile financial market

---

[8] A controversial vision that is certainly worth discussing is outlined by futurologist Erik Händeler, who sees a healthcare system based on the pillars of 'prevention', 'self-responsibility' and 'innovation' as the path to a more resilient and sustainable society (taken directly from Wellensiek, 2011, pp. 42–51).

system. In this context, Wink, states that forms of financing, e.g. for investments in innovation projects, that rely on the global financial market, which is already susceptible to crises, would contribute to a correspondingly high level of vulnerability (Wink, 2014). Others even discuss economic resilience as a 'decoupling' of regions from the developments of the crisis-prone global economy. However, decoupling theory remains controversial.[9] Overall, economic resilience includes a broad variety of aspects, such as economic and financial strength, low debt and relative robustness to the fluctuations of the global financial system, maybe some sort of 'independence', high innovative capacity, competitiveness, and economic diversity.

- *Ecology:* Preparedness for environmental hazards is typically a central topic of all above-mentioned studies. In the context of the sustainability discussion, particular importance is generally attached to independence from energy and resource imports with regard to the ecological resilience of societies. The studies by Triple Pundit (Cohen, 2011) and the ÖAR-Regionalberatung (Lukesch et al., 2010) consider greenhouse gas emissions in addition to self-sufficiency via renewable energies. The latter also cites criteria such as risk exposure to natural disasters. Analogously, the Pestel Institute considers the (forest) land use per inhabitant a resilience factor. The larger the forest area, the lower the exposure to extreme weather and the more independent the region is of energy raw materials and building materials, thus the more self-sufficient it is in setting up house and infrastructure buildings. The study additionally mentions agricultural land per capita and the share of organic farming as further indicators (Pestel-Institute, 2010).

- *Infrastructure (material/virtual):* The protection of critical infrastructures such as water, internet, energy supply and mobility, is of particular importance in the resilience discourse. How can the society cope with a prolonged and widespread power blackout? Could the supply of the population with (vital) goods be ensured? These and other questions are the focus of a resilient infrastructure. The focus of the ÖAR Regionalberatung and the Pestel Institute is, among other things, to investigate the extent of the population's 'mobility constraint'. Here, the Pestel Institute study considers a low share of interregional mobility (e.g. commuting to work) and a low share of private transport (which in turn is as-

---

[9] The central question here is whether the emerging economies, especially the BRIC countries (Russia, Brazil, China and India), will be able to partially decouple themselves from the development of the global economy in the future. This is countered by the antithesis that the economies are too intertwined for individual regions to be able to become independent of the global economy (Peña & Giné, 2009).

sociated with a high dependency on oil imports) as resilience-promoting for the region (Pestel Institute, 2010). These criteria are exemplary for a decentralised and diverse resilient urban development. The guiding principle of a decentralised urban design would be maintaining the functionality of the overall system, even if individual parts temporarily fail (Godschalk, 2002). It is noticeable that the studies do not explicitly consider the danger of cyber attacks and the importance of protecting virtual infrastructures. Typical approaches include digital control systems that are resilient to disruptions and hacker attacks, or backup systems that are intended to provide multiple protection against information loss (c.f. e.g. Petermann et al., 2010).

- *Preparedness:* An increasingly relevant topic in the resilience discourse, but not directly considered in any of the above-mentioned concepts, considers disaster preparedness of households and communities to ensure survival and self-sufficiency. In this regard, a higher proportion of 'prepper' households would contribute to a more reşilient society, at least in relation to external threats such as natural hazards or a war. Corresponding indicators could be the number and accessibility of bunkers (as mentioned, a society like Switzerland would have a globally leading position) and the average length of time that a household could survive without external supplies. Here, the German Federal Office of Civil Protection and Disaster Assistance (Bundesamt für Bevölkerungsschutz und Katastrophenhilfe [BBK]) recommends German households to stockpile supplies that will allow them to be self-sufficient for at least 10 days. A checklist can be found on the BBK website.
- *Collective mindset:* Another relevant resilience dimension, which is currently not very well developed, is concerned with the collective resilience mindset of a society. As it appears to be hard to measure, this dimension is typically not considered in the quantitatively researched approaches of the Pestel Institute (2010), the ÖAR-Regionalberatung (Lukesch et al., 2010) and Triple Pundit (Cohen, 2011). The criteria of the other two approaches, Grosvenor Report and Community Resilience, which emphasise a more qualitative perspective of collective resilience, seem to have at least an indirect link to the 'collective mindset'. These include 'governance', 'leadership', 'local knowledge' (Patel et al., 2017), a culture of learning (Grosvenor, 2014), and mental attitudes (Patel et al., 2017). What are the basic components of resilience culture and, in the broadest sense, of 'collective mindset'? A first criterion relates to mental health of the population, also in conjunction with spiritual practices which could serve as a preventive factor in addressing the trend of increasing mental illness (including depression and occupational burnout) and collectively traumatic events (like terrorist attacks). Secondly, a collective resilience culture could also imply the

aspect of a population's collective 'endurance' in the face of crises. Since not all societies have the same crisis experiences in terms of the nature and extent of the problems, it is difficult to make comparisons and draw appropriate conclusions. Is the collective crisis mentality of the Japanese population more resilient than that of Switzerland? It is noteworthy that crisis mentalities can significantly vary from culture to culture. The Japanese motto 'ganbaru' emphasises a different facet of resilience than the Thai 'mai pen rai', and is also quite different than the British 'Keep calm and carry on'. A third criterion relates to the social construction of resilience, i.e. how society perceives its own vulnerability in the face of concrete challenges. Does a society perceive the situation as a threat, as a challenge or even as a chance? How does it collectively cope with these challenges and how are they narrated in the cultural media? This aspect is pointed out by the IRS Leibniz Institute (Christmann et al., 2011). A fourth criterion, provided by the National Accounts of Well-Being, relates to 'social cohesion' and 'resilience and self-esteem' (Michaelson et al., 2009). However, as will be outlined elsewhere, these contributions to happiness research remain controversial, as concepts and perceptions of happiness outside the Western world differ significantly and are therefore not comparable. The resilience spirit of the Japanese population, for instance, which is based on a high degree of self-discipline influenced by a collectivist value structure and in large parts an eclectic Buddhist-Shintoist religious culture, differs considerably from the U.S.-American resilience spirit, which is characterised, among other things, by individual (in some cases also small-community) precautions, distrust of the state and a predominantly individualistic, Western value structure, which emphasises independence and liberty.

For various reasons, it is not possible to adequately capture these different resilience dimensions, especially the soft ones, with indicators. Despite the many limitations, however, an approximative assessment scheme that aims to capture societal resilience as comprehensively as possible should at least consider the following dimensions and according criteria (Table 2.1):

The model outlined here summarises the essential more or less quantifiable criteria that are relevant for a society to respond to a wide range of different crises. However, this approach proves to be very limited in at least three respects:

Firstly, as mentioned above, not all criteria can be translated into suitable indicators. The question of which society is the most resilient in the world according to the criteria formulated here cannot therefore be answered. However, according to the criteria, one could assume that Switzerland and the Scandinavian societies would probably have the highest scores or at least resources. Given the consider-

Table 2.1 Typical indicators to capture resilient societies

Resilience dimensions	Selected criteria
Psychic	Psychological resilience of the population (average resilience quotient) Meditation practice
Social	Public health Number of individuals at risk (including young early school leavers, child poverty, etc.) Social cohesion (including volunteering) Trust (interpersonal and institutional)
Economy	Diversity of the economy Savings rate debt Gold reserves Other financial resources
Ecology	Quality (soil, air, water) Renewable energies Renewable water resources Available forests Organic farming Land use/available forest land per capita Self-sufficiency through agriculture/organic farming
Infrastructure	Number of days before central utility power can be restored after a power/system outage Vulnerability of critical infrastructure and digital/automated control systems to disruption Cybersecurity (protection against hacker attacks and other disruptions from the internet) Expansion of local public and passenger transport (dependence on individual transport) Dependence on supra-regional mobility
Preparedness	Bunkers: Available number and accessibility of bunkers per household Precaution: Average number of days a household can survive without outside supply

able differences between societies, however, e.g. in terms of their exposure to certain crises, their experiences or their collective perception of threat, it is not possible to compare their overall resilience.

Secondly, while the above-mentioned criteria can be applied to a wide range of problems, it is unlikely that they can provide solutions to all problems that may arise in the future, such as a cyber war between the EU and Russia, climate wars or the next pandemic or the next refugee crisis. Similar limitations can also be found in the widely neglected area of man-made existential risks, such as the scenario of a 'grey goo' or a misguided super-intelligence, which are not so likely at present, but will pose by far the greatest threat in the future. These crisis scenarios are unlikely to be adequately answered by resilience alone (Bostrom, 2017), but require appropriate impetus from other societal concepts, such as the developed society and the sustainable society. This will be discussed in later chapters.

Thirdly, the model also proves to be relatively limited in the face of crisis bundles and bundled crises. Most of the above-mentioned dimensions, such as economy and preparedness, provide answers to solve problems, but in reality, societies are confronted with multiple, interrelated challenges and multi-dimensional crises occurring at the same time. Moreover, the diversity of contemporary interrelated problems is characterised above all by unpredictable change. Resilient societies are therefore required to develop their general collective capacity to deal with multiple unpredictable and highly complex situations. Complementary to these dimensions and indicators applying for 'mono-resilience' responses, further reflection concerning so-called 'multi-resilience' are therefore necessary. These considerations will be discussed in later chapters. Before that, an overview of the practical discourse on resilience will be given.

2.4 Two Practical Approaches for a Resilient Society

The practical discourse on resilience is dominated by two schools of thought that are characterised by a different understanding of resilience:

- *'Crisis management'*, based on the security discourse and a more static and stability-oriented resilience conception emphasising the value of 'permanence';
- *'Crisis transformation'*, based on an innovation-oriented discourse, including a more dynamic concept of resilience emphasising 'constant evolutionary change'.

2.4.1 Crisis Management

The approach to crisis and disaster management is characterised by a resilience concept that emphasises *'maintaining the ability to function'*. The concept dates back to the security discourse of the 1970s and the overarching discipline of 'disaster risk management'. To date, it continues to dominantly shape the current discourse on resilient societies. The relatively high public attention of this concept is largely a result of an increasing awareness of the vulnerability of highly technology-dependent modern societies and the resulting need to protect critical infrastructures.

Critical infrastructures include information technology and telecommunications, healthcare, energy supply, water supply, traffic and transport, and health care. They provide the population with (vital) goods and services through an interwoven, technical network, which is, however, highly vulnerable due to its internal complexity and interdependence. An impairment or even breakdown of critical infrastructures would have far-reaching consequences for the entire societal system. Some studies analyse the consequences of long-lasting and large-scale power outages. For modern societies, e.g. for the Federal Republic of Germany, studies 'have shown that, after only a few days in the affected area, it is no longer possible to ensure the nationwide and demand-oriented supply of the population with (vital) goods and services. [...] If this were to happen [...], the consequences would lead to a national catastrophe. This would not be "controllable" even by mobilizing all internal and external forces and resources, and could at best be mitigated' (Petermann et al., 2010, p. 237).

Critical infrastructure protection is linked to a wide range of other fields, i.e. not only urban and transport planning or political crisis areas such as terrorism, but also cyber security in the wake of increasing digitalisation and automation of control systems (e.g. in traffic control systems, industrial operations, power plants, automobiles, or smart meters in the home). The discussion about critical infrastructure protection therefore increasingly concerns the aspect of cyber security, including the protection of virtual space in general and digital control systems from hacker attacks and susceptibility to disruption in particular. The increasing importance of virtual space is reflected in the fact that it has become a relevant battlefield in every political, military and economic conflict, such as the current hybrid conflicts between the West and Russia and China. Virtual space is also an important object of high-level espionage as well as criminal activities, sometimes with considerable financial damage. At the heart of any cyber security discussion are at least two conflict areas between the state, the economy, and the civil society, based on a conflict of interests between 'security vs. freedom'. The protagonists of the first conflict area are the state and the private economy, which are concerned with the question of how cyberspace and the organisations, services, etc. that depend on it can be secured, without at the same time preventing the positive effects of liberalisation, globalisation, and privatisation through too much regulation. The second conflict area occurring between the state and the citizen, deals with the question of finding the right balance in the digital space between security (including authorisations of surveillance required by the police and intelligence services) and freedom (including civil rights, particularly the fundamental right to anonymity and informational self-determination). As mentioned above, cyber security has become increasingly relevant in recent years. One example of this is Operation Waking Shark, which was carried out in the United Kingdom to simulate cyber attacks on the financial sector in order to prepare against such incidents (Wilson, 2013).

Above all, the disaster management discourse is dominated by the protection of non-virtual technology-operated infrastructures of major cities, for example through earthquake-resistant architectures, robust electricity and telecommunications networks, and heat supply. Typical of the entire resilience concept of crisis management is that strategy development is usually dominated by measures to be decided 'top-down' – whether in national policy or in international policy (e.g. the United Nations International Strategy for Disaster Reduction [UNISDR] or the United Nations Development Programme [UNDP]). For some years now, there has been a growing awareness of the need to make emergency planning more participatory and thus to actively involve businesses and civil society 'bottom-up'. With reference to current research on 'resilience communities', there is increasing attention to the crucial role of civil society in contributing to societal resilience. According to political scientist Claus Leggewie, these communities 'would not be coincidental in places where earthquakes, typhoons and other extreme events are particularly frequent, while the rate of self-organisation is comparatively low in societies that delegate risks to a large extent and entrust precaution to formal organisations and insurance companies (Leggewie, 2017).' At this point, Greg Bankoff, historian and expert in the study of 'disaster cultures', uses the example of the Philippines to show how inhabitants evolved their architecture and established concepts of informal self-help and neighbourhood help (Bankoff & Hilhorst, 2009). According to Leggewie, these examples may 'demonstrate how active risk management can efficiently respond to the evolved vulnerability of modern societies – learning from the "South" of the world (Leggewie, 2017).'

From an international perspective, it is striking that the discourse on the resilience concept of crisis management appears to be dominated by contributions from Anglo-Saxon countries, such as the Foundation for Resilient Societies (USA), the World Institute for Disaster Risk Management (USA), Loughborough University (UK), and from Asia. Since 2005, numerous initiatives have been launched in the US and in the UK. In the UK, for instance, a lot of information on topics such as crisis and disaster preparedness, emergency planning or business continuity has been developed and made available to the public on a dedicated website[10] to en-

---

[10] https://www.gov.uk/government/policies/improving-the-uks-ability-to-absorb-respond-to-and-recover-from-emergencies

courage information sharing and partnership building. In addition, so-called 'Regional Resilience Teams' were established in the UK to ensure more effective coordination of local emergency preparedness (CSS, 2009). In the US, as a consequence of the experience with Hurricane Katrina, the Department of Homeland Security (DHS) has developed an all-hazards approach to enhance resilience as the top priority. Similar to the UK, a website was set up to address disaster preparedness and crisis management. [11] Other measures include training from the Federal Emergency Management Agency (FEMA) on how to create highly resilient communities and establishing a Community Preparedness Division which focuses on the role of individuals in coping with disasters. The aspect of individual preparedness has a long tradition, particularly in the United States, dating back to the critical 1950s and 1960s during the Cold War and has come to the fore again since the terrorist attacks of September 11 and the devastation of Hurricane Katrina.

Examples from the Asian region, such as the Japanese Social Resilience Report 2010, show that not only disaster prevention, but also socio-economic crisis prevention – as a direct response to the global economic crisis – are taken into account. Japan is generally regarded as a particular source of inspiration. (PECC and JANCPEC, 2010). The country is characterised on the one hand by a high level of experience in the development of prevention programmes, broad-based training measures and damage reduction technologies, and on the other hand by cultural techniques that foster a spirit of resilience in the population in the sense of 'capacity to suffer' and a high level of discipline (I have already mentioned the 'Ganbaru' mentality at this point).[12] In contrast, the 'prepper' scene may be typical for Western countries, based on individual values and, especially in the Anglo-Saxon countries, on a great distrust of the state. As the capacity of governments is limited to support the population in the face of a blackout, preparedness is proving useful. In this context, it can be observed that prepper communities have increased in recent years (Duclos, 2012).

2.4.2 Crisis Transformation

This relatively new resilience concept does not refer to 'risk minimisation' and 'disaster management' (Sect. 2.4.1), but rather to evolutionary 'risk adaptation' and 'disaster transformation'. The focus here is on the study of factors of social and

---

[11] http://www.ready.gov/

[12] For a more in-depth description of collective crisis management mechanisms in Japan, see Sect. 6.4 of another publication of mine (Fathi, 2019a).

technological change that enable societies to coexist in an adaptive equilibrium with a rapidly changing environment. Therefore, the question is not on how critical infrastructures in modern societies can be protected or become more robust in order to maintain their functionality in the face of adversities; rather, it is about how societies as such must change in order to evolutionarily adapt to global challenges. New ways of thinking that help to grasp complexity and to promote social innovation are seen as essential.

The concept of crisis transformation implies a dynamic notion of 'resilience' that emphasises the aspect of adaptation and, in the broadest sense, constant evolutionary change. In the sense of the concepts 'Resilience 2.0', 'Transformational resilience' or 'bouncing forward', it would be less a matter of enduring crises or 'protecting' oneself from adversities (as is the case with crisis management), but rather of developing the affected system in a way that it can allow crises to occur, learn from them, and evolve with them. This principle has gained importance in various theoretical and practical disciplines, with reference to the original meaning of the term crisis as 'turning point' or 'decision'. The core thesis arising from this basic idea is, in the words of the science philosopher Sir Karl Popper: 'All life is problem solving'. In other words, every development – from personal knowledge-acquisition to collective evolution – is based on a process of 'learning from problems' (Popper, 2004).

Ideological pioneers of the crisis transformation concept can already be found in the mid-twentieth century, focusing on conflict management of societies.[13] Among the best-known approaches are, for example, the 'open society' approach of Karl Popper (2003) or Ralf Dahrendorf's research on conflict sociology (Dahrendorf, 1961). Popper's concept of an open society aims at releasing 'the critical competencies of humans' without violence. His notion is closely related to the governance form of democracy, which, however, is not understood in its original sense as 'rule by the majority', but as the possibility of voting out the government without violence (Popper, 2003). Above all, however, Dahrendorf's contribution laid the foundation for today's understanding of conflicts and crises in modern conflict research, which in turn is indirectly reflected in the resilience concept of crisis transformation described above. According to this, social conflicts and crises can be understood as unavoidable events, but not per se as threats. Rather, they are a central element of social coexistence, even a driving force of social change. On the premise that conflicts cannot be avoided anyway, a peaceful society is not char-

---

[13] 'Conflict' comes from Latin *confligere* and originally meant 'clash'. In this regard, a conflict can be defined as a clash between two or more parties pursuing different interests and resulting in behavioural tendencies contradicting each other.

2.4 Two Practical Approaches for a Resilient Society

acterised by the absence of conflicts, but by its capacity for conflict. The ability to deal with conflict means to manage conflicts in non-violent ways, to endure contradictions, even to overcome them dialectically, and to develop further as a society. Johan Galtung, one of the founding fathers of peace research, highlights the key competencies of empathy and creativity as particularly important for sustainable and optimal conflict resolution (he calls it: conflict transformation). Here, 'empathy' means the ability to understand the deeper needs of all involved conflict parties through open and non-violent communication. As such, unsatisfied needs are regarded as the deepest source of any conflict. 'Creativity' means the ability to think 'the new' and thus to create new solutions from originally contradicting opposites that optimally satisfy the needs of all involved parties.[14] The key to overcoming any interpersonal conflict and to developing corresponding capacities in social systems lies in the design of successful communication. In this sense, knowledge exchange and cooperation between a wide range of representatives from all social sectors and disciplines can help to identify conflict potential at an early stage and develop innovative solutions. As we will see, for the practice of societal resilience promotion, cross-sectoral communication is a relevant pillar for coping with different adversities and thus for multi-resilience.

Typical organisations that attempt to implement the concept of crisis transformation described here and specifically promote cross-sectoral knowledge generation include the US innovator network PopTech and the academic Stockholm Resilience Centre (2013). The primary aim of crisis transformation is to develop social and cultural innovations towards a decentralised and more independent structure of any kind of supply (e.g. energy, food). In concrete terms, the aim is to develop regions that are as independent and self-sufficient as possible, and to initiate systemic change 'bottom-up' towards a resilient society. Typical agents in this movement are civil society initiatives that specifically promote cultural techniques of regional management, decentralised energy production and self-sufficiency. Representative of such communities are, among others, the 'Transition Town' movement originally from England, or the Austrian 'zämma leaba' (translated from German: 'living together'). Here we find a significant overlap with the 'post-growth economy' approach in the sustainability discourse which will be presented in more detail below.

In summary, two concepts and according discourses on the resilient society can be contrasted as follows (Table 2.2):

---

[14] For a good introductory reading on this, see Galtung (1998).

Table 2.2 Two discourses on societal resilience at a glance

Discourse/concepts criteria	Crisis management	Crisis transformation
Motif	Making critical infrastructure more resilient	Making society as such more adaptive and robust
Approach	Security policy, civil protection, socio-economic measures 'top-down' Increasingly also: Involvement of the private sector and civil society in emergency preparedness	Promotion of civil society self-sufficiency 'bottom-up' Promoting innovation at all levels More effective knowledge sharing at all levels
Vision	The safe society	The adaptive society
Agent level (current implementation)	National policy, supranational policy	Civil society (including science)

2.5 Conclusion

The resilience discourse has moved beyond the psychological discipline and gained widespread public attention, as such, it is currently being actively pursued at several nested system levels. There is simultaneous talk of 'individual', 'organisational', 'urban' and now 'societal resilience', with each bigger level expected to be even far more complex. At the societal level, the concept of resilience is being traded as the 'new concept of sustainability', although, as we will see, both sustainability and resilience cannot substitute each other. In general, two 'resilience' terms can be distinguished, representing two ends of a spectrum of different definitions. At one end of the spectrum is a static understanding of resilience as the ability of a society to maintain its ability to function in the face of adversities. On the other side of the spectrum, resilience is understood as the capacity for adaptation, and constant evolutionary change and transformation. An underlying criterion of all understandings of resilience is the principle of decentralisation and the associated ability of regions (and smaller system units, such as households) to be self-sufficient. Overall, the resilience discourse is dominated by the Anglophone world (especially the USA).

There have been a few initiatives to measure societal resilience on the basis of indicators and compare countries. According to this approach, a 'generally resilient' society – i.e. a society that is equipped to deal with many different crisis scenarios – would have to include at least the dimensions of social, economy, ecology,

2.5 Conclusion

preparedness, infrastructure and collective mindset. This approach proves to be limited in at least two respects. First, the high context-dependency of resilience and vulnerability and the limited measurability of soft resilience factors make it almost impossible to measure resilience and compare societies' general resilience. Second, in practical terms, these dimensions have proven to be far from sufficient to prepare societies for 'as yet unimagined' crisis scenarios and existential risks. Third, these categories are also insufficient to respond to crisis bundles and bundled crises.

In practical discourse, at least two guiding concepts and corresponding fields of discourse can currently be distinguished, which slightly overlap: 'crisis management' (based on a static understanding of resilience), and 'crisis transformation" (based on a dynamic and transformative understanding of resilience). Unlike in the sustainability discourse, the agent level of the private sector as an active co-shaper of societal resilience currently (still) plays a relatively minor role in both concepts, especially in crisis transformation. It is also noticeable that the resilience discourse as a whole is far less emotionally charged than the sustainability debate, which will be described in more detail below. This is probably due to the fact that the political recommendations under the two guiding principles – crisis management and crisis transformation – do not fundamentally contradict each other and are driven independently by different agents. Crisis management is primarily driven 'top-down' by representatives from the public sector. In contrast and complementary to this, the relatively new model of crisis transformation seems to be largely determined by civil society acting 'bottom-up'.

A convergence of the two guiding concepts can be observed, based on two insights: Firstly, both share the insight that a top-down approach alone proves insufficient for promoting societal resilience and that there is no getting around the involvement of the other subsectors, especially the civil society. Secondly, the representatives of both concepts share the view that decentralised urban planning and regional self-sufficiency represent a central criterion for societal resilience to many adversity types.

With the increasing importance of civil society and the relatively new concept of crisis transformation largely driven by it, a development trend is emerging in which resilience is becoming a core concept of social innovation. Roland Benedikter makes the following observations (hereafter Fathi & Benedikter, 2013):

1. In the context of analytical stocktaking of social trends, civil society in turn becomes a concept for identifying resilience. Alongside 'precaution', 'energy' and 'infrastructure', social participation and the possibility of taking action 'bottom-up" are becoming quality criteria for assessing the resilience level of societies.

2. Civil society is becoming an increasingly important part of the security discourse, insofar as it forms networks of organised citizenship which, when needed, especially in the case of more profound crises, can and probably increasingly must not only complement publicly organised aid but also expand it and enrich it with missing aspects, as the Fukushima nuclear disaster, for example, has shown.
3. Modern social movements are characterised by initiating social processes 'bottom-up' and leading them to long-term success by means of numerous, usually small and often regionally limited, but continuously implemented steps – which is a characteristic of resilience as a social innovation factor.
4. Finally, social movements and civil society are becoming indispensable agents for overcoming complexity. This is especially true in the context of the democratisation of knowledge, which is necessary to cope with complexity. Civil society actors are becoming indispensable agents for the generation, distribution and exchange of knowledge in order to be able to adequately analyse and address the exponentially growing complexity of current and future societal challenges.

Overall, societal resilience promotion remains a multiple challenge for society as a whole, with several unresolved questions: What does 'general', or: 'multi-resilience' mean, as demonstrated by a coping capacity in the face of a variety of different crises, crisis bundles and bundled crises? How can a society prepare for and respond to unforeseeable, even unimagined new crises? How can a resilient society prepare for existential risks? These and other questions will be addressed in the following chapters.

Excursus: Two Approaches to a Cross-Disciplinary Resilience Model 3

Given that the highly complex global crisis bundles and bundle crises of our time cannot be adequately addressed with the knowledge of a single scientific discipline, a transdisciplinary conception is becoming increasingly important. As early as the 1970s, the French philosopher and pioneer of cross-disciplinary research, Edgar Morin, criticised that the dominant view in science was characterised by the reductionism of the respective disciplines, could not capture the complexity of the world, even distorted reality and ultimately led to global ignorance (Morin, 2010). Since the 1990s, there has been increasing discussion about 'multidisciplinarity', 'interdisciplinarity' and 'transdisciplinarity' (for an overview, see Sect. 3.1). Although more and more disciplines are involved in crisis and resilience research, there has been a lack of approaches that bring together the diverse knowledge of the disciplines into a 'big picture' perspective. However, building societal future preparedness that does justice to the complexity of today's multi-dimensional risks/crises and bundles of risks/crises is unlikely without a cross-disciplinary approach. Section 3.2 outlines the two most important approaches in the current complexity debate. They provide a theoretical basis for developing a concept of multi-resilience that can be applied to a wide variety of different risks/crises and risk/crisis dimensions.

© The Author(s), under exclusive license to Springer Fachmedien
Wiesbaden GmbH, part of Springer Nature 2022
K. Fathi, *Multi-Resilience - Development - Sustainability*,
https://doi.org/10.1007/978-3-658-37892-9_3

3.1 Multi-, Inter-, Transdisciplinarity: What Is It?

Overall, there is conceptual confusion regarding the exact definition of frequently used buzzwords such as 'multi', 'pluri', 'cross', 'inter', 'trans' disciplinarity. The term 'interdisciplinarity' implies by far the greatest conceptual variations. Researcher Julie Thompson-Klein attributes this to three reasons.

> First, there is a general uncertainty about the meaning of the term. Many fields were pronounced 'interdisciplinary' with no clear definition of what that meant. (…) Not surprisingly, then generalizations about the nature of interdisciplinarity emerged prematurely. (…) The second major reason for confusion stems from the widespread unfamiliarity with interdisciplinary scholarship. Given all the talk about interdisciplinarity, published work on the subject is used by a relatively small group of people. (…) The third and related reason for confusion is the lack of a unified body of discourse. Discussion of interdisciplinarity literally sprawls across general, professional, academic, governmental and industrial literatures. (Klein, 1990, p. 12 f.)

This impression is confirmed by Balsinger (2005) and Kocka (1987).

The most common distinction of forms (I also like to speak of 'traditions') of cross-disciplinary cooperation is: multidisciplinarity, interdisciplinarity and transdisciplinarity. Before that, let us briefly define what is meant by 'disciplinarity' in the first place.

Individual sciences are also referred to as scientific *disciplines* or *monodisciplines*. In academic discourse, a distinction is made between four scientific fields, which are further differentiated into individual disciplines: Humanities (history, psychology, cultural studies, sociology, etc.), natural sciences (e.g. physics, biology, chemistry, medicine, earth sciences), human sciences (e.g. ethnology, linguistics, also areas from psychology and sociology) and structural sciences (e.g. systems theory, mathematics) (Heidrun, 2004). According to Balsinger, the purpose of monodisciplines is to provide 'a necessary reduction of an epistemological knowledge whole', without which no cognitive achievement would be possible (Balsinger, 2005, p. 57). Following on from this, Carl Friedrich von Weizsäcker wrote that one 'cannot ask all questions at the same time' because then 'not a single question' can be answered (quoted in Hübenthal, 1991, p. 9). The consequence is that monodisciplines deliberately 'fade out' in order to focus only on those aspects that are considered relevant (von Hentig, 1987). Since the second half of the twentieth century, the plethora of individual disciplines has increased, leaving an impression of fragmentation today. In response, three traditions of cross-disciplinary cooperation and knowledge integration have developed.

3.1 Multi-, Inter-, Transdisciplinarity: What Is It?

The tradition of *multidisciplinarity* (also 'pluridisciplinarity') is understood as a concurrent investigation of a research object by monodisciplines working independently of each other, without any significant exchange taking place between them. A synthesis of perspectives only takes place in an additive manner by bringing together the results obtained separately. The pattern of cooperation corresponds to 'plan side by side – act side by side'. According to Mittelstrass, this form of cooperation appears to be 'intended-but-not-skilled' (Mittelstraß, 1998; Klein, 1990).

The term *'interdisciplinarity'* currently implies the greatest diversity and vagueness in terms of definition. One reason for this is that interdisciplinarity is often used in everyday language as a collective term for any interdisciplinary cooperation per se. Despite different definitions, it is characteristic, at least for a narrow concept of interdisciplinarity, that the exchange of knowledge is carried out 'between' (thus: inter) the disciplines and that the partial aspects are brought together. A typical product of interdisciplinary knowledge integration are 'hybrid disciplines', i.e. new multidisciplines that are made up of several monodisciplines – for example social economics or biochemistry (Frodeman et al., 2000).

Transdisciplinarity is understood as a scientific practice that brings together the knowledge of all (or at least) many disciplines. It goes 'beyond' (thus: trans) these disciplines, so to speak. This term first appeared in the OECD Conference of 1970 (OECD, 1972). So far, two main uses of the term can be distinguished:

- On the one hand, transdisciplinarity is understood as a procedure that links scientific and practical knowledge (Bergmann & Schramm, 2008). Wolf Reiner Wendt aptly summarises this claim in four points: First, it 'stands for a collaboration between university and industry in research.' Second, it 'incorporates an extra-scientific understanding of a problem into the scientific definition of a problem.' Third, transdisciplinarity 'involves social agents in concept development and research.' Fourth, it means that 'boundaries of knowledge domains are dissolved and also non-scientific sources are considered' (Wendt, 2003, p. 1).
- Another understanding of transdisciplinarity does not focus on the practical procedure, but on the epistemological integration of the knowledge of the monodisciplines involved. In this sense, representatives of different disciplines work with corresponding meta-principles or meta-categories that go 'beyond' their areas of expertise. This enables a systematic integration and cooperation of different monodisciplines without reducing them to one another (Stokols et al., 2008). Thompson Klein sums up this epistemological implementation of transdisciplinarity as follows:

> Whereas "interdisciplinarity" signifies the synthesis of two or more disciplines, establishing a new meta-level of discourse, "transdisciplinarity" signifies the interconnectedness of all aspects of reality, transcending the dynamics of a dialectical synthesis to grasp the total dynamics of reality as a whole. (Klein, 1990, p. 66)

In practice, both uses of the term are found in transdisciplinary projects: on the one hand, overcoming disciplinary boundaries in terms of content and theory, and on the other hand, practical and organisational cooperation between science and other sub-sectors of society, e.g. between science, civil society and politics (Nowotny et al., 2001; Lieven & Maasen, 2007).

Complexity expert Gitta Peyn aptly summarises the difference between interdisciplinarity and transdisciplinarity as follows:

> While "interdisciplinary" means that disciplines exchange with each other, "transdisciplinary" builds bridges, provides connecting models that themselves do not necessarily intervene in the respective disciplines, but can build on them. For example, modern systems theory has on the one hand entered the disciplines (e.g. in the natural sciences and social sciences) and in this way participates in interdisciplinary exchange, but at the same time it is an overarching model which cannot be clearly assigned to the natural sciences or the humanities (Peyn, 2019).

An increasing diversity of disciplines can currently be observed in resilience studies. The multidisciplinary approach currently appears to be the predominant form of cooperation – Rüdiger Wink's (2016) anthology '*Multidisziplinäre Perspektiven in der Resilienzforschung*' (in English: '*Multidisciplinary Perspectives in Resilience Studies*') is a good example of this. However, in view of the complexity of existing knowledge and of contemporary global crisis bundles and bundle crises, there is a need for a more profound integration of knowledge. A corresponding transdisciplinary approach, which would be necessary for this, has been lacking so far – both in practical and in epistemological terms. From a practical point of view, the question arises as to how cooperation between actors from different disciplines and sectors of society can be organised. In epistemological terms, the question is how the diversity of knowledge (implying perspectives from different disciplines and society sectors) can be integrated into a holistic complexity-adequate picture. This shortcoming has already been recognised in sustainability studies; therefore, this field has an explicit reference to transdisciplinary studies. The much-acclaimed work '*Klimawandel im deutschen Wissenschafts- und Hochschulsystem*' (translated: *Climate Change in the German Science and Higher Education System*) by Uwe Schneidewind, the president of the German Wuppertal Institute, and Mandy Singer-Brodowski (2013), can be regarded as representative of transdisciplinary knowledge integration in sustainability sciences.

In the context of this book, it is assumed that the analysis and practical design of a resilient society – and in the broadest sense of societal future preparedness – cannot avoid a transdisciplinary approach. Although transdisciplinarity has been discussed since the 1970s, the concept has so far been relatively little explored. As Hübenthal aptly notes in her dissertation:

> Especially between sciences from different fields, such as between the humanities and the natural sciences, it is hardly possible to find a common theoretical basis. How such a basis should look here and how an interdisciplinary discourse should take place [, is] not yet discussed (Hübenthal, 1991, p. 17).

3.2 Two Cross-Disciplinary Approaches to the Resilient Society

Currently, two approaches can be found in the complexity debate contributing to a transdisciplinary approach to knowledge diversity and complexity. Both offer a different approach to a complexity-adequate resilient society. I call these two approaches or traditions 'simplify' (Sect. 3.2.1) and 'complexify' (Sect. 3.2.2).

3.2.1 Simplify

The '*simplify*' approach aims to reduce complex phenomena to their constituent patterns and thereby making them understandable and manageable. In a certain sense, this approach represents a pragmatic reduction, which, however, should not be confused with the reductionism of the monodisciplines. The latter would be the case, if a representative of a monodiscipline, for example a psychologist or physicist, were to attempt to grasp and explain all complex phenomena and aspects of life, such as modern society, economic crises, love, life, etc., exclusively with the methodologies and concepts of his/her discipline. The attempt made by this representative would in all probability prove to be very limited and under-complex in its outcome. The situation is different with the simplify approach. It is true that here, too, many aspects of reality are deliberately excluded. But in contrast to the specialised monodisciplines, the claim is made to reduce the complexity of phenomena to constituent patterns that are valid in all disciplines. As a result, 'the whole' would be simplified in analysis and description, but still adequately captured in what essentially constitutes 'the whole'. Among the best-known simplify approaches at present are the theories of what is known as 'systems thinking'.

Systems thinking emerged in the 1920s as a critique of physics' hitherto unchallenged attempts to explain the world. The representatives of physics assumed that they could explain the world more or less completely by understanding its smallest components. The biologist and pioneer of general systems theory, Ludwig von Bertalanffy, on the other hand, criticised the fact that complex, living organisms could in no way be explained by the functioning of their smallest components, such as atoms. Rather, 'the whole (...) is more than the sum of its parts'. For instance, the organism is not merely a bunch of interconnected cells, but also has qualitatively new properties. Cells, in turn, are also made up of molecules, but also have qualitatively new properties compared to them, and so on. This paradigm shift ushered in the birth of systems thinking (Kneer & Nassehi, 2000).

Today, modern systems theory is considered a cross-disciplinary approach in which fundamental aspects and principles of systems are used to describe and explain complex phenomena. Systems cope with complexity by combining it into 'wholes', i.e. systems. Systems theory has gone through several phases of development up to its present state. According to Luhmann, there were essentially three phases, whereby he assigns the paradigm shift described here, 'the whole is more than the sum of its parts', to the first phase (Luhmann, 1993).

In the second phase, a fundamental distinction is made between system and environment. According to this, every system is an open entity and enters into an exchange relationship with its environment. In doing so, it delimits itself externally as a system from the environment and in turn represents an environment internally. At the same time, system and environment influence each other. This principle can be compared to a thermostat (temperature controller) which 'observes' the environment through its mode of operation. If the room temperature falls below the programmed setpoint, the thermostat counteracts this and causes the room temperature to rise. As soon as the room temperature has reached the setpoint, the thermostat causes the heating process to stop. Observer (thermostat) and environment (room temperature) influence each other. This principle was influenced by cybernetics, which was coined by Norbert Wiener and William Ross Ashby. Cybernetics is understood as the study of the 'control' of systems (Kneer & Nassehi, 2000). The second paradigm shift led to the insight that complexity does not only follow a linear causality, but is rather structured in a circular and interrelational way.

The decisive impetus for the next paradigm shift was provided by the biologist Humberto Maturana and, following on from him, Francisco Varela. They introduced the concept of autopoiesis, which is still central to systems theory today. Translated, autopoiesis means 'self-production' or 'self-generation'. According to this, every system interacts with its environment, but is also 'closed'. 'Closed' means that every system carries out operations through which it reproduces itself, in other words: 'keeps itself alive'. For example, the cell separates itself from the environment by its cell membrane and is therefore closed. At the same time, it produces the necessary components (e.g. proteins, nucleic acids, lipids) from its interaction with the environment in order to maintain itself as a system. Due to their closed nature, autopoietic systems refer exclusively to themselves – they are 'self-referential'. Thus, they do not know a direct input-output relationship in terms of how they are internally organised. Systems, such as the cell, exchange energy and matter with their environment, but the exchange is controlled and channelled by the system. In this process, the cell takes in from the environment only what it needs to produce its components in order to keep itself alive. In other words, the exchange with the environment is selective and is only made possible by the closed nature of the system (Maturana & Varela, 1982).

Another relevant aspect of the third paradigm shift is that systems can only be oriented from the outside, but not directly influenced. They internally organise themselves. According to Maturana/Varela and Luhmann, this becomes particularly clear in the distinction between 'trivial' (in other words: simple or complicated) systems and 'non-trivial' (in other words: complex and chaotic) systems. Trivial systems are relatively easy to control because a certain input from outside is always followed by a relatively predictable output. This is not the case with non-trivial systems (such as living, social, or psychological systems), because they have their own internal complexity, which, as described above, is only selectively coupled to their environment. This means that complex systems are not in direct contact with the environment – rather, through their operations, they construct an image of the 'internal system' from the environment of the 'external system' (Maturana & Varela, 1982). An intervention from outside, such as counselling or leadership, encounters a complex 'life of its own' of the system. Consultants or leaders can therefore only marvel what their intervention triggers in retrospect at (Willke, 2016). In other words, environmental events do not exert a determining influence on complex systems; rather, they only irritate the system. However, the system's own operations alone determine in which sense it processes the irritations from the environment. In terms of resilience, this means that not every system (be it an individual, an organisation or a society) reacts in the same way to an event or even feels the same stress. How it responds to the event depends in many cases on its internal components, e.g. its threat perception, its past experiences, its coping resources, etc.. Following these principles, William Ross Ashby coined one of the core principles of cybernetics, Ashby's Law, according to which any system can only cope with as much external complexity as its inherent complexity. The higher the inherent complexity of a system, the higher its ability to react flexibly to external events and thus control complexity. Hence, the higher the inherent complexity of a system, the higher its viability (Ashby, 1956).[1]

The founder of sociological systems theory, Niklas Luhmann, finally initiated the fourth development by transferring the autopoiesis principle from living systems to other types of systems. According to this, humans would not only be living systems, but also mental systems. Both systems perform self-referential operations in exchange with the environment in order to maintain themselves. But according to Luhmann, the components differ. For example, while the living system of a cell produces lipids, proteins, etc., the mental system of a person produces its own manifestation – namely consciousness, which in turn contains sensations and thoughts. Luhmann also considers social systems. According to him, the basic element of mental systems is consciousness, whereas the basic element of all social systems is 'communication'. Communication alone – not persons – constitutes social systems, because persons, such as the employees of an organisation, can only interact with each other through communication (Luhmann, 1993). The complexity of other dimensions, such as thoughts, are pragmatically excluded because they only exist in the social system when they transition into a communication event. In other words, anything that is not communicated in a social system, does not exist in the social system (Simon, 2014). Nevertheless, Luhmann's model presupposes that these very three types of systems – namely living, mental and social systems – are structurally coupled, even 'interpenetrating' each other. Social systems thus presuppose life and consciousness, but the crucial basic operation that constantly reproduces the social system is communication (Luhmann, 1993).

Figure 3.1 shows a simplified representation of a system with all typical components for the description of this system:

In the context of resilience, there are at least two systemic models that describe aspects of resilience and are applicable to all system levels (e.g. organisations, societies) and different contexts (e.g. ecological, economic, social problems).

---

[1] Strictly speaking, Ashby's law states that a system's ability to master the complexity of its environment depends above all on its having at least as much complexity or diversity in terms of its possibilities for action (the technical term for this diversity of action is "variety"). Ashby summarised his law with the words "variety can destroy variety" (1956), which was later positively reformulated for the management context by the founder of management cybernetics, Stafford Beer, as "variety absorbs variety" (Beer, 1974). This principle is still considered valid today in the complexity debate and thus has far-reaching implications for a fundamental understanding of general resilience or multi-resilience.

3.2 Two Cross-Disciplinary Approaches to the Resilient Society

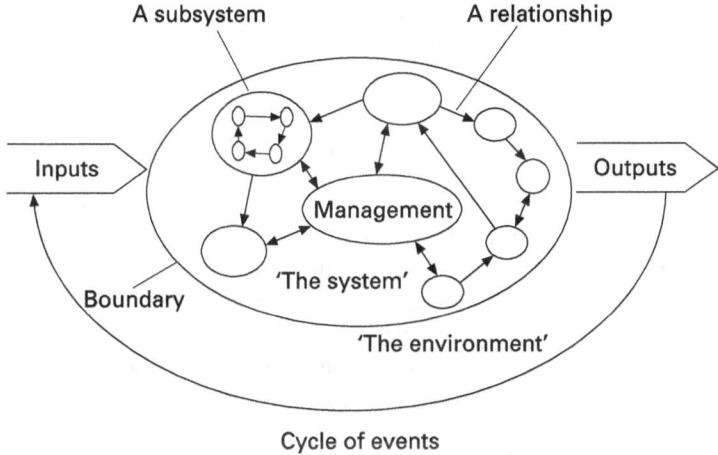

Fig. 3.1 Typical components of a system. (Jackson, 2003, p. 6)

3.2.1.1 The Panarchy Model of Adaptive Systems

The 'Panarchy Model of Adaptive Cycles' developed in the 1970s by Holling and Grunderson (2002) is one of the few approaches that explains resilient processes on a multi-contextual level. This model, which originates from systems thinking, can be applied to individual, organisational or even societal systems, as well as to completely different types of systems and crises. It illustrates that in a system, such as a society, the relationship between adaptation and stability can constantly change:

1. *Exploration or accumulation phase:* The accumulation phase is characterised by expansion and growth of the system. Typically, this phase is signed with 'r', which originally stands for population growth in a logistic equation or for rapidly reproducing species in ecology (Holling & Grunderson, 2002). A social system, such as a city, expands in this phase through high population growth and accumulates physical, cultural and social capital. The resilience of the system is high, as failure of the system still results in relatively few costs (Schnur, 2013).
2. *Maintenance or conservation phase:* The conservation phase is marked with 'K', named after the mathematical abbreviation for the maximum achievable population. In this phase, the former growth reaches its limits (Holling & Grunderson, 2002). What has been achieved is conserved, the self-preservation

effort increases, and innovation declines. Routines become established. The system becomes increasingly less flexible, which increases its overall susceptibility to crises (Schnur, 2013).

3. *Releasing phase*: This phase is typically denoted by the Greek Ω, the 'end'. Once a crisis has occurred, the releasing phase comes into effect. The system begins to release the capital that was once tied up; dysfunctional structures (this is also referred to as 'connectivity') are dismantled. In the words of the innovation pioneer Joseph Schumpeter, this ideally leads to 'creative destruction' (Schumpeter et al., 2006) and, in the transition to the next phase, an intense period of development in which the resilience of the system begins to increase again (Holling & Grunderson, 2002).

4. *Reorganisation phase*: In this phase, marked with 'a' for 'beginning', an adaptation to the new environmental conditions takes place. The 'structural potential' increases again, while the 'connectivity' (here for example the forms of regulation of a system) remains weak (Schnur, 2013). At the same time, it is also a phase of increased uncertainty, because new paths have to be taken; at the same time, resilience continues to increase until it reaches the level of the first phase again. The cycle begins again, albeit under different conditions (Holling & Grunderson, 2002; Fathi, 2016a) (Fig. 3.2).

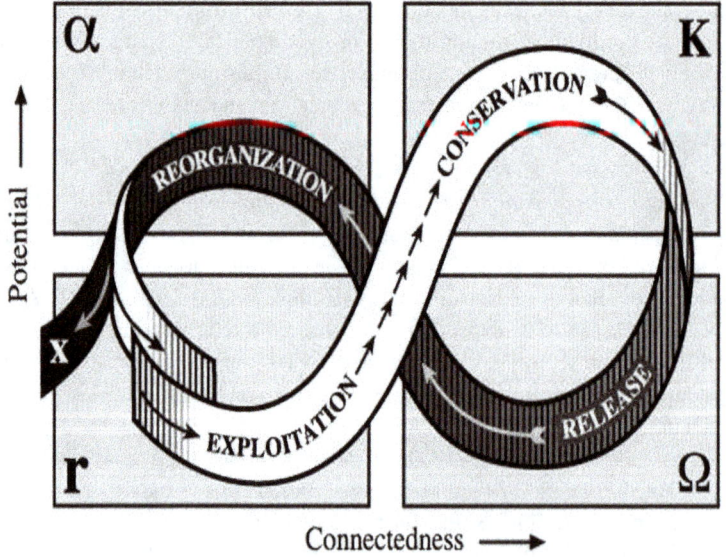

Fig. 3.2 The Panarchy Model of adaptive cycles. (Holling & Grunderson, 2002)

3.2 Two Cross-Disciplinary Approaches to the Resilient Society

The Panarchy Model of Adaptive Cycles presents resilience as an evolutionary process and provides a systemic rationale for the Resilience 2.0 concept presented above (cf. Fig. 2.1). According to this model, the system affected by a crisis does not bounce back to an 'original normal state', as would be the case with a tumbler according to the old concept of Resilience 1.0. Rather, crises usually require the affected system to adapt to new conditions and 'bounce forward' to a new normal state. Similarly, in cybernetics an affected system can achieve several states of equilibrium after a disturbance, which is called a 'poly-stable system' (Malik, 1992).

Given that systemic changes can occur through abrupt shocks (e.g. political upheaval, coronavirus) or through gradual changes (e.g. climate change), Holling/Grunderson also assume that adaptation cycles are nested within other adaptation cycles. For example, an economic region is at once 'a subsystem within a global economic system and a super-system within which individuals, households, firms, local governments, and organizations act and interact' (Pendall et al., 2010, p. 78). Subsystems and overall systems are linked by two different functions, which Holling et al. refer to as 'revolt' and 'remember' (Holling et al., 2002). Subsystems can pressurise or change overall systems via innovations. This would be a 'revolution'-function from stage Ω (reorganisation) to stage K (conservation). With reard to multi-resilient societies, this aspect could illustrate spillover effects from the local level to the overall societal level, for example in the context of social innovations or protest actions by civil-society participants or disruptive technological innovations by private sector organisations.

By contrast, overall stable systems can govern subsystems. This function is called 'memorising' or 'remembering'. It happens from stage K (conservation) to stage α (growth). A typical example of multi-resilience would be the revitalisation of banks by the superordinate system 'government' during the subprime crisis (2007–2009).

In this respect, the term 'panarchy' reflects the ambivalent relationship between superordinate systems and subordinate subsystems, which leads to positive and negative cascades of change (Holling & Gunderson, 2002). It is no coincidence that 'panarchy' refers to 'Pan', the Greek god of nature who, depending on circumstances, time and context, can equally cause well-being and its opposite, 'panic' (Holling et al., 2002).

3.2.1.2 The Cybernetic Model of Viable Systems

Complementary to the Panarchy Model, the *Viable Systems Model* (VSM) does not describe the exterior evolutionary process of social systems in the face of crises, but instead focuses on the inner mechanisms and communication structures of these systems which could contribute to transformative resilience. The Viable Systems Model stems from the *cybernetic* tradition, which was introduced by Norbert Wiener in the 1950s. The Viable System Model (VSM) was first formulated in 1959 by Stafford Beer in his book *'Cybernetics and Management'* (Beer, 1959). Cybernetics can be briefly understood as the scientific study of how humans, animals and machines control and communicate with each other. Ashby's Law is inherent to the VSM and describes the core principles of a system's viability in complex environments, which is used here, to some extent, identically to resilience. Central to any viability is the ability of a system to master complexity. According to Ashby's Law, here the parameter for measuring complexity is the variety within the system. Variety reflects the number of possible states of a system in coping with the complexity of the environment. The higher its internal complexity, i.e. the number of states it can take, the higher the system's ability to process the information from the environment and deal with corresponding adversities.

As highlighted here in Ashby's Law: The bigger the variety of action of a system controlling another system, the greater the number of disturbances in the control process it can absorb (Beer, 1974, 1985). Another important principle of viable systems is that of 'homeostasis'. This term describes that complex systems maintain stability in their internal environment through self-regulation while dealing with the unpredictable external environment.

Overall, the VSM serves as a reference model to describe, diagnose and manage complex social systems. Here, according to the British systems theorist Michael Jackson, the VSM

> integrates the findings of around 50 years of work in the academic discipline of organization theory. And it goes beyond organization theory by incorporating those findings into an applicable management tool that can be used to recommend very specific improvements in the design and functioning of organizations (Jackson, 2003, p. 107).

An important practical conclusion for the implementation of multi-resilience is to find a balance between decentralised, autonomous decision-making by local units and effective, comprehensive (centralised, homogenised and combined) knowledge distribution and 'bundling' between units (Malik, 1992). The contribution of the VSM for resilience research lies in its generality, which can be applied to all types of systems and subsystems of organisations and societies to make them function better. In terms of multi-resilience, it provides us with information about the aspects of a system that are essential for viability (Jackson, 2003).

According to the VSM, a viable system is composed of five interacting subsystems, pictured in the following Fig. 3.3 (hereafter Malik, 1992; Jackson, 2003; Beer, 1985):

3.2 Two Cross-Disciplinary Approaches to the Resilient Society

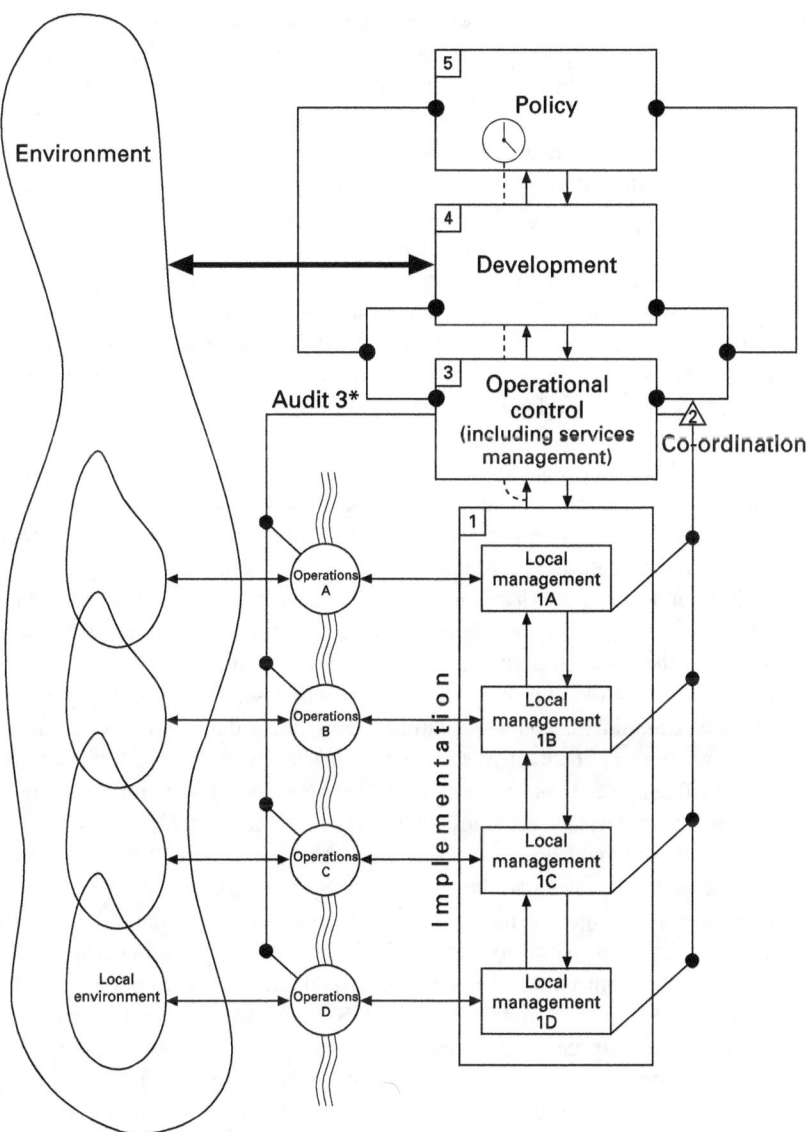

Fig. 3.3 The Viable System Model. (Jackson, 2003, p. 92)

- *System 1:* This system is representative of the local operating units of a viable system, which are represented by circles A-D. From them one can see what the viable system does. In the words of Stafford Beer: 'The purpose of a system is what it does. And what the viable system does is done by System One' (Beer, 1985, p. 128) In an organisation, system 1 corresponds most closely to quasi-autonomous divisions, with the circles representing the leadership of each division. In a national state, the circles 1A-D might represent the leaderships of all state activities, such as health care, police, etc. (Malik, 1992). Characteristically, System 1 is 'recursive', which means that each System 1 is an autonomous sub-system and includes a VSM with all five systems in itself. According to the principle of viability, each System 1 needs to be as autonomous as possible in order to deal with its environment in the most efficient ways. Thus, they should develop their own policy, development, operational control, coordination and implementation functions.
- *System 2:* Each individual System 1 has, in principle, complete decision-making autonomy. However, since they are part of a more comprehensive system, their scope of behaviour is limited for the benefit of the larger whole and the other systems 1, which are also part of the comprehensive system. Accordingly, the core function of System 2 is to harmonise and coordinate the behaviours of all Systems 1A-D. In this regard, System 2 represents the information channels and bodies that allow the primary activities in System 1 to communicate between each other. Here, practically all forms of coordination instruments are used, for example conferences, committees, planning control systems, but also the informal communication relationships between Systems 1A-D. In summary, System 2 fulfils a coordinating function and ensures harmony between the parts of System 1. This allows System 3 to monitor the activities within System 1 (Malik, 1992).
- *System 3:* Although System 2 coordinates the interaction of Systems 1, this alone cannot guarantee that the coordinated Systems 1 together have a greater effect than the sum of the individual activities. This is only ensured by System 3. One of its main tasks is to draw up guidelines and specifications that restrict the principle of freedom of behaviour of Systems 1 while incorporating information from Systems 4 and 5. In addition, System 3 checks their performance, quality, conformance to regulations, and maintenance of System 1. In doing so, it is responsible for the optimal allocation of resources to Systems 1 and monitoring the scheduled use of these resources. Its key role is thus the operational control of System 1 and the management of services. In order to perform this task well in an environment characterised by constant change, System 3 has several direct communication links upwards to Systems 4 and 5 and downwards to Systems 1 and 2. In addition to a vertical axis of command running from the

top via System 3 to the bottom, direct communication channels also exist from the bottom to the top through which System 3 receives information about the efforts and pressures of Systems 1 and 2 (Malik, 1992). In organisational systems, this task is usually performed by the human resources management (Jackson, 2003). From an overall system perspective, the activities of Systems 1, 2, 3 are directed inwards and focus primarily on maintaining internal stability. They make up what Beer calls the 'autonomic management' of an organisation (Malik, 1992).

- *System 4:* The viability of a system also depends on information about the system environment. Receiving, processing and forwarding this environmental information to Systems 3 and 5 are core tasks of System 4. As such, system 4 is responsible for integrating internal information provided by System 3 and external information of the system's total environment. If the organisation is to match the complexity of the environment it is dealing with, it needs a model of that environment that enables predictions about its likely future state. System 4 develops and delivers this model and communicates the information to System 3 when rapid action is required or to System 5 when it has longer-term implications (Jackson, 2003). System 4 exhibits certain functions of staffs, but goes far beyond this advisory information processing, which is why it is also directly on the vertical command axis. In general, activities such as strategy, planning, research and development, marketing, and public relations are home to System 4 (Malik, 1992). This could also mean that scenario planning, strategic forecasting and other methods of anticipatory governance are important to imagine unpredictable adversities in the future. This is a key activity for the resilience of a given system operating in complex environments – be it an organisation dealing with a highly competitive environment and disruptive innovations, or a state dealing with geostrategic or world economic developments (Jackson, 2003).
- *System 5:* The balancing of internal and external equilibrium results from the interaction of Systems 3 (representing internal demands) and 4 (representing external demands) and is monitored and influenced by System 5. If Systems 3 and 4 cannot agree on a common course, System 5 makes the final decision. If System 5 agrees with the decision of Systems 3 and 4, no intervention is needed. System 5 thus forms the top decision-making level with the main tasks of drawing logical conclusions and observing Systems 3 and 4. It formulates a strategy based on the information received from System 4 and communicates it to System 3 for implementation by the divisions. According to Malik, along the vertical axis of command, a concise characterisation of the main function of each system can be summarised as follows:

- System 1: What is happening here and now?
- System 2: Coordination of Systems 1
- System 3: What will happen soon?
- System 4: What could happen in the context of vaguely identifiable development trends?
- System 5: What should happen considering all the information? (Fig. 3.3)

From a cybernetic perspective, a viable system is characterised by its ability to adapt to external and internal changes by sensibly utilising information from the environment and operating accordingly.

Historical Case Example: Project Cybersyn
The Cybersyn ('cybernetic synergy') project outlined below is one of the best-known applications of VSM principles. It was an attempt during Salvador Allende's Chilean government (1970–1973) to control the socialist planned economy in real-time through computers. Essentially, it was a teletype network that involved equipping Chile's main nationalised factories with a computer into which current data could be entered. The data included various information, such as capacity utilisation, raw material, energy requirements, or workers' sick leave. This data were to be transmitted via telephone lines to a central computer in Santiago, and the Cyberstride software was to correlate it. If the target values were not met, the programme would sound an alarm – similar to a thermostat. Stafford Beer himself is considered the main architect of the system. Cybersyn took a year to build, but was never fully completed. The biggest use of the system ctook place in October 1972, when the powerful Association of Hauliers called a nationwide strike, resulting in some 50,000 hauliers to block the streets of Santiago. Through the teleprinters, the government was able to coordinate the transport of food into the city with only 200 government-owned trucks. Remarkably, the actual Cybersyn core programme was not used at all, but the telex network through which the logisticians exchanged capacities and routes among themselves. However, before the effectiveness of the VSM could be tested in the long term, the project ended abruptly with Augusto Pinochet's military coup and Allende's suicide. (Medina, 2011). In later times, Fernando Flores, then Minister of Finance, and Raúl Espejo, then technical director of the group, saw the effective communication pattern of the project as one of the first beginnings of today's internet, although this was not the intention. *The Guardian* called the project 'a sort of socialist internet, decades ahead of its time'. Today it is called the 'socialist internet' (Beckett, 2003).

The contribution of the VSM for (multi-)resilience research lies in its generality, which can be applied to all types of systems and subsystems of organisations and societies so that they can persist and work more effectively in crisis situations.

In summary, what do these two 'simplifying' models, the Panarchy of Adaptive Cycles Model and the Viable Systems Model, teach in terms of a more multi-resilient response to crisis bundles and bundle crises?

One major practical consequence for more multi-resilient societies can be derived from both models: Societal subsystems should be granted autonomy so that they can absorb some of the environmental variety that could otherwise overwhelm reasoned management on governance levels. In turn, this implies a governance approach based on coordination and audit instead of exclusive hierarchical control. In other words: multi-resilience implies multi-level governance. Multi-level governance in this perspective is meant to empower a multipolarity of forces to become (diverse and pluralistic) agents in a more resilient social structure in continuous development. The superordinate governance level should be open enough to spill-over-effects from the bottom (e.g. technical or social innovations generated on the local and regional levels) to enable evolutionary development 'from below' in crisis situations. On the other hand, it should also have the capacity to support and maybe also regulate top-down, if crisis situations require it.

3.2.2 Complexify

The *complexify* or *integral* tradition is characterised by the fact that it attempts to grasp complex phenomena in their multidimensionality. In contrast to *simplify*, it does not attempt to break down complexity into underlying patterns. Rather, it tries to categorise reality in terms of non-reducible categories. Here it assumes that different disciplines have different scopes that cannot be reduced to one another.

A typical contribution in the tradition of complexify is '*Rethinking Social Theory*' by the social scientist Roger Sibeon (2004), in which he names criteria that are indispensable for an integrative methodology. Among other things, he defines 'four deadly sins' of complexity management that are committed in current discourses in the humanities and that can be prevented with an adequate meta-methodology:

1. *Reductionism:* Attributing complexities of social life to a single explanation, such as 'rational choice', 'risk society', 'patriarchy', etc.

2. *Reification:* Ascribing agency to entities that are not actors or agents. As will be shown in more detail below, a distinction should at least be made between agents and structures.
3. *Essentialism:* Assuming homogeneity of social phenomena and ascribing more uniformity to social categories such as 'men', 'working class', 'the law', etc. than they actually exhibit.
4. *Functional teleology:* Circular reasoning that amounts to explaining causes of social phenomena by their outcome. An example would be to explain the existence of the institution 'marriage' by the fact that marriage is a system with a self-preservation interest (Sibeon, 2004).

Besides Sibeon's work, integral metatheories have increasingly been referred to by different social sciences and by practitioners. They are the result of the current challenge of dealing with the increasing complexity of disciplines and methodologies responding to more complex environments and aiming to complement rather than contradict each other. To date, a variety of models can be found, such as those of conflict workers John Paul Lederach (2003) and Johan Galtung (1998, 2008), the social scientists Roger Sibeon (2004) and Derek Layder (1997), or even renowned science philosopher Sir Karl Popper (2004). All of their approaches have one typical aspect in common: they use multi-dimensional schematics, assuming that social phenomena include correlating facets that cannot be analysed in a mutually exclusive manner.

Typical tools of a knowledge integration intended in the context of *complexify* are models that make it possible to categorise and combine the knowledge and methodologies of different disciplines. The typical procedure for this consists of sifting through as much scientific material as possible, critically examining it, and formulating statements on a general basis. The two most popular models are the quadrant model, which is presented in more detail in the following subchapters, and the development level model, which will also be referred to in a later chapter.

The quadrant model is based on the assumption that there are several fundamental perspectives on the world that cannot be reduced to one another and that are each epistemologically encoded in a different 'language'. Usually, these schematics consider variations of the variables 'micro versus macro' and 'subjective versus objective'. The result is usually an integrative matrix of three or four social dimensions which cannot be reduced to their individual parts. Typical examples of models that distinguish three epistemological dimensions are Karl Popper's three 'worlds' (World 1 [I], World 2 [It] and World 3 [We]) (Popper, 2004) or Jürgen Habermas' three validity claims (truth [objects], sincerity [subjects], justice [intersubjectivity]) (Habermas, 1981). Each of these dimensions brings a distinct

3.2 Two Cross-Disciplinary Approaches to the Resilient Society

methodological approach to the object of investigation. Therefore, they cannot be separated; they are part of a methodological pluralism and thus, in the ideal case, form a transdisciplinary map of the object of investigation.

In the following, I draft a brief, not exhaustive, outline of a typical 'four quadrants' model which can be widely applied in transdisciplinary knowledge integration:

- *Individual-objective / It-perspective / Empirical truth:* This perspective follows the validity criterion of 'empirical truth' and refers to all factors that are measurable or at least objectively observable. This quadrant, coded in the 'it' language, is captured using quantitative, positivist, and empirical methods (Popper, 2004; Sibeon, 2004; Layder, 1997). In the context of resilience, relevant indicators are observable resilience behaviour and physical resilience factors of individuals as well as, for example, social and economic data.
- *Individual-subjective/Ego-perspective/Truthfulness:* This perspective contains data that can only be sensed and experienced in one's own subjectivity. These include intentions, feelings and thoughts (Popper, 2004) and self-identity (Sibeon, 2004; Layder, 1997). These aspects are not accessed through empirical observation of an objectively verifiable truth (e.g. whether it is raining outside), but through interpretation of subjective experience (e.g. how one feels about the rain). Associated with this quadrant are qualitative methodologies encoded in 'I' language: From psychoanalysis, depth psychology, phenomenology, to meditative practices. Typically, this quadrant covers much of what is discussed in the context of psychological resilience, such as the psychological resilience factors.
- *Collective-subjective/We-perspective/Cultural meaning:* The data of the I-perspective is always embedded in an intersubjective context that provides a collectively shared meaning. The corresponding validity claim is 'cultural fit'. Here, the focus is on mutual understanding, based on a shared notion of 'right', 'wrong', 'normal', etc. This 'we' perspective clarifies that agents can agree on a shared context of meaning (Popper, 2004). Typically, this quadrant includes content such as morality, ethics, language, and shared symbols (Sibeon, 2004; Layder, 1997), or values or culture in general (Lederach, 2003). Epistemologically, this dimension is accessed through constructivist theories, such as social psychology or qualitative sociology, among others. Regarding the resilient society, this dimension typically refers to any above-mentioned aspects of collective mindset or resilience culture, including narratives, degree of trust or mottos.

- *Collective-objective/Its-perspective/Functionality:* While the It-perspective relates to directly measurable data and facts, such as economic key figures, neurological activities or observable behaviours, the underlying pattern of these data and facts is addressed in the Its-perspective. This can be, for example, big data algorithms that help predict how the observed system will 'behave' next after evaluating all the data and facts. Another typical instrument that stands for this epistemological dimension is the systemic feedback loop, which explains how a conflict reproduces itself through an interplay of the behaviours of the members of a social system. Here, the focus of consideration is not on culture, meaning and values (this would be part of the we-perspective), but rather on the external functionality of the phenomena, i.e. whether they have a stabilising or destabilising effect on the system 'society'. Typical approaches that have their focus on this dimension can be found in systems thinking (Fathi, 2019a). Furthermore, this dimension also allows for the consideration and analysis of societal structures and institutions (Lederach, 2003).

Graphically, the main contents of the four quadrants can be summarised as follows (Table 3.1):

It stands to reason that the PESTLE factors are mostly related to the two right-hand quadrants. This acronym stands for *Political, Economic, Sociological, Technological, Legal, Environmental* and represents the typical factors of external environmental analysis. Each of these areas could in turn be analysed in the light of the four quadrants.

Table 3.1 Four quadrants

	Subjective	Objective
Individual	**First-person perspective:** Motives, feelings, thoughts, inner drives to act	**It-perspective:** Measurable (numbers, facts), observable (behaviour)
	Methodology: Qualitative inquiry, psychotherapy, phenomenology, meditation	Methodology: Quantitative methods, based on positivism, empiricism, behaviourism
	Validity criterion: Truthfulness	Validity criterion: Truth
Collective	**We-perspective:** Shared meanings (culture, norms, values, ethics, etc.)	**Its-perspective:** Structures, systems, patterns, feedback loops
	Methodology: Social psychology, qualitative sociology, etc.	Methodology: Systems theory, structuralism, big data algorithms
	Validity criterion: Cultural significance	Validity criterion: Function/functionality

Based on Lederach (2003), Sibeon (2004, pp. 108–110), Layder (1997, pp. 2–4)

3.3 Conclusion

Given the complexity of today's crises, the conception of societal resilience cannot avoid a cross-disciplinary approach. Strictly speaking, the resilient society requires a transdisciplinary approach, i.e. an approach that helps grasp and contextualise the knowledge and contributions of different monodisciplines. Compared to multi- and interdisciplinarity, the transdisciplinary approach is characterised by the highest degree of integration and might therefore be mostly relevant for a complexity-adequate conception of (multi-)resilience. Overall, two transdisciplinary approaches can be distinguished that are of importance for our later considerations regarding a multi-resilient society: I call them *simplify* and *complexify*.

The *simplify* approach, typically represented by systems thinking, epistemologically manages complexity by breaking it down to its 'system-preserving' essential features and operations. Thus, a social system (such as a society) is defined by the operation 'communication'. The other transdisciplinary approach, *complexify*, is often also referred to as 'integral' and is characterised by metacategories in order to integrate the diversity of knowledge and perspectives and avoid reductionisms. From this perspective, the resilient society emerges as a multi-dimensional entity – none of these dimensions would be neglected in the course of resilience promotion. What this can mean in concrete terms is illustrated in the next chapter.

The *complexify* approach has the advantage of being able to consider knowledge and viewpoints from different disciplines (e.g. psychology from the top-left quadrant and economics from the top-right quadrant) in a complementary context. This enables an appropriate multi-perspective view of the crises and protective factors of our time. The disadvantage is that hardly any practical recommendations for action result from this approach and that it does not provide an answer on how to deal with non-knowledge and unpredictability. At this point, the *simplify* approach is likely to provide more added value. Here, complex systems, such as crises or societies, are assumed to be 'black boxes'. From this perspective, there is a conscious acceptance of non-knowledge and at the same time a method of experimental approach to gain knowledge about the complex system. This aspect, which is also of central importance for the resilient society, will also be examined in more detail in the next chapter. Based on the two transdisciplinary approaches, the essential components of a multi-resilient society will be outlined.

Five Principles of a Multi-resilient Society 4

What characterises a society that is resilient in the face of multiple crises – be it an economic crisis, a natural disaster, cyberterrorism, the refugee crisis, or an interdependent combination of all of them? In this chapter I outline general principles that foster capacities for a 'multi-resilient' society.

4.1 General Principles for a Multi-resilient Society

The resilient society of the twenty-first century is likely to face at least three key challenges:

1. In an interdependent world, multi-dimensional problems and risks and bundles of them cannot be adequately captured from the limited viewpoint of a single discipline or sector of society. Consequently, a single protective factor or problem-solving approach that addresses only one dimension, be it political, economic or social, will usually not suffice. Accordingly, this requires integrated approaches that take into account the resources of agents from different sectors of society and disciplines and enhance collective intelligence.
2. The interdependence and complexity of today's modern societies are accompanied by the fact that protective factors can turn out to be risk factors and vice versa. How can vulnerabilities be reduced and protective factors increased?
3. Contemporary (and maybe future) challenges are mostly characterised by VUCA, which means: volatility, unpredictability, complexity and ambiguity.

They therefore involve a high degree of unknowns. How can we deal with problems which cannot be predicted, or even imagined?

Appropriate answers to these challenges may emerge by changing the level of abstraction and on the basis of a transdisciplinary approach. In their much-discussed book '*Why Things Bounce Back*' (2012) resilience researchers Andrew Zolli and Ann Marie Healy made a valuable pioneering contribution to the question of whether and to what extent there are 'patterns' or 'universal or general principles' that characterise resilient systems across different disciplines, problem contexts and system levels. According to their analysis, 'patterns of resilience' are characterised primarily by (1) redundancy, (2) close-meshed feedback mechanisms, (3) decoupling of subsystems, (4) the ability to reorganise dynamically, (5) constructive error culture ('the ability to fail gracefully'), (6) strong social cohesion, and (7) people who make connections ('translational leaders'). The plausibility of Zolli/Healy's analysis can hardly be disputed and is largely transferable to these reflections. At the same time, it leaves some questions unanswered, since the examples listed by Zolli/Healy refer almost exclusively to resilient communities, which are a fundamental component of the resilient society, but are not identical with it. What does the 'tight feedback mechanisms' pattern mean for society? Secondly, the Zolli/Healy study focuses almost exclusively on physical threats. As such, the pattern 'redundancy' indeed turns out to be necessary as a preparation for a variety of threats (such as in the sense of stockpiling, economic reserves, and back-up systems). More subtle threats such as the refugee crisis, as well as the aspect of the social and cultural construction of threat in the collective perception of a society, are neglected in Zolli/Healy's study and are not covered by patterns such as 'redundancy'. For the considerations in this book, however, the question of general principles that foster the capacity of a society to manage the complexity of crisis bundles and bundle crises is particularly important. Thirdly, as with the discourse on 'real resilience factors' described above, it is striking that the patterns identified by Zolli/Healy tend to correspond to observed behavioural characteristics of resilient systems.

Based on the evaluation of the crisis types presented above (see Chap. 2) and the contributions of both transdisciplinary traditions (see Chap. 3), at least five general principles of a multi-resilient society can be derived. They do not replace contemporary approaches on resilient society, but provide a complementary, general orientation on dimensions to foster multi-resilience:

- Principle 1: The multi-resilient society is fostered by resilient citizens.
- Principle 2: The multi-resilient society is comfortable with unknowns.

4.1 General Principles for a Multi-resilient Society

- Principle 3: The multi-resilient society is based on decoupling and knowledge integration of its subsystems.
- Principle 4: The multi-resilient society makes collectively intelligent decisions.
- Principle 5: The multi-resilient society is based on a strong learning culture.

It is obvious that the five principles are interrelated. At the same time, it should be emphasised that they cannot be reduced to one another. Each of these principles has a different focus in the quadrant model presented above. The model makes it clear that, regardless of the nature of the crisis, it is important for a society to keep at least these four dimensions in mind. As shown above, these dimensions can include a variety of protective factors. In the context of the following considerations, it is of particular interest which general criteria are relevant for a multi-resilient society.

- *I-perspective (subjective):* This quadrant represents the core area of psychological resilience of individuals. Various indicators, especially from comparative happiness research, are often used in connection with societal resilience. These make it possible to capture social and emotional well-being, perceived happiness in life, resilience and self-esteem, as collected, for example, in the 'National Accounts on Well-being' of the New Economics Foundation (NEF) (Michaelson et al., 2009). Another approach might be to identify the average 'resilience quotient', which was developed by the founder of positive psychology and dominant resilience researcher Martin Seligman (e.g. Reivich & Shatté, 2003). These and other indicators illustrate the potential for social conflict in a society (as will be shown in more detail in later chapters). However, they prove to have only very limited transferability to non-Western societies as well as limited applicability to all types of crises. Within the framework of Principle 1 *'The multi-resilient society is fostered resilient citizens'*, it is necessary to investigate which general criteria contribute to higher overall psychological resilience – regardless of culture and problem type.
- *It-perspective (objective):* This dimension encompasses a wide range of empirically measurable data and facts that provide information on, among other things, the self-sufficiency of the population (stockpiling), physical protection facilities (e.g. bunkers), economic resilience (e.g. productivity, diversity of sectors and low debt) and the robustness of technical infrastructures. From a multi- or cross-crisis perspective, it would be important to consider how effective a society and its subsystems are in their general problem-solving capacity, especially in the face of crises and crisis bundles that cannot be predicted. This question is examined in the context of Principle 2, *'The multi-resilient society is confident with unknowns'*.

- *Its-perspective (collective-objective):* This typical systemic perspective reveals how viable the overall system is, especially with regard to its communication structures. The systemic sociologist Niklas Luhmann considered communication to be the basic element of social systems (and thus of societies) and consequently calls such systems 'communication systems'. Communication can be problem-solving, functional or dysfunctional. In this respect, the type and quality of communication is decisive for how well a society succeeds in using its collective resources and its 'swarm intelligence' to cope with complexity. I address these very complex considerations within the framework of Principle 3 *'The multi-resilient society is based on decoupling and knowledge integration of its subsystems'* and Principle 4 *'The multi-resilient society makes collectively intelligent decisions'*.
- *We-perspective (collective-subjective):* This dimension reflects the risk and protective factors of a society, which result from its collective perception. Typical country-comparative indicators pointing to this resilience dimension include interpersonal trust and tolerance of a population. Both are identified by the Gallup World Poll and the World Values Survey (Medrano, 2012). Another indicator would be 'trust and belonging' and 'supportive relationships', surveyed by the nef (Michaelson et al., 2009). As noted above, the context of collective mindset includes many aspects which cannot be captured with indicator-based analysis. One example is how societies collectively perceive threats, as studied by the IRS Leibniz Institute. The collective resilience mindset is also likely to vary strongly from culture to culture. Resilience Principle 5, *'The multi-resilient society is based on a strong learning culture'*, discusses the general criteria that characterise a resilient culture – regardless of cultural differences (Table 4.1).

4.1.1 Principle 1: The Multi-resilient Society Is Fostered by Resilient Citizens

It can hardly be disputed that resilient citizens contribute to a resilient society. However, the question of what makes people generally resilient in the face of various and multiple challenges, remains unexplored. The concept of individual resilience has undergone enormous change since its introduction in the middle of the twentieth century. At the time of Emmy Werner's pioneering studies, resilience primarily referred to a particular quality of individuals. The research mostly focused on children who maintained their mental health in conditions that would

4.1 General Principles for a Multi-resilient Society

Table 4.1 Resilience dimensions of a society

	Subjective	Objective
Individual	**I-perspective:** Inner attitude, regulation of own feelings, thoughts	**It-perspective:** Socio-economic data, external behaviour and practices
	Manifestation of resilience: Psychological resilience, mental toughness	Manifestation of resilience: Coping strategies & methods, economic resilience, technological resilience etc.
	Criteria: Resilience quotient, psychological well-being	Criteria: Economic performance, Gini-Coefficient, employment rate, problem-solving behaviour, redundancies, etc.
Collective	**We-perspective:** Culture, norms, values, narratives, etc.	**Its-perspective:** Structures, systems, patterns, feedback loops
	Manifestation of resilience: Cultural resilience	Manifestation of resilience: Systemic viability of the social system, resilience-promoting structures and institutions
	Criteria: Trust, belonging, relationships, collective threat perception	Criteria: Complexity management capacity of the overall system and its subsystems, effective knowledge management within and between the subsystems

Inspired by Fathi, 2013, pp. 41–74; Galtung, 2008; Lederach, 2003; Sibeon, 2004, pp. 108–110; Layder, 1997, pp. 2–4

have broken most people (e.g. poverty, refugee situations, mental illnesses of the parents etc.). Mental resilience is not only beneficial in extreme or potentially traumatic situations. In recent decades, the term has been expanded to other contexts. Accordingly, mental resilience is not only seen as beneficial for coping with extreme situations, but also for stressors in 'normal' everyday life. What characterises such resilience, which relates to different and diverse problems, and how can it be systematically promoted?

An orientation still widely used in resilience coaching today is the assumption of so-called 'resilience factors'. In the German-speaking world, the following 'seven pillars of resilience' are usually mentioned (in the following Wellensiek, 2011):

- An *optimistic* outlook on life;
- *Acceptance:* The ability to accept things that cannot be changed and to reflect on one's own options for action;
- *Solution-orientation:* Not getting bogged down in problems, but looking for ways out;

- *Leaving the victim role:* not feeling at the mercy of crisis situations;
- *Taking responsibility* for one's own feelings of suffering and taking active steps to change one's situation;
- *Network orientation:* The ability to ask people for help and to activate contacts who can provide support in specific situations;
- *Planning for the future:* The ability to think in the long term and to orient one's own actions and thinking accordingly.

In response to this, Denis Mourlane (2017) emphasises that these seven pillars are a misrepresentation of the research of Reivich and Shatté (2003), who, together with the founder of positive psychology, Martin Seligman, had set up the world's largest resilience research and promotion programme. According to them, the factors of acceptance, leaving the victim role or network orientation were particularly misnamed. Although it is true that highly resilient people accept setbacks more quickly, see themselves less often as victims and seek support from other people, these are 'only' behaviours that are based on the seven resilience factors defined by Reivich and Shatté. Mourlane therefore writes of the seven 'real' resilience factors that can be measured with the help of the programme based on Reivich and Shatté (Resilience Factor Inventory). These 'real' resilience factors include (in the following Mourlane, 2017; Reivich & Shatté, 2003):

1. *Emotion control:* The ability to remain calm under pressure.
2. *Impulse control:* The ability to control one's own behaviour in pressure situations, not giving in to initial impulses and following through.
3. *Causal analysis:* The ability to analyse problems accurately and to correctly assess reasons for successes and failures.
4. *Empathy:* The ability to put oneself in another person's psychological and emotional shoes based on observed behaviour.
5. *Realistic optimism:* The inner attitude that things can and will change for the better, and the ability to see a sense of purpose and something positive even in very difficult situations.
6. *Self-efficacy:* The desire to take on challenges and the belief that one can change things through one's own actions.
7. *Goal orientation (Reaching-Out):* The ability to set new goals and to pursue and implement them, regardless of adversities and the opinion of others.

A direct comparison suggests that the factors of both resilience models largely coincide or are at least interrelated. For example, emotion control requires an attitude of acceptance, and only empathy makes it possible to develop support networks. Self-efficacy beliefs lead to an abandonment of the victim role and

4.1 General Principles for a Multi-resilient Society

planning for the future presupposes goal orientation. Finally, the ability of causal analysis correlates directly with the factor of personal responsibility.

Even beyond resilience studies, there are strikingly many similarities with the factors of other approaches that deal with psychological resilience. The best-known approach outside resilience studies, which is nevertheless usually discussed in conjunction with psychological resilience, is the concept of 'salutogenesis', which was coined by the Israeli-American medical sociologist Aaron Antonovsky. This approach translates as *'origins of health'* and focuses on factors that support human health and well-being, rather than on factors that cause disease (pathogenesis). Originally, his model addressed the relationship between stress, health, and coping through a study of Holocaust survivors. Antonovsky found that some survivors were able to thrive later in life, apparently due to powerful health-promoting factors (Antonovski, 1997). In line with these findings, Antonovsky developed the Sense of Coherence (SOC) model. It forms the basis of psychological resilience in one's own way of life and describes a basic feeling of confidence in one's own abilities and the 'coherence' between one's own perception of the world and the events experienced. The SOC is based on three components that partly complement the resilience factor model and partly overlap with it:

- *Understandability:* The ability to understand and make sense of the environment.
- *Feasibility:* The conviction to actively cope with situations and problems.
- *Meaningfulness:* The confidence that life has a deeper meaning (Antonovski, 1997).

Another concept that describes aspects of personal resilience was introduced in 1979 by Suzanne C. Kobasa under the name *'Hardiness'*. Hardiness translates as 'resilience' and refers to personality traits that are able to protect people from illnesses despite great stress and critical life events (Kobasa, 1979). The focus here is on the individual's ability to cope with stress factors ('stressors'). Essentially, hardiness contains three further components that complement and confirm the above-mentioned resilience and SOC factors:

- *Commitment* (engagement and self-commitment) means curiosity about life. This refers to a person's effort to identify with and commit to everything he or she does or encounters. At this point there is an overlap with 'impulse control' from the resilience model.
- *Control* can be understood as the opposite of helplessness. People with a high level of control do not experience events as something alien overwhelming them, because they recognise various options available to them. This factor appears to correspond with the 'feasibility' factor of salutogenesis.

- Finally, *challenge* means that change is not perceived as a threat, but as a positive opportunity for new experiences and an incentive for further growth (cf. Kobasa, 1979). This might correspond with the resilience factor 'realistic optimism'.

A third concept is the coping model developed by Charles S. Carver. It was popularised by him at the end of the 1980s and describes the way of dealing with a life event or period that is perceived as significant and difficult. Essentially, coping distinguishes between adaptive and maladaptive coping strategies. According to this, a resilient person is characterised by coping strategies that are highly adaptive to the given circumstances. A psychological test to measure coping with stress was developed in 1989 under the name COPE by Charles Carver. It measures 14 different coping strategies (Knoll et al., 2005). The strategies are classified as follows:

Effective coping strategies
- Proactive avoidance of stress
- Active elimination of the causes
- Seeking support from friends
- Relaxation
- Think positive
- Humour

Ambivalent coping strategies
- Distraction
- Accept fate
- Refuge in religion

Ineffective or risky coping strategies
- Denial of stress
- Use of alcohol and drugs
- To give up
- Self-Reproach

A comparison of the models mentioned here yields the following correlations and mutual complements (Table 4.2):

The multiple correlations point to several, indeed 'universal' factors that characterise resilient individuals in different challenging situations. This is supported by the fact that these models have been researched in different crisis contexts – e.g. stress and burnout (coping), life crises (hardiness), coping with traumatising events

4.1 General Principles for a Multi-resilient Society

Table 4.2 Comparison of different models and strategies for promoting individual resilience

Resilience	Salutogenesis	Coping (effective and ambivalent strategies)	Hardiness
Impulse control			Commitment/ Self-commitment
Causal analysis	Understandability		
Self-efficacy	Feasibility	Active elimination of the causes	Control
		Proactive avoidance of stress	
Realistic optimism	Meaning	Refuge in religion	
		To think positive	Challenge
Empathy		Search for support	
Emotion control		Accept fate	
		Relaxation	
		Distraction	
		Humour	

and illnesses (salutogenesis), self-development under adverse circumstances (resilience). To some extent, these factors can be seen as a narrow manifestation of multi-resilience at the system level of individuals. 'Narrow' in this context means that these criteria might be applicable to various and different problem contexts. What remains unexplored, however, is what makes us resilient to multiple threats occurring simultaneously. In this context of multi-resilience, other factors could play a role, such as decision-making skills, including the ability to prioritise effectively, as well as cognitive and intuitive skills to cope with complexity. These considerations need to be explored in more depth in later chapters.

Another critical factor limiting the universality of the psychological factors mentioned above is the high context-dependence of resilience. In this context, the second wave of resilience studies points out that a protective or resilience factor can also become a risk factor in certain contexts (e.g. optimism or empathy) and that resilience varies in the respective situation. It has often been observed that people prove to be relatively resilient in one life situation, but are relatively quickly overwhelmed in another (O'Dougherty et al., 2013). Despite these limitations, this does not change the fact that resilience can be developed systematically and that this often applies to different (and possibly multiple) adversities.

Studies show that resilience in people is to some extent an innate characteristic. However, resilience can always be learned to a significant degree. Recent studies even suggest that people whose innate resilience is relatively low have a greater potential to learn resilience (Drath, 2014).

In the current practice of individual resilience promotion we can find a multitude of methodological approaches. The tools of cognitive behavioural therapy and positive psychology have proven particularly effective in working with stressful beliefs. An important finding is that stressful emotions (such as anger or fear) do not result directly from observable adversities, but rather from stressful beliefs and evaluations of these events. This is evidenced by the fact that people who face similar or almost identical adversities do not feel the same way because of their different evaluations of these situations. An employee with low self-esteem will possibly feel stressed and discouraged by his or her superior's criticism of his work. The source of this stressful feeling would not be the criticism itself, but the judgement 'I am a failure'. By contrast, an employee with a belief system reflecting high self-confidence, would possibly feel less stressed and maybe even encouraged to do better next time. Cognitive resilience techniques enable systematic questioning and reinterpretation of stressful beliefs (e.g. Reivich & Shatté, 2003). Another approach to increasing one's own resilience is to directly release one's own stressful feelings, for example with breathing techniques from yoga (e.g., Bönsel, 2012). Approaches that have a long-term effect, but at the same time require continuous practice, can be found in various traditions of meditation or mindfulness training (e.g. Lutz et al., 2004). Other approaches of resilience coaching focus on reflecting and developing one's own resources. This would mean, for example, systematically reflecting on all stressors and all resilience-promoting resources in one's own life and deriving appropriate measures to reduce stressors and increase resources (e.g., Wellensiek, 2011).

In summary, the dimensions of individual resilience and resilience promotion can be outlined as follows (Table 4.3):

Within the framework of this resilience principle, resilience promotion of the citizens is essential for any resilience promotion of societies. Similar to Daniel Goleman's call for fostering emotional intelligence in primary and secondary school curricula (Goleman, 2001), individual resilience promotion could also be a component of an education policy for systematic competence development of the population.

Two basic forms of resilience promotion that are applicable to almost all types of crises are likely to be, on the one hand, ongoing meditation practice[1] and, on the other, 'learning to learn' or 'lifelong learning'.

---

[1] Although, as already mentioned, numerous studies indicate positive effects in the treatment of various mental and even physical illnesses, it should be noted that the practice of meditation can also have negative effects in certain constellations. For example, it is contraindicated in schizophrenia, borderline syndrome, psychoses and drug-induced psychoses, bipolar disorder and severe narcissistic personality disorder. There is still a need for research here to find out the appropriate methods for people in different life situations.

4.1 General Principles for a Multi-resilient Society

Table 4.3 Dimensions of individual resilience and resilience-promoting disciplines. (Inspired by Fathi, 2016b)

Resilience dimensions	Disciplines that promote resilience
Beliefs/Attitude: Cognitive resilience factors, e.g. reinterpreting setbacks, realistic optimism, etc.	Cognitive behavioural psychology/ positive psychology
Intuition/Mindfulness: Impulse control, calmness, being able to let go of negative thoughts and feelings	Meditation/Yoga
Resources: Personal short- and long-term strategies, own stress limits	Resource-oriented approaches/ salutogenesis
Own networks: Friends, experts, supporters, etc.	Resource-oriented approaches
Values/Sense	Resource-oriented approaches/ salutogenesis
Social skills: Being able to set boundaries, conflict management skills, empathy, communication skills, etc.	Empathy training, conflict competence, appreciative communication, etc.

The many resilience-promoting effects of meditation are supported by the fact that an increasing number of studies demonstrate at least the following effects after only 8 weeks of daily meditation practice (which is, for example, part of the 'Mindfulness Based Stress Reduction' program):

- Increasing one's own mindfulness.
- Deepening of body awareness and thus an improvement of one's own health-promoting behaviour.
- Awareness of own (potentially stressful) thoughts and feelings.
- Change in dealing with feelings, especially difficult feelings.[2]

Studies have shown that long-term meditation can lead to positive changes that can be detected in the entire brain structure (Barinaga, 2003; Carter et al., 2005; Davidson et al., 2003; Lazar et al., 2005; Lutz et al., 2004). The research results were summarised in a meta-study in 2010 (Bohlmeijer et al., 2010) and 2011 (Fjorback et al., 2011).

Following on from this, the principle of 'learning to learn' emphasises the continuous (lifelong) pursuit of development, but also self-observation and recurrent reflection on one's own individual abilities, which in turn has indirect effects on the learning capacity and performance enhancement of the surrounding systems, e.g.

---

[2] A list of all clinical trials and meta-studies since the 2000s can be found on the MBSR Association website: http://www.mbsr-verband.de/mbsr-mbct/forschung.html

organisations and society. This characteristic corresponds to the 'personal mastery' which is one of five collective resilience factors of the 'Learning organisation' popularised by Peter Senge in his famous book *'The Fifth Discipline'* (Senge, 1996). It is not surprising that this aspect of 'learning' and evolutionary adaptation to one's own development is associated with the rise and resilience of Japan and China. This will be discussed in more detail in later chapters.

4.1.2 Principle 2: The Multi-resilient Society Is Comfortable with Unknowns

Collective problem-solving competence does not only require the integration of existing knowledge, but also the ability to effectively deal with unknowns. In this context, the current complexity discourse distinguishes between simple, complicated, complex and chaotic problems. The *Cynefin model which was* developed in the late 1990s by Welsh complexity researcher Dave Snowden provides a framework which integrates corresponding strategies to ensure complexity-adequate decision-making and problem-solving. *Cynefin* is a Welsh word that literally translates as 'habitat', which means that we all have multiple pasts which we can only be partially aware of: cultural, religious, geographical, tribal, etc. (Snowden, 2000). The four domains of knowing or not knowing can be represented as follows:

Simple (known knowns): This domain represents the 'known knowns' referring to phenomena whose relationship between cause and effect is clear (Snowden, 2000): if you do X, expect Y. The corresponding strategy in such situations is 'sense–categorise–respond', meaning: establish the facts ('sense'), categorise, then respond by following the rule or applying best practice. Snowden and Boone offer the example of loan-payment processing: An employee identifies the problem (e.g. a borrower has paid less than required), categorises it (reviews the loan documents), and responds (follows the terms of the loan). For simple problems, clearly identifiable best practices can usually be used (Snowden & Boone, 2007).

Complicated *(known unknowns):* Complicated phenomena are characterised by 'known unknowns'. Here, the relationship between cause and effect requires analysis by experts with specialised knowledge, such as engineers, surgeons, intelligence analysts, lawyers, to ensure a refined judgement. Artificial intelligence can also serve well here. Characteristically, there is no 'one right way' here and it is advisable to be open to as many solution options as possible, which can also be identified relatively quickly. This is where 'good practices' come into play. The framework recommends 'sense–analyse–respond': assess the facts, analyse, and apply the appropriate good operating practice (Snowden, 2000).

4.1 General Principles for a Multi-resilient Society

Complex (unknown knowns): Complex phenomena can only be explained in their relationship between cause and effect in retrospect, but not in advance. Unlike complicated systems (e.g. cars or industrial plants), complex systems (e.g. people, corporate cultures, ecosystems or markets) are more than just the sum of their parts. A 'take-it-apart-and-see-how-it-works approach' does not work, because one's very actions change the situation in unpredictable ways. Typically, problem-solving processes are openly designed and require a high degree of trial and error and dialogical knowledge exchange. As there are no pre-defined right answers, and solutions emerge out of the open process, Snowden calls the corresponding problem-solving practices 'emergent practices' (Snowden, 2000). The general approach is always: probe (experiment), sense, respond. As they write 'instructive patterns (...) can emerge (...) if the leader conducts experiments that are safe to fail.' Cynefin calls this process 'probe–sense–respond'. According to Snowden and Boone, typical tools in dealing with complex challenges therefore include: democratic large group dialogues, ritualised debates and/or agile problem solving in mixed teams (Snowden & Boone, 2007).

Chaotic (unknown unknowns): With chaotic phenomena, there is no discernible relationship between cause and effect, even in retrospect. Since such events as the September 11 attacks require immediate action, there is no time for experimentation and step-by-step trial and error (Snowden & Boone, 2007). The general approach would be: 'act–sense–respond', which means: Action as the first and only way to respond appropriately and to establish order; *sense* where stability lies; *respond* to turn the chaotic into the complex (Snowden, 2000). Another aspect of chaotic problems is that, similar to crises, they typically require new ways. Thus, corresponding problem-solving approaches would be so-called 'innovative practices'. For organisations facing chaotic crisis situations, Snowden and Boone recommend to simultaneously assign teams to different tasks. One team would concentrate on acute crisis management, while the other would closely monitor the current situation and try out new practices, and evaluate whether they prove to be better. This, according to Snowden, is to ensure that action is taken and at the same time no opportunities are missed (Snowden & Boone, 2007).

A typical example for organisations that work effectively in chaotic situations are so-called 'High Reliability Organisations (HRO)', which are a natural model for resilient and multi-resilient organisations.

Five Meta Practices of High Reliability Organisations (HROs) Against Chaotic Phenomena

The concept of HROs originated in the work of a research group at the University of California at Berkeley, which in the mid-1980s looked at the analysis of organisations with a high-risk potential. They studied organisations that operated daily under difficult conditions with unpredictable risks and at the same time were affected by far fewer accidents and incidents (and could therefore operate 'reliably') than would be statistically expected. These included organisations involved in electricity generation and supply, air traffic control, nuclear-powered aircraft carriers and nuclear power stations. Inspired by this research, scholars Karl Weick and Kathleen Sutcliffe analysed other organisations, such as hospital emergency rooms, accident investigation teams, wildland firefighting teams, and hostage negotiation teams (Weick & Sutcliffe, 2001). In analysing HROs, it was found that they were characterised by particular features of mindful collective action. This is achieved through the interplay of several meta-practices (in the following Weick & Sutcliffe, 2001):

- *Focus on errors:* Members of HROs constantly question their own abilities and expectations, thereby remaining alert and sensitive to possible sources of error (Weick & Sutcliffe, 2001). On aircraft carriers, for example, the landing officers brief the pilots on the deck in the final seconds of flight. In doing so, the officers admittedly expect that experienced pilots do not panic during the landing approach. At the same time, they know from experience that pilots can lose their orientation during night landings. The officers are sensitive to signs of tension in the pilots' voices that indicate that their expectation is wrong and that the night landing could fail.
- *Sensitivity to the current state:* Compared to conventional teams and organisations, HRO teams usually react situationally. In order to be able to identify operational disruptions at an early stage and prevent far-reaching effects, they strive to have a comprehensive knowledge of current operational processes. This requires that employees report signs of operational disruptions at an early stage and that all employees are informed about current events. HROs therefore promote information transparency and encourage communication. If, for example, anomalies occur in nuclear power plants, the various departments are immediately informed about

(continued)

4.1 General Principles for a Multi-resilient Society

continued

the current situation. In meetings and briefings, the understanding of all employees for the overall system and the complex interrelationships is promoted.

- *Aversion to simplistic interpretations:* To raise their awareness, the HRO teams strive for complex and comprehensive perceptions. They reject simplistic interpretations because they cannot adequately represent the complex environments in which they operate. To counter simplifications, HROs systematically encourage diversity of perspectives. For instance, when incidents occur on aircraft carriers or in nuclear power plants, members of different departments form a team to assess the situation and possible risks from different angles and develop different options. This makes it less likely that something will be overlooked.
- *Striving for flexibility:* HRO teams must be prepared for both expected and highly rare extreme scenarios (such as a military operation or a nuclear power plant accident). Since each disaster scenario is unique, there is usually a lack of established routines and standardised detailed knowledge. As a consequence, HRO teams keep themselves fit with regular simulations and intensive training for different disaster scenarios. In doing so, they do not commit themselves to a single correct solution. Another central component is that redundancies are created in the work processes and solutions are available in surplus. Most importantly, this involves alternative communication channels beyond formalised paths, such as informal networks and knowledge communities. This practice may seem inefficient and at times confusing in normal situations, but it increases the group's ability to improvise in times of crisis.
- *Flexible forms of decision-making:* In order to be able to react adequately to disturbances, decisions must be made where the situation is best known. HROs therefore allow for alternatives to hierarchical decision-making paths. Flight officers, for example, are used to strictly hierarchical decision-making paths, but are trained to disagree with the commanding captain in extreme situations. This means that in the event of technical problems during a landing attempt on an aircraft carrier, the head of the flying squadron, who is best able to assess the skills and behaviour of his pilots, is allowed to pre-empt the superior officers in the tower, contradict them in case of doubt and decide how the aircraft should land.

Snowden also distinguishes a fifth domain of *'disorder'*. This is a lack of knowledge about what kind of causality exists or in which of the above-mentioned four domains the current problem can be appropriately classified. Often, it can be observed that decision-makers confuse the domains with one another, for example by forcing chaotic situations into the 'Simple problems' domain. In this context, Snowden and Boone emphasise that decision-makers should beware of over-simplifying, 'entrained thinking' (being blind to new ways of thinking), or of becoming complacent (Snowden & Boone, 2007). Furthermore, it is important to keep in mind that phenomena and associated solution approaches may change over time. Innovative practices that have proven effective become 'best practices' and the once seemingly chaotic problem may transform into a 'simple problem' as more knowledge is gained about it (Snowden, 2000). This shows, as Malik confirms, that it is ultimately the information at hand that determines whether a problem is simple or complicated (Malik, 1992). If there is enough information to fully grasp and predict the problem, a linear, systematic approach using good or best practices is possible. If the number of unknowns is too high, a more open, cyclical-evolutionary approach is required, which is a typical component of emerging and innovative practices. In order to prevent disorder, Snowden and Boone recommend that leaders provide a communication channel, anonymous if necessary, so that dissenters (e.g. within a workforce) can warn against complacency and against decisions that are not appropriate to the situation (Snowden and Boone, 2007).

Moreover, it is noteworthy that events can encompass all domains simultaneously, which requires a correspondingly multi-resilient approach. Snowden and Boone give the example of the 1993 Brown's Chicken massacre in Palatine, Illinois, where robbers murdered seven employees. Deputy Police Chief Walt Gasior had to act immediately to contain the initial panic (*chaotic*), while keeping the department running (*simple*), calling in experts (*complicated*), and maintaining community confidence in the weeks that followed (*complex*) (Snowden and Boone, 2007).

> **Different Types of Problems and Solution Practices Using the Example of Refugee Accommodation (Taken from Fathi, 2016a)**
>
> In the context of the refugee crisis of the years 2014–2016, the German media increasingly reported on massive challenges in the accommodation of refugees. Challenges such as the extremely short-term influx of new residents or mass brawls among them involved several problem types and solution approaches at once, which can be assigned to the different domains of the Cynefin model.
>
> Typically, simple problems are about maintaining normal operations. Catering has to be purchased, provided and managed and the professional staff responsible for this does this through well-rehearsed routines that are similar in all refugee facilities and have proven to be a standard in terms of best practices. Complicated problems require specialist advice and expertise, such as on technical issues (e.g. repairs) or on security issues (e.g. emergency exits), and provide good practices.
>
> These two phenomena are often closely linked to complex or/and chaotic phenomena. Thus, even in the face of for example, inter-ethnic conflicts (complex problem) or a sudden massive influx (chaotic problem), normal day-to-day business must be maintained (best practices) and special expertise (e.g. what does a conflict management system for refugee shelters entail?) is required (good practices). In the context of complex challenges, which include, for example, unpredictable conflicts between residents, approaches to solutions could arise through the involvement of unofficial authorities of the respective ethnic groups (emerging practices). Unlike expert advice, the solution is not known at the outset and emerges from the situation. A chaotic challenge, such as a lack of space due to a short-term influx, requires crisis management and improvisation – also by drawing on existing resources and developing new solutions. In one case I know of, the nearby military barracks was spontaneously called upon to set up tents when several busloads of new residents arrived one night without warning. As a follow-up, it was considered to organise a scenario workshop with actors and stakeholders from different areas and integrate their different perspectives on the issue of the 'Refugee Shelter 2020'. The insights from the resulting alternative future scenarios would help with strategic planning and the development of innovative practices (innovative practices) (Fig. 4.1).
>
> *(continued)*

continued

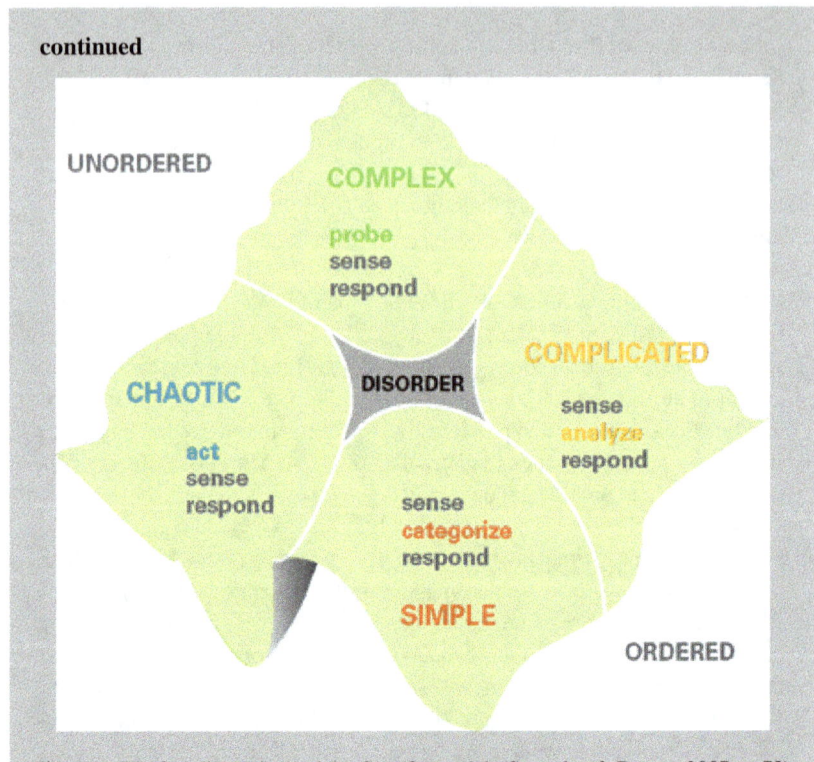

Fig. 4.1 The five dimensions of the Cynefin model. (Snowden & Boone, 2007, p. 73)

To summarise: Dealing with complex phenomena, chaotic phenomena, or with a combination of them requires dealing with a number of unknowns. How can we manage the unexpected in a resilient way? A simple response to this is: learning. Only from learning processes – as exemplified by the Cynefin model – can emergent and innovative problem solutions emerge. In general, different types and levels of learning can be distinguished. In the following, I will discuss two popular learning approaches that are relevant in the context of resilient complexity management.

First, *learning loops* approach, formulated by Gregory Bateson (1987) and popularised by Chris Argyris and Donald Schön, distinguishes between three levels of learning that differ in the depth of reflection (Argyris & Schön, 2008). These are (1) single-loop learning (adaptation learning), (2) double-loop learning (change learning), and (3) deutero-learning (learning to learn). *Single-loop learning,* also referred to as 'doing things better' or 'adaptation learning', represents an optimisation of one's actions without changing the existing framework. The focus is therefore on the efficiency of actions. *Double-loop learning,* also referred to as 'doing better things' or 'change learning', goes deeper. Here it is a question of realigning one's own actions and strategies to the changing conditions. This involves reflecting on the conditions and expanding one's own potential for action in the sense of Ashby's Law, described above (Argyris & Schön, 2008). *Deutero-learning (triple-loop-learning),* also referred to as 'learning to learn' or 'process learning', analyses and questions the previous learning processes and institutional embeddings. This leads to the optimisation of one's own learning behaviour for maximum flexibility in the face of unexpected events (Argyris & Schön, 2008). This form of learning requires a high degree of self- and process observation and is of central importance for promoting multi-resilience of individual and collective systems.

The second learning approach was popularised by Claus Otto Scharmer and is part of his widely acknowledged *U Theory*. Here, he distinguishes between 'learning from the past' and 'learning from the future' (in the following Scharmer, 2009). The former involves problem solving on the basis of knowledge that we have accumulated from past experiences ('downloading'). Typically, this results in good and best practices, but these answers usually prove insufficient, when it comes to thinking about new problems in innovative ways. Scharmer calls this approach 'learning from the future'. Assumptions from past experiences are systematically released ('Letting go') and the user(s) opens up in the here and now for the potentials that lie in possible futures ('Letting come'). In the further process, it is important to concretise these potentials that often emerge abstractly in one's own intuition ('Crystallising') and to test them practically via prototypes ('Prototyping'). Which dialogue formats could be suitable for 'learning from the future'? Two approaches are outlined below:

- *Scenario planning:* This method organises knowledge in future-oriented problem-solving processes. Participants in scenario workshops jointly develop models of possible futures in a communication process which is structured through several phases. The goal is not to predict the future (what will happen?), but to reflect on possible futures (what could happen?). In this way, it is possible to strategically prepare for unforeseeable scenarios (Fathi, 2016a; Gabriel, 2013). In a typical scenario workshop, the participants first clarify the guiding question of the workshop. In a further step, factors that have a significant influence on this issue are identified and integrated with a view to their future development – usually with the help of computers and cross-impact-matrix-tables. The results of this integration are scenarios, i.e. images of consistent worlds. The scenario technique requires a heterogeneous group of participants and is therefore par-

ticularly suitable for integrating cross-disciplinary perspectives. It is usually applied to anticipate problems that are characterised by high complexity and uncertainty and require a long-term orientation.
- *System simulation:* Experience building in organisations and societies is often associated with a high expenditure of resources. For this reason, more and more attempts are being made to simulate 'real' experience building. Examples of this are case-based learning, which has been used for centuries in legal training, and research knowledge generated in simulation experiments in the natural sciences (Willke, 1998, p. 70). A special form of computer-based simulations are micro worlds. Originally developed as programmed learning environments for children, these systems are designed to give the user the feeling of living in the simulation in order to gain a deeper understanding of the aspects of the real world that are replicated by the micro world. This tool is increasingly being used in the development of organisational learning and transformation processes. In this context, micro worlds are simulations that are created on the basis of pre-existing mental models of organisational members (Senge, 1996). The general advantage of all these simulation tools is that complex problem situations can be represented and problem solutions can be tested in a time- and space-saving manner. At the same time, however, it is also apparent, similar to scenario planning that it is limited by the fact that it is currently not yet possible to represent complex realities in all their facets. This gap will probably narrow in the future with the development of augmented reality and immersive reality applications.

Application Example: Two Approaches to 'Learning from the Future' in the Refugee Crisis (From Fathi, 2016a)
In the context of the refugee crisis, at least a short-term and a long-term approach would be conceivable. Both would contribute to 'preventive resilience', i.e. preparation for unpredictable developments. In principle, both approaches outlined below can also be applied to other crisis phenomena.

Formats such as scenario workshops would be conceivable in the short term and without major expenditure of time and money. This idea is not fundamentally new, it originates from 'strategic wargaming' and served the military during the Cold War as a tool to strategically prepare for different scenarios. Today, this method is increasingly being used in companies outside Europe. It could also be used as a strategic tool for a more resilient refugee policy. Within a two-day workshop, the knowledge of different agents (e.g.

(continued)

continued
from refugee initiatives, security personnel, the political administration, entrepreneurs, etc.) could be brought together and influencing factors and resulting scenarios could be discussed with the help of special moderation techniques. This would not (!) be about predicting the future, but about systematically pooling knowledge and preparing for possible futures. With regard to the refugee crisis, scenario workshops could anticipate future bundle crises like 'Multi-resilient refugee politics in the Federal Republic of Germany in 2030 in the face of climate change and international wars'. The approach corresponds to Nassim Taleb's proposal to anticipate unexpected chaotic problems of high impact (which he calls 'black swans') with methods of counterfactual thinking. Counterfactual thinking involves an examination of conceivable futures, which enables the strategic development of a higher general level of responsiveness to unexpected crisis situations (Taleb, 2008).

In the long term, further formats would be conceivable – here, too, preferably with heterogeneous groups – in which, for example, socio-economic and integration policy frameworks are rethought and tested in real experiments. So far there seems to be no welfare regime in the world that simultaneously fulfils all three requirements for social stability and high resilience: First, ensuring the greatest possible social equality; second, the broadest possible access to work; third, all this at low financial and bureaucratic cost (for more on this, see Chap. 8). Similar considerations apply to the various approaches to integration policy. Current discussions include the Anglo-Saxon multicultural model, the Dutch 'tolerance model', the 'promote and demand' approach in Switzerland and the assimilation model in France. Apart from the fact that historically developed socio-economic structures can hardly be transferred from one country to another, I am not yet aware of any sufficiently successful integration policy approach that could serve as an inspirational blueprint in the refugee crisis. All this suggests that it is not enough to look for 'best practices' or 'good practices' and thus to learn from the past, but it is necessary to open up to completely new approaches and thus 'to learn from the future'.

There is no shortage of formats that could contribute to 'rethinking' and link valuable experiential knowledge from different areas of society and test new approaches in innovation labs and real experiments: whether Design Thinking, Scharmer's U-process or the Collaboratory format (a neologism from 'Collaboration' and 'Laboratory') – all these and other approaches could serve to systematically and creatively use the collective intelligence of groups and – appropriately moderated over a longer period of time – of societies.

In summary, securing the future of a society within the framework of the principle outlined here requires,

- distinguishing between problem types of varying complexity and unpredictability and associated problem solutions, and
- addressing problems that are associated with a large number of unknowns (which is the classic domain of resilience research and practice) in a way of permanent learning, including self-observation and process observation in action ('deutero learning') and 'learning from the future' (by means of e.g. counterfactual thinking and scenario planning as well as innovation support).

4.1.3 Principle 3: The Multi-resilient Society Is Based on a Fine Balance of Decoupling and Knowledge Integration of Its Subsystems

The multi-resilient society is based on decoupling and knowledge integration of its inherent subsystems.

The former means that complex systems (such as modern societies) become more vulnerable the more they are interconnected, as is the case with power grids or in the global financial system, for example. Even a selective disruption, e.g. in a power grid spanning several countries, can trigger a severe chain reaction that leads to the collapse of the entire system (see also Zolli & Healy, 2013). A multi-resilient society should therefore be characterised by decoupling and independence of its inherent subsystems. As outlined above, this might include the independence of societal subsystems (such as counties, communities, or cities) from any external supply of energy, food etc. In this context, corresponding 'top-down' (e.g. decentralized urban planning, earthquake-resistant architectures) and 'bottom-up' (e.g. Transition Towns, to promote regional self-sufficiency) initiatives have already been discussed.

So far neglected and of considerable importance is a discourse located at the level of households and private communities influenced by the 'prepper' or 'survivalist' movement. In this social movement, individuals or groups proactively prepare for emergencies, including natural disasters, as well as disruptions to social, political, or economic order. Among the most popular forums are the American Preppers Network and Survivalblog. Out of doomsday opportunism, preppers and survivalists (both terms can be used largely synonymously) also pursue the goal of extensive independence from the outside world in order to be

4.1 General Principles for a Multi-resilient Society

able to bridge an emergency for an extended period of time (if not forever) without outside help. This includes the creation of a secret or defensible retreat, haven, or bug-out location (BOL) as well as the stockpiling of water (e.g. with water canisters), water-purification equipment, non-perishable food, clothing, firewood, seed, defensive or hunting weapons, ammunition, agricultural equipment, and medical supplies as well as the corresponding exchange of knowledge in survival techniques via internet forums. The scene, which originated in the conservative Christian right milieu of the USA, has now established itself via the internet as a subculture encompassing millions of people worldwide, especially in the Anglo-Saxon countries and Europe. During the ongoing COVID-19 pandemic, which was declared a Public Health Emergency of International Concern by the WHO in early 2020 (WHO, 2020), survivalism has received renewed interest, even by those who are not traditionally considered preppers (Garrett, 2020). Today, the prepper scene encompasses all social milieus – not only paranoid conspiracy theorists feel addressed, but also urban 'bourgeois bohemians', isolationist populists, such as environmentalists or even 'people who simply want to know what to do in the event of a failure of the water or electricity supply' (Duclos, 2012). A commonality between the Transition Town discourse and the survivalism/prepper discourse is that both are primarily driven by civil society initiatives and that they have emerged from a distrust of the state in terms of its ability to provide security. An essential difference is that the prepper/survivalism discourse is primarily determined by the motive of securing one's own survival, whereas the Transition Towns discourse deals with alternative, post-growth forms of living together. Moreover, as will be described in more detail in later chapters, the Transition Town initiative also tries to contribute to a more sustainable society by triggering change bottom-up.

In sum, promoting self-sufficiency is an essential feature of resilient societies. Ideally, however, this characteristic does not only extend to the level of household communities and municipalities. In line with the principle of subsidiarity, it should also encompass cities, counties, countries, states and, beyond that, communities of states (e.g. the EU) and supranational organisations (e.g. the UN). Self-sufficiency is, however, only one facet of the principle of subsidiarity.

The other characteristic implies close-meshed knowledge integration through networking between the decoupled and self-sufficient subsystems. The resulting 'balancing act' between decoupling and autarky on the one hand and close-meshed networking and knowledge management on the other is captured in two concepts: the subsidiarity principle and the cybernetic VSM model outlined above. The so-called 'principle of subsidiarity' (from Latin *subsidium* = 'help, reserve') establishes a precisely defined order of priority for social measures, whereby the larger social unit in each case should only intervene in a regulatory manner if the smaller unit is unable to do so. In other words, problem-solving should, as far as possible, only be undertaken by the individual, the smallest group or the lowest system level that is directly affected by a particular problem. Only if this is not or hardly possible, or if the added value of cooperation is obvious and it meets with general approval, should higher system levels intervene in a supportive way (Calliess, 1999). Both the subsidiarity principle and the cybernetic model of viable systems imply that each subsystem is largely autonomous and independent in its decisions and is self-sufficient and responsible for its own viability; at the same time, the system as a whole benefits from a distribution and integration of the different knowledge of all subsystems.

Originally, the approach of understanding society as a whole as the interplay of several functional areas or subsystems goes back to the founder of anthroposophy, Rudolf Steiner, who was active at the beginning of the twentieth century. With his model of social threefolding he distinguished three functional areas, which he regarded as autonomous and of equal rank, but different in their purpose and function:

- *Spiritual life,* which includes education, science, religion and culture – compared to the other areas, this area has a moral and culture-creating function;
- *Legal life,* which includes laws, rules and agreements of society and, in the broadest sense, politics – this area has a regulating, governing function of and within society;
- *Economic life,* which includes the production, trade and consumption of goods and services – this sector has a supplying and production function (Schmelzer, 1991).

Following this model, the model of social tripartism is increasingly referred to, emphasising the need for collaboration and knowledge integration between at least these three sectors. The Philippine sociologist and holder of the Alternative Nobel Prize Nicanor Perlas is probably one of the best-known representatives (Perlas, 2000). Also, since the early 2010s the EU Committee of the Regions et al. (2016) has invoked a 'triple helix' and more recently a 'quadruple helix' approach. The former distinguishes between policy, private sector and academia, and has in recent years mainly been applied by local and national innovation initiatives. The quadruple helix also distinguishes a fourth domain: civil society. It was used as a reference approach for the preparation and implementation of the 'Research and Innovation Strategies for Smart Specialisation'. Both, triple and quadruple helix approaches basically assume that only collaborative knowledge networking of all subsystems can contribute to regional development and to higher societal sustain-

4.1 General Principles for a Multi-resilient Society

ability and resilience. The EU Commission here explicitly refers to a transdisciplinary understanding of science (EU Committee of the Regions et al., 2016). Why is cross-sectoral knowledge integration so important for building societal future preparedness?

According to the observation of many contemporary analyses, it might currently be the only known answer to dealing sustainably with the highly complex risk potential of (post-)modern societies. Among the most influential contemporary analyses are Anthony Gidden's theory of structuration and the image of the 'side-effect society' (Nebenfolgengesellschaft) (Giddens, 1995), Ulrich Beck's theory of reflexive modernity and the image of the 'risk society' (Risikogesellschaft) (Beck, 2007), and the sociological systems theory according to Niklas Luhmann (1993). All of these and other approaches share the observation that

- (post-)modern societies are functionally differentiated, i.e. include functional subsystems or sectors (such as the economy, the media, science, politics), which, similar to the division of labour in companies, ensure an effective and efficient processing of information of events.
- The respective sectors of (post-)modern societies generate side effects, e.g. when attempts are made to manipulate complex systems such as the financial system, the ecosystem, etc. In the worst case, the side-effects can lead to human-induced global crises. This risk potential is inherent in the system, as the scope of the consequences could often not be recognised and adequately managed by the respective sectors (Beck 2007; Giddens, 1995).

Furthermore, the interdependence theory points out that the increasing political interdependence of modern societies can lead to a reciprocal import of problems from other states. The resulting increase in the number of stakeholders leads to a constellation which is all the more difficult to manage (Spindler, 2006). Accordingly, effective complexity management would not only require decoupling, decentralisation and system autonomy, but also the opposite, the collaborative networking and knowledge sharing of all agents/stakeholders from all society sectors (e.g. civil society, science, public sector, private sector) and at all levels of subsidiarity (e.g. communities, counties, states, supranational institutions). Similar to the EU, the idea of a subsidiary multi-level governance framework would integrate aspects of networking/integration (e.g. supranational law) as well as system autonomy (e.g. national sovereignty). On the global level, the ideal would be a so-called 'global governance' ensuring a world domestic policy (Nuscheler et al., 2007; Radermacher & Beyers, 2011).

However, the concrete implementation, especially the networking/integration aspect, is still largely underdeveloped. In this context, the Corona crisis has highlighted the vulnerability and fragility of global supply chains, leading to an increased shift towards selective self-sufficiency and decoupling. However, according to observers such as Richard Haass, President of the U.S. Council on Foreign Relations, or John Allen, President of the Brookings Institution, the trend might have also led to a reduced willingness or commitment to tackle regional or global problems, including climate change (Allen et al., 2020). In contrast, the aspect of integration – both between nation states and between stakeholders representing different society sectors – continues to be neglected.

The multi-resilient society would at least require an increased dialogue and knowledge exchange between the different society sectors (including civil society, science, the private sector and politics). In this respect, as will be outlined in Chap. 9, there are some initiatives that build 'build bridges' between these sectors. Moreover, given the fact that all sectors follow different functional logics, facilitators with a strong competence in shaping complex multi-stakeholder communication processes could make a significant contribution. They could moderate or even mediate the often quite conflictual communication between the stakeholders. Depending on the situation, the role of such an expert would be that of a moderator and mediator on the one hand, and knowledge broker on the other (see Sect. 9.1). In the broadest sense, the role of these experts corresponds to that of the 'translational leader' mentioned by resilience researchers Zolli and Healy (Zolli & Healy, 2013).

How can knowledge be distributed and integrated at all levels of subsidiarity in a multi-resilient society? A proven approach in organisational knowledge management is the communication forum. These are meetings fixed in time and space with the primary aim of exchanging experiences. Examples of this are weekly departmental meetings or monthly project meetings. Ideally, the discussions held here enable an 'integration of complex interrelationships' and 'broaden the view of the overall system' (Probst & Büchel, 1994, p. 148). Communication forums have therefore always been considered 'cross-discipline, cross-gender, cross-cognitive' (Hoffmann & Patton, 1996, p. 25). In communication forums, but also in other conceivable instruments of knowledge exchange, dialogue represents the central and most desirable form of conversation (Schein, 1993; Senge, 1996). Dialogue is to be sharply distinguished from debate, which is more about competing with the other and imposing one's own opinion. Dialogue, on the other hand, is about broadening one's own perspective and learning together. In this context, Peter Senge refers to the physicist David Bohm, who, according to Senge, sees thinking 'as a largely collective phenomenon':

> As with electrons, we must view thinking as a systemic phenomenon produced by our mutual interactions and discourses.

4.1 General Principles for a Multi-resilient Society

According to Bohm, dialogue originally meant '*a free flow of meaning between people, as in a stream flowing between two banks.*' In dialogue, Bohm argues, the group gains access to a larger

> reservoir of shared meaning that is not accessible to the individual. [...] The purpose of dialogue is to go beyond the limits of individual understanding. [...] A new form of thinking emerges, based on the development of a common sense [...] People are no longer in opposition to each other, nor can they be said to interact. Rather, they participate in this reservoir of shared meaning, which can constantly evolve and change (Bohm in: Senge, 1996, p. 292 f.).

In a dialogue, a group explores difficult complex issues from many different views. Individuals do not commit to their opinions, but they openly share their assumptions. This leads to the fact that the participants are able to unhinderedly explore and bring to the surface the full richness of experience and thought, but reaching far beyond individual opinions (Senge, 1996, p. 293).

Where does dialogue-oriented knowledge management take place? In the context of the multi-resilient society, think tanks or learning laboratories have proven to be suitable: Think tanks or competence centres represent places where knowledge can be pooled across disciplines, divisions and society sectors, and new ideas can be developed. The topic-related gathering of experts enables collective learning as well as development processes that would not have been possible for individuals. In the context of multi-resilient anticipation and strategy planning in the face of current future crisis bundles or bundle crises, a high diversity of experts representing as many disciplines and society sectors as possible, as well as a suitable dialogue format for integrating this diverse knowledge – such as the scenario method – would be important.

Unlike think tanks, learning laboratories are more likely simulations in which working groups experiment with alternative solutions to problems. In this way, mistakes can be made in a protected space, allowing for new insights and ultimately innovative solutions to problems to be developed and tested (Rehäuser & Krcmar, 1996). Overall, different types can be distinguished, depending on the focus and agents involved, e.g. Grassroot Labs (e.g. Trial and Error Kulturlabor or Vétomat), Coworking Labs (e.g. Open Design City or Webworker Berlin), company-owned Labs (e.g. Telecom School of Transformation or UFA Lab), or university/research-related Labs (e.g. the gameslab of HTW Berlin or the Berlin Knowledge and Innovation Communities) (SenWTF und Landesinitiative Projekt Zukunft, 2013). In addition, there are also mindlabs, living labs (Kleibrink & Schmidt, 2015) and real-life labs ('Reallabore') (e.g. Transition Towns) (Schneidewind & Singer-Brodowski, 2013). The latter are among the most frequently used instruments in the context of resilience and sustainability.

The list of measures presented here does not claim to be exhaustive. Knowledge networking within society requires not only the involvement of key individuals and the most important decision-makers, but is to be understood in the broadest sense as a multi-dimensional knowledge distribution and change process for society as a whole. Effective knowledge networking and integration in society therefore requires, above all, corresponding dialogue and learning competencies of those involved, which are developed within the framework of Principle 1. At the same time, collective competence is needed to use the combined knowledge and to incorporate it into good decisions in a time-efficient manner. But what are the conditional factors for time-efficient and good decisions in collective contexts? This question is addressed in the next principle.

4.1.4 Principle 4: The Multi-resilient Society Makes Collectively Intelligent Decisions

What are 'good' decisions? Good decisions have been made, when the intended goal is achieved. Good decision-making in social systems, such as organisations or societies, is ensured by high collective intelligence. A system with high collective intelligence is typically characterised by enhanced problem-solving capacities that cannot be attributed to the individual members of the social system; rather, they result from the interaction of these members. Aristotle is considered the earliest pioneer of collective intelligence. According to his so-called 'summation thesis', the decision of a larger group of people is better than that of a few experts (Aristoteles, 2006). Since then, the phenomenon of collective intelligence has been explored in a variety of disciplines and fields.

One of the best-known metaphors for collective intelligence or 'swarm intelligence' is the ant colony. At the beginning of the twentieth century, the biologist Morton Wheeler demonstrated that ants, which individually have only a very limited behavioural repertoire, achieve results in self-organising interaction that far exceed the abilities of individual ants (Feynman, 1992; Wheeler, 1911). In addition to ants, other insect states, such as bees and termites, also exhibit distinct collective intelligence (Hölldobler & Wilson, 2009). To date, the concept of collective intelligence has been increasingly applied to human social systems. Many observers view the internet as an information infrastructure that facilitates the coordination of decentrally distributed human knowledge and the aggregation of human intelligence (Rheingold, 2003). One of the most prominent researchers in this field is the Belgian Francis Heylighen. The research goals of his 'Global Brain' project, which he co-founded, include nothing less than the development of algorithms that are

4.1 General Principles for a Multi-resilient Society

intended to further the World Wide Web into a self-organising, learning network with collective intelligence in the sense of a 'Global Brain' (Heylighen, 2007).

What are the conditioning factors for collective intelligence and how can it be further developed with regard to societal multi-resilience? What does collective-intelligent decision-making imply, even under time pressure and information scarcity?

Various experiments from psychological and group dynamics research have shown that people can make better decisions together, but also worse decisions than individuals. Poor collective decision-making or 'swarm stupidity' is, according to current knowledge, primarily a consequence of herd behaviour. In his popular work '*The Wisdom of the Crowds*' (2004), sociologist James Surowiecki summarises the current state of knowledge from a multitude of studies and experiments in four points. According to this, swarm-stupid behaviour is characterised by at least one of the following errors:

- *Homogeneity:* This includes suppression of diversity of opinion.
- *Emotional factors:* Among these, Surowiecki counts emotional actions that develop their own dynamics in the group and override conscious thinking. Examples are the herd instinct, peer pressure and, in extreme cases, mass hysteria.
- *Imitation:* Sometimes decision-makers simply copy past decisions on the basis of a so-called 'information cascade' characterised by a number of people making the same decision in a sequential fashion.
- *Centralism:* Agents at the higher level of the hierarchy do not draw on the knowledge of those working on lower levels which are directly and locally concerned with the challenges. Surowiecki explains the accident of the space shuttle Columbia in 2003 primarily on the basis of this principle.
- *Lack of transparency:* As Surowiecki stresses, the US community was unable to prevent the September 11, 2001 attack because information from one sub-agency was probably not passed on to another. He points out that groups work best when they choose their own work and get the information they need themselves.

These points are essentially confirmed by Günther Dueck's recently published work '*Schwarmdumm*' ('*Swarm stupid*') (Dueck, 2018).

Other studies confirming the reverse suggest that swarm intelligence results from the diversity of the group and its ability to integrate this diversity of perspectives through successful communication. One of the most famous studies in this context comes from Anita Woolley (Carnegie Mellon University, Pittsburgh) and Thomas Malone (MIT Sloan School of Management in Cambridge, Massachusetts). They observed teams with different compositions tackling complex problems. The result: mixed teams performed significantly better than homogeneous teams. Another key finding was that many smart people with high intelligence quotients do not automatically make a group smarter. What matters most, is the diversity of perspectives and the group's ability to integrate this knowledge through empathetic communication (Woodley & Malone, 2011; Woolley et al., 2010). Confirming these and other research findings, James Surowiecki summarises the conditioning factors for collective intelligence as follows (Surowiecki, 2004):

1. *Diversity of opinion:* Each person should have private information.
2. *Independence of opinion:* Each person's opinion is respected and is not determined by the views of the others.
3. *Decentralisation:* People are able to specialise and draw on local knowledge.
4. *Aggregation:* (Dialogic) mechanisms are in place to turn individual judgements into a group opinion and collective decision.
5. *Trust*: Each person trusts the collective group to be fair.

In summary, the collective intelligence of a system depends on its ability to manage its communication in a way that integrates its inherent diversity of perspectives. This will be much easier to realise in small groups and organisations than in cities or societies, where communication processes can develop a wave-like, unpredictable and largely uncontrollable dynamic due to the influence of social media.[3] Regardless of the system level, however, whenever decision-makers come together, the crucial challenge is how to design a communication process that ensures time-efficient decision-making involving all perspectives concerned. A variety of guiding principles can be found here in organisational discourses that aim to increase the collective intelligence of social systems. Among these, the *'agile organisation'*, the *'learning organisation'* (Senge, 1996), and the *'teal organisation'* (Laloux, 2014), or the *HRO* (Weick & Sutcliffe, 2001) might be the most popular approaches and the most relevant for the topic of multi-resilience. All of these approaches broadly share the insight that diversity of perspectives is fundamental, but that mere democratic decision-making does not always prove to be optimal. As a disadvantage

---

[3] For more details, see Fathi & Osswald, 2017.

4.1 General Principles for a Multi-resilient Society

of democratic decision-making, experts criticise that the communication processes can be too time-consuming due to endless discussions (if unanimity is sought) and too unsatisfactory in the outcome due to a majority decision. In contrast to the consensus principle typical of democratic decision-making, some complexity managers prefer the so-called 'consent principle' for sociocratic decision-making. Consent is defined as 'no objections'. Members discussing an idea in consent-based governance commonly ask themselves if it is 'good enough for now, safe enough to try'. If this is not the case, there is an objection, which leads to a search for an acceptable adaptation of the original proposal in order to obtain consent. In other words: 'By consensus, I must convince you that I am in the right; by consent, you ask whether you can live with the decision' (Bockelbrink et al. 2022). The advantage of consent-based decision-making is that it allows for clear roles and responsibilities as well as time-efficient, collectively committed decisions (for more on this, see Laloux, 2014).

As an example for an integrative framework which incorporates many ideas and topics of the current, cross-disciplinary debate on decision-making, I will outline the so-called K-i-E method, which was invented by agile coach and consultant Richard Graf. In his major work *'Die neue Entscheidungskultur – mit gemeinsam getragenen Entscheidungen zum Erfolg' (The New Decision Culture – Success through Shared Decisions)* (2018), he presents a decision-making method that has received little attention so far, but that has proven itself in practice and could in principle be applied at all system levels. He calls it 'K-i-E' – the abbreviations stand for 'cognition' (in German: 'Kognition'), 'intuition' and 'emotion'. This implies the assumption, also held by other decision researchers and psychologists, that people base their decisions on two psychological systems that have different qualities: the emotion system (including intuition) and the cognition system. The former is characterised by largely unconscious, relatively fast, effortless, intuitive decision-making and is almost identical to Daniel Kahneman's 'fast thinking' (Kahneman, 2012). The latter is characterised by conscious, and relatively effortful, slow decision-making, almost identical to Kahneman's 'slow thinking' (Kahneman, 2012). Since both decision-making systems can come to correct conclusions on their own, but also to completely wrong ones – as the influential emotion researchers Daniel Kahneman (2012) and Gerd Gigerenzer (2007) independently point out – and also influence each other, it is advisable to integrate and address the impulses of both systems. Following the research of Benjamin Libet and the blindsight experiment, Richard Graf assumes that the decision impulses of the two systems express themselves differently in the conscious mind. These peculiarities can essentially be summarised as follows:

- The impulses of the largely unconscious emotion system emerge significantly faster than those of the cognition system and before they are conscious. This has been confirmed by Benjamin Libet's experiments (Libet, 1985). In reference to that, Graf states that any impulses that occur within 350 milliseconds are pure intuition or emotion. Everything that comes after that, from 550 milliseconds at the lates, is superimposed by influences from the cognitive system (Graf, 2018).[4]
- The information processing of the emotional system and the cognitive system is very different. As research experience with the phenomenon of blindsight suggests (see Weiskrantz, 1986), the emotion system can be specifically activated by yes/no questions and by scaling questions. The cognitive system, on the other hand, responds very well to open-ended questions (Graf, 2018).[5]

According to Graf, taking into account the peculiarities of the emotional and cognitive systems is the key to designing integrated decision-making processes. The core of his method is the so-called K-i-E scale as a standardized evaluation system for

---

[4] As early as the end of the 1970s, Libet measured in his series of experiments the time interval between a nerve activity in the brain that precedes a certain hand movement and the consciousness of the associated decision to act. He came to the remarkable conclusion that the point in time at which the voluntary decision became conscious was in every case clearly after the point in time at which the initiating nerve activity in the brain had already begun (Libet, 1985). What does this mean? Every conscious decision is preceded by an unconscious impulse, within less than 350 milliseconds (Graf, 2018). Libet thereby concludes that humans have 'no free will' with regard to their actions, but they do have a 'free won't', I.e. the possibility of a conscious veto as soon as the decision-maker becomes aware of the impulse (Libet, 1985).

[5] The term 'blindsight' was coined by Lawrence Weiskrantz, who examined the eyesight of test subjects who could not consciously perceive their surroundings, let alone describe them, even though their sense of sight was intact. As their visual centre in the brain was damaged and thus could not produce conscious visual impressions, they were also unable to deal with open-ended questions such as 'What do you see?'. They could, however, correctly identify simple representations that could be answered with 'yes/no', e.g. 'Is that a red playing card?' or 'Is that an O or X?', in the vast majority of cases. In doing so, they denied having perceived the information, but could somehow 'feel' what might be correct. Apparently, the processing of the visual information took place without conscious perception (Weiskrantz, 1986), but on the basis of their intuition. According to Graf, this suggests that the stimulus also reaches the emotional system, whereas defective processing in the cognitive system cannot construct a visual image. Thus, Graf emphasises that the emotion system, which operates in the unconscious, expresses itself in a 'go/no-go' impulse. In contrast to the largely unconscious emotion system, the conscious cognition system works more slowly, but in a more detailed and goal-oriented way. According to Graf, the cognitive system can be activated very well by open-ended questions such as 'What do you see?' (Graf, 2018).

4.1 General Principles for a Multi-resilient Society

all types of decision-making processes to which intuition responds. Similar to the traffic light model commonly used in coaching and facilitation, which distinguishes between 'green' (Go), 'orange' (Maybe) and 'red' (Not-Go), the K-i-E scale is based on a spectrum ranking from 1 to 10, with the same three ranges of meaning. The larger range represented by scale values 1–5 represents various gradations of 'no'/'no go'; scale values 6–7 represent 'maybe' or 'it depends'; and the upper scale values 8–10 represent gradations of a clear 'yes' or 'go'. The No-range is proportionately larger because risk aversion and associated 'no' responses take on a relatively high importance in our archaic subconscious. Stages 6–7 ('Maybe' or 'It depends') still correspond to 'no' in the broadest sense, but they usually involve solutions that can be accessed through the cognitive decision-making system (Graf, 2018).

Self-Exercise: Making Coherent Individual Decisions with the K-I-E Scale

> The purpose of this short self-exercise is to become more aware of the 'very first' impulses coming from the emotion system / intuition. In a situation where you have to decide between two or more options (e.g., 'which T-shirt should I wear?'), it would be recommended to ask your intuition (the very first impulse) for each option: 'Should I wear the black T-shirt? Should I wear the grey T-shirt? Should I wear no T-shirt/something else?'. If your emotion system responds (within the first milliseconds) to one option with a clear 'yes', the decision is made.
>
> Sometimes, however, several options are good enough. In this case, Graf would recommend to use scaling questions and the corresponding 10-point scale. To do this, you can modify the questions, e.g. 'How much do I like this T-shirt?'. The option with the highest score might be the best one. However, Graf points also out, that a score between 1 and 7 ('Not Go' would be 1–5 and 'Maybe' would be 6 and 7) could still be accompanied by an unsatisfactory feeling. Furthermore, in the range of 6–7 the tendency would be more or less a clear 'no': If you are in the 6 and 7 range, Graf recommends asking an open question that addresses your cognitive system. He recommends always to formulate open questions in a solution- or resource-oriented way, such as: 'What would be need to grade this option with a score of 8 or higher? How can I make this happen?'. Then ask the initial scale question again. If your intuition is still at a value less than or equal to 7, the decision should be considered as a 'no' (Graf, 2018).

This process which systematically integrates cognition and intuition can also be applied to collective decisions. The typical challenge here is that as the size of the group increases, so does the diversity of perspectives and thus the inherent complexity of the group. At the same time, the number of decisions decreases in most groups because a jointly supported decision that reflects the perspectives represented in the group requires correspondingly costly communication processes. The advantage of a scale-question based approach, by making use of a traffic light model or the K-i-E scale described here, is that it provides a 'standardised language' that can map the intuitions and viewpoints of group members very quickly. The procedure is essentially based on several steps that characteristically alternate between a collective activation of the emotion system and the cognition system. The following is an illustration of such a decision-making procedure using the K-i-E method as an example:

1. As a pre-step, the group should commit to this decision-making process. This can be made transparent within seconds by asking the scale question, for example: 'On a scale from 1-10, how much do you agree to work with the K-i-E in the following decision-making process?' If there is any response between 1 and 7, the process should already be stopped and should not be proceeded.
2. When starting the first step of the actual decision-making process, it is recommended to first address the emotion system by making use of the scale question, e.g.: 'To what extent will this project be successful?' This makes it possible to gain a transparent overview of the participants' positions within a very short time. For the query the participants can write their number down or they can use a numbered set of cards – the card numbers would correspond to the ten K-i-E score values. If all participants had a value of 1–5 (no) or 8–10 (yes), there would already be a uniform result in terms of 'no go' (leave the project) or 'go' (go for it). Mostly, however, a mixed result is the case.
3. In this case, Graf recommends to shift to a resource-oriented question: 'What do you think would be needed to get closer to the functional "go" range (8-10) on the K-i-E scale?' This follow-up question is tailored to the cognition system and allows for solution-focused dialogue. In doing so, the solution-oriented phrasing of the question is intended to prevent the discussion from drifting into time-consuming lamenting and problematising. An essential aspect of solution-oriented questions is that those who have concerns can no longer dwell on their criticisms, but are held constructively accountable by having to put their safety concerns into action. In this step, Graf recommends not to start with participants who are already in a 'no go' range (1–5), but rather with those whose positions are in the 'maybe' range (6–7).

4.1 General Principles for a Multi-resilient Society

4. After the solution-oriented discussion with the moderate worriers should have brought to light a number of resources and improvement measures, the third step can again be a scale query with all participants, targeting their intuition. It is very likely that the values of all worriers will have changed towards the functional 'go' range and in most cases even converged. If there is still no joint decision tendency, the cycle can be continued with an alternation of questioning the emotional system (using a scaling question with the K-i-E scale) and the cognitive system (using a resource-oriented question) (Graf, 2018).

According to Graf, the step cycle described in step 2 and 3 should not be repeated more than three times. If, even after the third cycle, the discussion produces a result in which the gaps between participants' scores are two or more points, it is unlikely that a collectively supported decision will be reached. In such cases, Graf recommends either voting collectively 'no' for the project, or authorising one person to make the final decision for all participants. This procedure essentially mirrors the principle of the consultative decision and consent. But here, too, according to Graf, it is important that the group has previously agreed on this very procedure in a commitment process (explained in the pre-step above). This is the only way to ensure that collective decisions are sustainable.

In addition to the commitment process, the K-i-E facilitation applications can also be used to inquire about or establish prioritisations ('How high do you rate the urgency and importance?'). Regardless of the different query contexts, the following applies: The group decision always begins with a scale-based query that appeals to intuition. If an indeterminate picture emerges after the query, there is a switch to a solution-oriented resource question that addresses the cognitive system. The purpose of the process is to come to a mutually shared decision. This can even occur if the participants too far apart in their assessments and evaluations and therefore cannot agree on a common action. In such a case there is agreement in disagreement, and the group can, for example, decide to abandon the project (Graf, 2018). A simplified version of this process, which can be applied to individual or collective decisions, would be based on the method of systemic consensing. Similar to K-i-E, it works with a scale question 'On a scale of 1-10, where 10 is the highest resistance to a decision option: Where is your/your resistance the highest?' The option that has the lowest score, i.e. the least resistance, would be considered the best (Maiwald, 2018).

The purpose of the moderation applications, such as the K-i-E method or systemic consensing, is to arrive at jointly supported (and thus highly collectively intelligent) decisions in a time-efficient and solution-oriented process. In principle, these methods could be applied at all system levels and in a variety of complex and chaotic situations. However, they require at least basic communication skills among all participants, a specially trained moderator, and the ability for cultural adaptation (by the participants/parties involved?).

Overall, one could critically ask whether decision-making procedures that are currently being tested at the level of groups and organisations work for modern societies. It can certainly be assumed that these tools can be applied in teams with societally relevant decision-makers, such as political committees and strategy meetings or cross-disciplinary think tanks. However, with regard to the society as a whole, it is necessary to consider how the methods and approaches outlined above can be integrated within the framework of cross-sectoral communication in such a way that they make communication more efficient, more transparent, simply 'smarter' for society as a whole. If this is the case, the population as a whole could also become more collectively intelligent and thereby more capable of managing complexity.

4.1.5 Principle 5: The Multi-resilient Society Has a Strong Learning Culture

If resilient individuals are characterised by psychological resistance factors (see Principle 1), it can be assumed that the collective psyche of societies also represents an essential dimension of collective resilience. This can be described as a 'resilience culture' or a 'collective resilience spirit'. Sect. 3.2.2 has already outlined different research approaches in this area. These different contributions reveal several complementary components that characterise a society's resilience culture. They can be summarised as follows:

Resilience-promoting virtues: It should be obvious that there is a reciprocal influence relationship between the resilience-promoting virtues of a society and the resilience behaviour of the individuals belonging to it. The influence of resilient individuals on the resilience of the system as a whole was reflected in conjunction with Principle 1 and will not be further considered in the following. The extent to which the behaviour of individuals is determined by their cultural imprint is primarily the subject of comparative cultural and intercultural research. Among other definitions, Geert Hofstede emphasises culture as 'a collective programming of the mind' (Hofstede, 1994). From this perspective, the resilience of communities

4.1 General Principles for a Multi-resilient Society

and societies is manifested in collectively lived virtues.[6] One of the most researched areas of collective resilience at present relates to specific ethnic groups, such as the above-mentioned Vietnamese Boat People in the 1970s, who demonstrate high levels of academic resilience despite discrimination by the majority society. An example for the society level I mentioned is the Japanese perseverance mentality (gambaru) in the face of diverse natural hazards.

Collective perception: Similar to individuals, societies also judge adversities collectively, which also reveals their collective self-image and worldview. Does the society perceive the problem as a threat or as an opportunity? Closely related to the assessment of events and one's own vulnerability is the question of identity. What in the collective self-concept should be preserved at all costs and what is allowed to be let go and change? This question does not only touch upon the typical conflicts of multi-culturalism and integration politics, it can be also applied to coping with past traumatic events. Depending on the prevailing understanding of resilience, at least two resilience strategies can be distinguished: First, the static resilience concept ('firm as a rock', resilience 1.0) which focuses on protecting the 'self-core' and protecting it from stressors that compromise the integrity of this core. Second, the evolutionary resilience concept (Resilience 2.0) which emphasises adaptation, change and development and also considers far-reaching changes to one's self in order to evolve with the crisis.

Cohesion: The ability of communities and societies to cope with any kind of crisis requires a correspondingly strong social cohesion, which is essential for the emergence of any collective self. As mentioned above, comparative indicator-based research cites several factors that provide information on the social cohesion of societies. These include mutual trust, trust in institutions and prosocial behaviour. If these factors are strong, the stability – and, in the broadest sense, the cohesion – of a society is likely to be high. Political practice shows that social cohesion can be developed even in societies with relatively poor indicator scores. A typically used nationalist strategy, for example, could be to try to distract the population from internal social problems (such as corruption, criminality, inequality, etc.) by establishing a demarcation from an external enemy. This strategy of shifting the focus away from problems, as can currently be observed in Russia, for example, can be seen as a cultural coping strategy, but is unlikely to be sustainable in the long term.

---

[6] However, and this is the justified criticism of transcultural studies and corresponding disciplines, it may be questioned whether cultures can be seen as homogeneous spaces. In reality, there are various contexts which form belonging and identity. In this regard, a Japanese engineer who is passionate about liberal ideas and the Hippie movement, would have more in common with a Saudi Arabian engineer sharing exactly the same passions than with a compatriot who considers himself a right-wing conservative and has totally different passions.

The dimensions mentioned here – cultural virtues, collective perception and social cohesion – have a decisive influence on the resilience culture of a society. As a rule, these cultural dimensions are shaped at all levels of subsidiarity (see Principle 3) of the (multi-)resilient society.

> **Case Study China: Resilience Virtues, Threat Perception, Cohesion**
> In their widely acclaimed analysis *'The Chinese – Psychogram of a World Power'* the authors Baron and Yin-Baron (2018) outline the cultural characteristics of China. With an age of around 5000 years, China represents the oldest civilisation alive today. It is therefore particularly suitable as a case study for cultural resilience. In the light of the cultural dimensions presented above, China's resilience profile could be briefly drafted as follows:
>
> - *Resilience Virtues*: Similar to the Vietnamese Boat People, many Chinese at home and abroad are above the global average in their academic performance. Discipline, diligence and above all constant learning and self-improvement are writ large in Chinese culture, in keeping with Confucian moral teachings. Whereas almost 100 years ago the famous sociologist Max Weber located the cause of China's 'backwardness' in this very Confucian philosophy, today Baron and Yin-Baron (2018) attribute the relatively high learning ability of the Chinese to this very influence.[7]
> - *Assessment and identity*: Many Chinese study abroad, but tend to return to the motherland. Although Baron and Yin-Baron (2018) assume that the youngest generation of Chinese students is more than ever influenced by the West, both authors assume that the sense of belonging to the Chinese community and culture is very strong. According to Baron and Yin-Baron (2018), a distinctive feature of the Chinese sense of belonging and collective self-image is the identification with a culture instead of a nation. In terms of China's collective resilience, this leads to a certain flexibility, since Chinese identity is not bound by national borders. At the same time Baron and Yin-Baron (2018) point out that China's perception of its own
>
> *(continued)*

---

[7] Nevertheless, they admit that there were certainly phases of stagnation in the course of historical development, which they attribute to an orthodox interpretation by those in power at the time. This shows that one and the same culture can inhibit or promote resilience and development.

4.2 Conclusion

> **continued**
> vulnerability is mainly characterised by an avoidance of social instability (if necessary, with repressive violence). This is due to the long historical experience in which the country has experienced multiple divisions and conquests.
>
> - *Sense of belonging*: According to Baron and Yin-Baron (2018), the sense of belonging for most Chinese is focused on the circle of the extended family and, in the broadest sense, to personal friends and acquaintances. Outside this nucleus, Chinese interactions tend to be characterised by distrust. In some ways, this leads to high resilience at the subsidiarity level of households and private communities, but at higher system levels it could lead to less cohesion, engagement and thus to higher vulnerability. Contrary to widespread assumptions, identification with a Chinese state is relatively weak in China (unlike e.g. Japan). According to Baron and Yin-Baron (2018), this is also a possible explanation for the fact that the Chinese state has been militarily defeated and conquered by much smaller and technologically less developed states on several occasions in history. Nevertheless, China remains the oldest living civilisation.

Regardless of their case-specific differences, the above dimensions can be found in the (multi-)resilience culture of every society. An underlying challenge for all of them is to maintain a delicate balance between permanence and change. The aspect of permanence emphasises the unchanging core of values that characterises the collective self. Moreover, it establishes which coping practices have stood the test of time, or in other words: What can remain as it is. The aspect of change, by contrast, opens up new coping practices. This can (but does not have to) go hand in hand with a changed self-image in times of crisis.

Although the balance between permanence and change may vary in the culture of each society and depending on the context, all resilience cultures have one common criterion: They all place a central value on learning as part of their cultural self-image. In a resilience culture, this is reflected in the promotion of two types of learning: On the one hand, the learning of established and proven resilience practices (best and good practices), and on the other hand, the development of new resilience practices (emerging or innovative practices). The former has been paraphrased as 'learning from the past', the latter corresponds to 'learning from the future', as mentioned above. To illustrate this, two short case studies are presented below:

Case Study Cuba: Best Practice Resilience Strategies as Central Components of Education Policy

The case of the so-called 'Cuban Way' illustrates the importance of education and skills development in fostering a resilient population. Cuba is regularly exposed to natural disasters. Between 1996 and 2002, the island was hit by six severe hurricanes that killed 16 people, a minimum of the total 665 deaths in the affected neighbouring countries. In some cases, e.g. Hurricane Charlie, fewer people died in Cuba (4 people) than in Florida (30 people), even though both countries were affected in a similar way and even though Cuba is relatively poor compared to Florida and lacks technical resources. The International Secretariat for Disaster Reduction (ISDR), the UN body that focuses on disaster reduction, points to Cuba's resilience-building education policies. Disaster preparedness and prevention are considered central components of the curriculum in schools, universities and workplaces. Citizens are continuously informed and trained from an early age onwards (Mohideen, 2010).

Case Study Netherlands: Innovative Practices and Transformative Resilience in Response to Flood Disasters

The Netherlands was and still is threatened by flood disasters. Until the 1990s, the resilience strategy focused on building dikes, which reflected the idea of 'keeping the water away'. The new resilience strategy of 'leven met water' provides a classic example of crisis transformation – in other words, the use of new opportunities arising from a crisis. Since then, floating architecture has dominated the cityscape of the Netherlands (Leven met water, n.d.; Christmann et al., 2011).

4.2 Conclusion

How can societies achieve greater overall resilience – multi-resilience – in the face of the unpredictable and complex variety of crises? According to the present analysis, the implementation of at least five principles contributes to an increase of societal multi-resilience:

4.2 Conclusion

- Principle 1: Promote resilient individuals
- Principle 2: Dealing confidently with unknowns
- Principle 3: Decoupling and knowledge integration of social sectors
- Principle 4: Collectively intelligent decision-making
- Principle 5: Strong learning culture

The first principle highlights the essential dimensions of a resilient society based on the four-quadrant model. The four-quadrant model originates from the Complexify tradition (see Chap. 3) and provides a rough location of the different dimensions of protection and corresponding indicators that are related to the resilience of a society in different crisis situations. The following Principles 2–5 can also be assigned to one of these quadrants in terms of their focus (Table 4.4).

Principles 1–5 do not replace the diverse other contents of Table 4.1, which may differ in their details from society to society. Rather, the principles listed here represent generally applicable orientation criteria that are independent of cultural manifestations and can be applied to as many different types of crises as possible. In the light of the five principles, it becomes clear that the multi-resilience of a society is primarily related to its ability to deal with complexity. Dealing with com-

Table 4.4 Core principles of a multi-resilient society

	Subjective	Objective
Individual	Principle 1: Psychological resilience	Principle 2: Being comfortable with unknowns
	Equanimity	Learning from the past (best and good practices)
	7 resilience factors	Learning from the future (emerging and innovative practices)
	Lifelong learning	
Collective	Principle 5: Resilience culture	Principle 3: Decoupling and knowledge integration of subsystems
	Learning and education as a central virtue	Self-sufficiency/autarky of the subsystems
	Balance between stability and change	Multi-level- and multi-sectoral knowledge management
		Principle 4: Collectively intelligent decision-making
		Conditional factors of collective intelligence (diversity of opinion, etc.)
		Time-efficient joint decision-making (e.g. K-i-E method, sociocracy, systemic consensing)

plexity requires the ability to deal confidently with events that cannot be predicted and therefore imply a high number of unknowns. Learning has been identified as a necessary core competence in this context. This means, on the one hand, the systematic integration of knowledge about best practices for problem-solving (learning from the past) and, on the other hand, the modification and innovation of new practices (learning from the future). Secondly, this presupposes extensive independence and collectively intelligent decision-making capacity of all societal subsystems and, at the same time, mutual collaboration and knowledge integration.

At present, no society can be observed that has implemented these principles and established full multi-resilience. However, there are a number of initiatives in different society sectors that provide important impulses and contribute to societal multi-resilience. These include, for example, Transition Towns and similar initiatives, platforms of transdisciplinary knowledge exchange (e.g. the European School of Governance), open-access organisations promoting mental resilience (e.g. Art of Living), and prepper initiatives. What all these and other initiatives have in common is that they target and are already implementing small-scale change (especially at the subsidiarity levels of households, communities and organisations). Multi-resilience development, encompassing society as a whole or even the supranational level and involving all system levels, has yet to be achieved. Principle 5 – the development of resilience cultures – may prove to be the most difficult to implement, as cultures change relatively slowly and can at best be indirectly influenced by the implementation of the other principles.

Is multi-resilience even sufficient to secure the future of societies in the twenty-first century? Not necessarily. It is true that an increased ability to learn and deal with complexity can prove useful in almost any crisis and in the face of bundles. However, this will not be enough to cope with man-made existential crises. To use the words of disaster expert Nick Bostroms:

> Our approach to existential risks cannot be one of trial-and-error. There is no opportunity to learn from mistakes. The reactive approach – see what happens, limit damages, and learn from experience – is unworkable. Rather, we must take a proactive approach. This requires foresight to anticipate new types of threats and a willingness to take decisive preventive action and to bear the costs (moral and economic) of such actions. (Bostrom, 2002, p. 5)

The multi-resilience concept presented here also takes into account a proactive component – but it takes the occurrence of the disaster as a given and focuses only on protecting the affected system. However, this will prove insufficient in the case of man-made existential crisis scenarios, for example, a superintelligence or nanotechnology out of control. There is also a need for proactive influence on the

4.2 Conclusion

environment and thus prevention of the occurrence of the crisis and mindful reflection of the consequences of societal action. Securing the future of society will also have to take into account and integrate other guiding concepts.

Several guiding concepts are currently dominating the discussion. These are, above all, the 'sustainable society', the 'developed society' and, since the turn of the millennium, the 'resilient society'. Sometimes they are used vaguely, and not infrequently they have to serve as buzz words in the PR strategies of governments, companies and civil society organisations. Nonetheless, each of these different concepts is likely to make an indispensable contribution to an integrative discussion of how to secure the future of societies. The model of the developed society focuses on how a society can promote its inherent potentials to the greatest possible extent and satisfy its needs accordingly in the best possible way. By contrast, the sustainable society focuses on the side-effects that result from societal development processes and actions and how these side-effects can be avoided. Finally, the concept of a (multi-)resilient society focuses on the question of how society can develop in the face of unavoidable side-effects.

Part II

Future Preparedness in the Twenty-First Century: Development, Sustainability, Resilience

The Developed Society 5

The guiding concept of a 'developed society' is characterised by various meanings. The following chapter provides a basic understanding of the inner and outer context of societal development (Sect. 5.1). This is followed by a more in-depth look at four discourses that contribute to the multifaceted concept of a developed society (Sect. 5.2). To what extent is it possible to measure and to compare the development of a society? In addressing this question, indicators of the four discourses (Sects. 5.3 and 5.4) are finally brought together.

5.1 The Inner and Outer Context of Societal Development

The developed society represents by far the oldest guiding concept. In his famous book '*Who Rules the World?*', British historian Ian Morris defines social development as the ability of a society to come to terms with itself and the world (Morris, 2011). Development can be examined from an inner and outer perspective.

In a philosophical context, development can be understood as evolving oneself through 'folding-out' one's own (in-folded) abilities to master one's own problems (Nuscheler, 2006, p. 226). Transferred to the society context, a developed society is characterised by its ability to provide its own population with essential goods and services, thus enabling them to live in dignity and unfold their potentials. Here, an important purpose is to create an efficient national economy that enables the broadest possible use of the society's own (especially human) productive forces. This goes hand in hand with the goal of achieving the lowest possible poverty level, which correlates with the broadest possible capital-forming middle class. This, in turn, would provide incentives for the government to establish structures such as the rule of law, protection of private property, social services, and educational institutions to optimise the productivity of the population. In turn, it can tax these generated benefits and thereby finance public services. This process typically describes the development cycle of a modern or modernising society. Conversely, a poorly developed society is characterised by a structure in which a rich elite enriches itself from the unprocessed resources of its country and in which many human productive forces are mostly unused. Instead, capital is invested in unproductive luxury goods and a police/military apparatus designed to maintain the status quo (Zakaria, 2004; Nuscheler, 2006).

If we can assume that 'development' follows a goal or is at least driven by a purpose, the notion is indirectly related to 'success'. Transferred to the context of society, one might ask: 'What is the fundamental purpose of any society? What do societies generally try to achieve?' From these questions it is possible to derive generalisable criteria for 'success', which might apply to all societies, Let us consult sociology on this: according to the definition of Ferdinand Tönnies, who analytically introduced the concept of society into the establishing discipline of sociology in the nineteenth century, society is understood as a special form of mutual intentional 'affirmation' of people. In other words, people come together and profess a common affiliation. According to Tönnies, this mutual affirmation is a means to jointly achieve one's individual needs and goals (Tönnies, 2005). This thesis is confirmed by Talcott Parson's perspective of structural functionalism. According to this, a society is only formed by individuals, if it is able to satisfy human needs permanently through certain social institutions and structures. From the sociological perspective sketched here, individuals and society are in an interdependent relationship and 'reproduce' each other. On the one hand, society supports individuals in satisfying different needs; on the other hand, it thrives on holding these individuals together – and this cohesion is cemented by manifold aspects such as norms, values, laws, ideologies, etc. that are lived and observed together (Parsons, 1971).

The goal of satisfying the (basic) needs of the population therefore appears to be a central aspect of a developed society. Models of human needs are widely accepted in different disciplines, ranging from conflict studies (Galtung, 2008) and international conflict work (cf. Sarvodaya) to developmental psychology (Maslow, 1954), economics (Drekonja-Kornat, 2001), humanistic psychology (Rosenberg, 2009) or international development cooperation (Steinberg, 1985). Apart from some differences in detail, all needs concepts assume that all people, regardless of time and culture, have needs. Among these are material needs such as food, health, shelter, security, etc., but also intangible needs such as identity, freedom, creativity,

5.1 The Inner and Outer Context of Societal Development

education, meaning, in the broadest sense spirituality and so on. Accordingly, a society fulfils its *raison d'être* most optimally by ensuring that the needs of the population are satisfied through appropriate social, economic, political and cultural framework conditions.

Peace and conflict studies teach that social conflicts and crises always result from unsatisfied and frustrated needs. As such, needs can be regarded as the deepest motivations and as universal components of every human being. At this point conflict management distinguishes between 'needs' on the one hand and 'strategies' or 'positions' on the other. Unmet needs are the deepest root of any conflict, and they are a universal component of any human being. As soon as one or more needs are not met, it leads to frustration and can result in conflict. Strategies or positions can be understood as means to fulfil these needs. A conflict results from a contradiction between opposing strategies/positions represented by different conflict parties (cf. Galtung, 1998, 2008).

Consequently, societal success or development prevents social conflicts and ensures cohesion within a society. One area of central importance to avoiding social conflicts is socio-economic well-being. To conclude, the success and the development of a society is characterised by its ability to provide its population with the broadest possible satisfaction of its (basic) needs.

In order to develop the necessary resources to maintain and increase its welfare, a society's core interest is mainly to increase its *economic performance* which is usually measured in terms of economic growth. Accordingly, growth is now considered one of the most important political goals of any society in today's system of global capitalism. Against this background, the term 'competitiveness', which originated in business administration, has been applied to national economies for some years. This is not without controversy. In the business context, 'competitiveness' means the ability of companies to sell their range of goods or services at a profit in the relevant markets and to secure or enlarge their market share. In the context of national economies, competitiveness could also refer to the ability to permanently generate more income and welfare than other countries.

In this context, the military and power political context would view war as, in the popular words of Carl von Clausewitz, the 'continuation of politics by other means' (von Clausewitz, 2010) as the preferred strategy for establishing national unity and securing economic growth and access to territorial resources. The context of competition here would be measured by military strength and 'hard power' in general. With the establishment of international law and the United Nations after World War II, the importance of this context may have diminished, although the ongoing Russo-Ukrainian War has led to a significant shift in priorities among European countries toward strengthening their military capacities.

Another context, perhaps still the most important at present, is largely influenced by the ongoing trend of progressing globalisation (let alone current attempts of the international community to isolate Russia) and the development of knowledge-based economies. The context of competition would here mainly occur between locations for investments and innovations. Accordingly, in the discourse on the future viability and thus the success of today's knowledge societies, the 'war of talents' is being emphasised more than ever. This includes the aspect of 'soft power' and ability of a society to attract direct investment and a highly qualified and inventive workforce, thereby ensuring its innovative capacity and economic growth potential. According to the latest Global Competitiveness Index[1] from 2019 the ten economies with the highest growth opportunities were dominated by East Asian countries, partly Anglo-Saxon countries and Northern European countries. Singapore ranked first, followed by the USA (2), Hong Kong (3), the Netherlands (4), Switzerland (5), Japan (6), Germany (7), Sweden (8), UK (9), and Denmark (10) (WEF, 2019a).

The competition discourse is closely linked to an outer perspective that emphasises objective and tangible criteria of societal development. In this context, the geostrategic dimension is probably one of the historically most far-reaching. At any time in history, communities, and later states, have strived for an expansion of their own geostrategic influence, be it military, technological, economic, geostrategic, political, or cultural. In the case of so-called 'empires', this development goal might have been achieved to the highest degree. In a cultural comparison, however, peace and conflict researcher Johan Galtung distinguishes between expansionist and non-expansionist powers. The former include, for example, the Islamic empires from the sixth to sixteenth centuries and the Western colonial empires from the seventeenth century onwards, or the USA today, all of which had a strong sense of mission (be it of Islamic, Christian, or Western values) and a desire to shape other states in favour of their culture. Representative of a non-expansionist power was, for example, the Chinese empire, which by the fifteenth century had enough resources to usurp the world. Nevertheless, it decided (quite possibly on the basis of an arrogant assessment) to withdraw into itself and leave the non-Chinese 'barbarians' to their own devices. (Galtung, 1998; Baron & Yin-Baron, 2018).

In today's empire discussion, there are several models that describe the complex global situation. Depending on the model, it is assumed that one (USA [e.g. Münkler, 2007] or Europe [e.g. Posener, 2007]), two ([USA and Europe [e.g.

---

[1] This ranking, compiled by the World Economic Forum (WEF), compares 144 economies with regard to their level of growth opportunities. The index takes into account data on infrastructure, health, education, labour market efficiency, goods market efficiency, level of technological development, etc.

5.1 The Inner and Outer Context of Societal Development

Bollmann, 2006]) or even three (USA, Europe and China [e.g., Khanna, 2009]) postcolonial empires dominate. If societal development is dependent on international power and influence, we can observe that the determining criteria are shifting as globalisation progresses. For a long time, the attainment of 'hard power' was considered the most important development goal, especially in the context of military strength. Since the end of the Cold War, this has become increasingly less important, partly because classic interstate wars have decreased over the last decades. Today, military activities mostly take place in the area of international counter-terrorism and above all in the context of domestic 'counter-rebel' operations (HIIK; AKUF, 2012). The ongoing Russia-Ukraine War, however, might be an exception from that. Compared to 'hard power', economic strength and above all 'soft power' (Nye, 2004, 2011), i.e. cultural and religious appeal as well as dominance in science and technology, are becoming increasingly decisive today. Although military power and geostrategic considerations have received increased attention due to an increase of the world military expenditure up to 1.9 trillion US$ (SIPRI, 2021), the following chapters will mostly focus on soft power. However, geopolitical reality today illustrates that any dimensions, such as economic growth, technological innovation, but also geostrategic considerations are intertwined. Currently, the most impressive examples can be found in Chinese foreign political initiatives, such as the Belt and Road Initiative (BRI) or the current construction of the world's largest solar energy park.

Case Example: China's Belt and Road Initiative (BRI)

The Belt and Road Initiative (BRI), or One Belt One Road (OBOR), was officially announced by the Chinese government in 2013. It involves infrastructure development and investments in 70–150 countries and the cooperation with international organisations in Europe, Asia, the Middle East and the Americas. The project is scheduled for completion in 2049 (Godehardt & Kohlenberg, 2017), which coincides with the 100th anniversary of the People's Republic of China. While the Chinese government calls the initiative 'a bid to enhance regional connectivity and embrace a brighter future' (China.org.cn, 2015), critical observers assess it as a push for a China-centred global trade network or even a complex strategy to become the next lone superpower by 'absorbing' Europe (Wo-Lap, 2016; Arafeh, 2018; overall cf. Fathi, 2021a).

Case Example: China's Global Energy Initiatives

> In March 2022, China officially announced plans to build solar and wind parks with a capacity of around 450 gigawatts. This capacity is equivalent to the output of 450 smaller nuclear power plants and more than four times as much electricity as the failed Desertec project that was supposed supply for Europe. As with the EU's plans in the Sahara, China's gigaparks are to be built in desert regions, especially in the Gobi desert. Since renewable energies are considered to be one of the largest global markets of the future, China's initiative might be driven by economic interests. The initiative might also be motivated by geopolitical, power political, and even resilience political considerations. Those who consistently develop such power grids can hope for a higher security of supply, falling energy costs and additional sources of income from patents and technology exports. From a resilience political perspective, power grids are considered an extremely critical infrastructure, because after all, a stable energy supply is nothing less than the basis for a flourishing economy (Schulz, 2022). In the long term, China plans to expand this initiative into a global power grid. These plans are only made possible by a new technology in which China is increasingly taking the lead: so-called ultra-high voltage (UHV) technology, in which 800,000 volts are conducted via direct current cables. This technology allows for electricity to be transported over very long distances with comparatively low losses. The Chinese government has been setting the course for global market leadership for more than 10 years (Schulz, 2018).

The competition for soft power (and to a certain extent also hard power) is evident in the current 'competition of the moderns' (Jaques, 2012). Until recently, the concepts of 'modernity' as a desirable stage of social development were defined and advanced almost exclusively by the West. After Japan, modern societies are now also emerging in other regions of the non-Western world, for example China or India, which stand for different 'ideas of the good life'. While Western concepts associate 'the good' or 'the desirable' with individuality, person-centred freedom, human rights and democracy ('All different, all equal'), for new economic powers such as China or Singapore 'modernity' tends to mean collectivism, unity, harmony and stability ('The mountain does not move'). This competition will intensify in the future with the rise of non-Western states (Jaques, 2012).

Overall, it can be summarised that the concept of a developed society essentially includes aspects such as the satisfaction of the population's basic needs through social welfare, which in turn is based on social cohesion as well as economic performance and and the population's ability to create value. Although armed conflicts are unlikely to disappear, soft power in the sense of cultural attractiveness, economic performance and technological innovation could become the most important criteria in the development discourse of the twenty-first century (at least in the first half).

5.2 The Developed Society in the Light of Four Discourses

The previous considerations show that societal development relates to different aspects which are analysed in different disciplines. The following sections focus on four discourses which currently seem to be the most influential on the concept of the developed society:

- The *Development Policy* discourse measures the development of a society on the basis of material, particularly economic, indicators. This framework leads to a fundamental distinction between the First World (developed countries), the Second World (transition countries), and the Third World (Third World countries). An alternative distinction is that between modern (First World, partially Second World) and traditional societies (particularly Third World) (Sect. 5.2.1).
- *Happiness Research* criticises the one-sided focus of development studies on material, above all economic, indicators. This discipline measures the development of a society by the degree of subjective well-being of the population, which in turn provides a variety of indirect conclusions about social cohesion, conflict potential, and thus the extent to which basic needs are met and how effective general policies are (Sect. 5.2.3).
- The *Developmental psychology* discourse evaluates the development of a society primarily according to the degree of complexity of the prevailing worldview – this includes, for instance, the cognitive and moral lines of development (Sect. 5.2.4).

5.2.1 Contributions from Development Studies and Modernity Research: Development and Modernity as an International Orientation Framework for Social Development

Confirming the above-mentioned theses, the notions 'poverty' and 'modernity' are largely considered as orientation criteria that determine the developed society. The corresponding research field is called 'development studies'. It deals with the causes, aspects and consequences of underdevelopment and development of societies. There are many synonyms for the term 'Third World country', such as 'Third World' or 'Fourth World'. The German Federal Ministry for Economic Cooperation and Development (BMZ) uses the English-language term 'LDC' (Least Developed Countries). It can be criticised that the notion of 'developed countries' suggests that these countries have completed the evolution process and that they don't need to develop further. As we will see in later chapters, this assumption is by far not adequate.

In the context of development studies, development is largely defined as the opposite of poverty. Here, development can be understood as

> the independent unfolding of the productive forces to provide the whole of society with essential material as well as liveable cultural goods and services within the framework of a social and political order that grants all members of society equal opportunities, allows them to participate in political decisions and to share in the jointly generated prosperity (Nuscheler, 2006, p. 245).

Poverty, by contrast, means the

> inability to lead a life that meets economic, social, and other standards of human well-being (Nuscheler, 2006, p. 149 f.).

In other words, poverty implies that the individual's opportunities to develop are limited. This affects economic (income, consumption), human (access to education, health, clean water, safe housing), political (human rights, political participation), and socio-cultural (participation in social life) capabilities. It also affects equal opportunities between genders and other social and cultural groups.

Both concepts, poverty avoidance and development promotion, are essential assessment criteria. Since the 1950s there have been so-called 'lists of characteristics' naming the central development problems. Although this is controversial within development studies, the approach provides a general orientation on the criteria that constitute a developed society:

- *Economic criteria:* The concept of value added is central to national accounting. Other and related characteristics include average income per capita, equality of income and wealth distribution (Gini-Coefficient), employment rate, foreign debt, savings and investment. Another model is the economical three- or four-sector model that divides economies into several sectors of activity: extraction of raw materials (primary), manufacturing (secondary), and service indus-

5.2 The Developed Society in the Light of Four Discourses

tries (tertiary) (Fisher, 1939). The fourth sector refers to post-industrial or knowledge-based economies, as an increasing proportion of economic activity is not directly related to material goods. This includes information technology, media, research and development, and information-based services (Economic Activity, 2017).

- *Ecological criteria:* Many Third World countries are particularly affected by ecological problems. According to studies by the World Watch Institute and the UN Environment Programme, 90% of the global species extinction, deforestation and soil erosion takes place in Third World countries. Environmental crises therefore hit them particularly hard, as natural resources are among their most important wealth and basis of existence (Nuscheler, 2006).
- *Demographic criteria:* Most Third World countries are characterised by a relatively high birth rate, which is mainly due to a low integration of women into the labour market. As infant and child mortality rate have declined with the improvement of medical care in recent decades, population growth has increased uncontrollably. A resulting factor is the so-called 'Youth Bulge' (youth surplus).[2] In this context, empirical studies show that the combination of a strong youth bulge and a weak absorption capacity of the labour market can lead to social protests (Nuscheler, 2006), such as the Arab Revolution 2011 (Fathi & Karolewski, 2015).
- *Public health criteria:* Current research differentiates between typical diseases in poor countries and so-called 'diseases of affluence', previously called 'diseases of rich people'. Diseases of poverty tend to be largely infectious diseases or the result of poor living conditions, such as tuberculosis, malaria, and intestinal diseases. In contrast, diseases of affluence mainly include chronic, non-communicable (non-transmissible) diseases and other physical health conditions for which societal conditions and personal lifestyle related to economic development are considered an important risk factor. These include type 2 diabetes, asthma, coronary heart disease, cerebrovascular disease, hypertension, peripheral vascular disease, obesity, cancer, some types of allergies (Ezzati et al., 2005), depression and other mental health conditions (Luthar, 2003).
- *Socio-cultural criteria:* Poor countries are more often characterised by a form of interplay of social, cultural and religious behaviours in which human

---

[2] According to the definition of Graham Fuller, who introduced the term in 1995, a youth bulge exists wherever 15–24-year-olds make up at least 20% of the total population, or 0–15-year-olds at least 30% (Fuller, 2003).

development potential is inhibited. The most widespread characteristic is inadequate education and high rates of analphabetism, especially among women. Another feature is an inhibition of reinvestment through treasury accumulation by the upper class (Nuscheler, 2006).

- *Political criteria:* A number of characteristics are cited with regard to the political problems of Third World countries. These include comparatively high levels of general corruption and a high degree of instability and inefficiency of political institutions and a resulting low acceptance level in the population. Other criteria cited by development researchers include political structures and human rights violations that might have an inhibiting effect on the population's individual potential for technological and business innovation. The most serious criteria are political instability and even state collapse (Nuscheler, 2006).

To date, a large number of indicators have been established that allow countries to be compared with regard to the above-mentioned criteria. Among other things, they provide information on economic performance (GDP, GNP, per capita income, etc.),[3] perceived corruption (CPI, measured by Transition Network 2015) poverty (MPI, measured by OPHI, n.d.) or human development. The latter, i.e. the United Nations Human Development Index (HDI), is one of the leading country-comparative indicators of well-being and correlates per capita income, education and life expectancy. Another set of indicators with completely different criteria is provided by the historian Ian Morris. His approach measures societal development by four characteristics: energy yield (e.g. calories/person extracted from the environment), urbanisation (size of the largest city), warfare capacity (e.g. number of troops, firepower of weapons), and information technologies (available means to exchange and process information) (Morris, 2011). None of these various indices is free from criticism, yet together they provide a widely accepted guide for defining differences in development between individual societies.

We can provisionally state: Development is considered an internationally recognised and comparable benchmark for assessing a society's success in satisfying needs and developing individual and collective potential. According to the indicators mentioned above, the societies of the West and increasingly also of some East Asian countries perform best in international comparisons. At this point, at least two important points of criticism are raised in the debate. First, the results from development studies presented here could give rise to the Eurocentric assumption of a 'superiority of Western culture'. This alone is debatable with regard to the question of the causes of development and the currently emerging

---

[3] Relatively up-to-date statistical data on all countries of the world, from the field of economics and beyond, are provided, among others, by the CIA World Factbook and the World Bank.

5.2 The Developed Society in the Light of Four Discourses

'competition of the moderns'.[4] Second, all studies and indices used in development

---

[4] Depending on whether the causes are located outside or inside society, a distinction is made between exogenous and endogenous theories. It is noticeable that each of these theories has explanatory gaps – including the culturalist theory, which assumes the 'superiority' of cultures (it belongs to the endogenous theories). An overview of the discourse shows that endogenous and exogenous theories might complement rather than contradict each other.

The *exogenous theories* assume that the cause for the lower level of development of Third World countries are to be found in their exploitation by the developed industrial countries. They emphasise the dependence of Third World countries on developed countries, and the causes that appear to lie in Third World countries (such as corruption) are regarded as consequences of this dependence. Accordingly, most arguments that make reference to the external factors of (under)development run under the umbrella of the so-called 'dependency theories". Among other things, they provided important insights into unequal trade conditions in which the world market tends to be shaped in favour of developed industrial societies – these insights have been taken up in the sustainability debate. At the same time, the greatest weakness in the explanatory scheme of the exogenous theories is that they can neither explain the industrial development of the 'white colonies" (South Africa, Canada, Australia), nor the rise of Japan to the top of the world's most powerful economies since the late 1960s, nor the rapid development of the Asian Tiger and Panther states into industrial and post-industrial societies. All of them have been former colonies (Nuscheler, 2006, pp. 212–218).

Endogenous theories point to internal system factors driving developmental change. Here, at least two types of theories can be identified. Firstly, theories that argue with the climatic and geographical framework conditions of a society and, secondly, theories that cite culture and mentality as an essential development factor.

The first, i.e. the geographical perspective, assumes that a country's (un)favourable geographical location is the cause of its development situation. This manifests itself, among other things, in a landlocked location with high transport costs or a climate characterised by long periods of drought and/or deadly tropical diseases (Sachs, 2005). Closely related to the geographical perspective is the so-called 'resource curse theory", which assumes that societies that are heavily dependent on the export of mineral and fossil raw materials generally have lower development incentives than resource-poor countries. Possible reasons are: a lower competitiveness of the remaining economic sectors, state misuse of revenues from the raw materials sector, lack of investment in education and thus the absence of an efficient middle class that contributes significantly to economic growth and, conversely, can demand democratic participation and the rule of law from the state leadership (e.g. cf. Zakaria, 2004). The resource curse theory reaches its limits when it comes to explaining the very high level of development of resource-rich Western societies, e.g. the USA, Canada or Norway.

The other major perspective among the *endogenous theories* is determined by the so-called 'modernisation theories' and the culturalist theories, according to which the causes of social development would lie in its interior, especially in its culture and in its degree of social-psychological development. An important inspiration was given by sociologist Max Weber's analysis, '*The Protestant Ethic and the Spirit of Capitalism*', published in 1905, in which he attempted to trace the industrial-capitalist revolution back to the ethic contained in Protestantism and Calvinism. At the same time, Weber concluded that other cultural areas (even the Catholic south of Europe) would not possess the capacity for capitalist-industrial development – an assertion that, from today's perspective, does not hold true. Typical culturalist references to the alleged 'otherworldliness' and thus 'hostility to development' of Buddhist and Hinduist or Muslim culture/religion can hardly explain the current dynamic development processes in countries such as Thailand, Malaysia or Indonesia (Nuscheler, 2006).

studies turn out to be little or not at all suitable for assessing the development potential of high-ranking post-industrial societies, as assessment criteria such as poverty, literacy and life expectancy mainly concern less developed countries or even 'non-modern' societies.

The development stage of the modern industrial society is therefore regarded as a critical threshold in this context. From an economic perspective, it is emphasised that the modern industrial society is associated with an increase in material prosperity, which results from its value creation processes based on the division of labour, technology and mass production (Slater & Tonkiss, 2001). In this context, in the field of development cooperation, the degree of modernity is used as the benchmark for designating a 'developed society'[5] (Nuscheler, 2006). The basic distinction between modern and traditional cultures is also frequently referred to in cultural comparative anthropology. A common metaphorical description is the distinction between 'cold' and 'hot societies' going back to Claude Lévi-Strauss (Treichel & Mayer, 2011) or Raymond Dasmann's differentiation between 'ecosystem' and 'biosphere people' (Dasmann, 1988).[6]

These and other perspectives are part of a discourse on modernisation that goes back to the nineteenth century, and which primarily identifies the particularities of modernity in terms of the following features:

- *Democracy:* In his famous article *'The Open Society and its Enemies'*, the science philosopher Karl Popper propagates the model of the 'open society', which is characterised by releasing 'the critical faculties of man' and by democratic decision-making. In this context, Popper sees the special potential for coping with complexity not so much in the rule of the majority, but in the possibility of being able to vote out the elected government without violence,

---

[5] The assumptions and goals of development cooperation can certainly be criticised. Activists such as the Big Mountain Aktionsgruppe e. V. (1993) or Vandana Shiva (2005), for example, point out that the goals and assumptions of development cooperation stem from a one-sided market economy-oriented perspective. The classification of traditional, subsistence-oriented societies as 'underdeveloped" would fail to recognise the importance of securing largely independent, socially and ecologically sustainable modes of existence (Shiva, 2005).

[6] The 'colder' a society is, the stronger its efforts to preserve its traditional cultural characteristics (Treichel & Mayer, 2011). As 'ecosystem people' they are characterised by subsistence farming and a low sphere of influence on nature (Dasmann, 1988). By contrast, 'hot' societies have a greater drive for profound and rapid modernisation (Treichel & Mayer, 2011). Typically, they are characterised by a relatively large influence on the biosphere (hence Dasmann's designation of 'biosphere people") and attempt to adapt nature to humans.

5.2 The Developed Society in the Light of Four Discourses

thus institutionalising a trial-and-error process (Popper, 2003, 2004). Hence, the contribution of democracy consists in the institutionalisation of non-violent conflict resolution.

- *Middle class/participation:* As mentioned elsewhere, many theories of modernity assume a broad middle class and high civil participation as a central feature of modernity. According to the sociologist Seymour Martin Lipset, the emergence of modern societies is primarily tied to the emergence of a prosperous middle class, which in turn fosters a democratic form of government. Put simply, according to Lipset, there is the following causal chain: economic development → rising levels of education → development of rational and tolerant attitudes and behaviour among citizens → democratisation of the middle class → emergence of civil associations demanding to participate in politics (Lipset, 1960). An affluent population would have better opportunities to participate materially in the affairs of state through taxes and would thus be empowered in their right to participate. Conversely, the government would have an important incentive to ensure rule of law and overall legal certainty (Zakaria, 2004). Analogously, Tatu Vanhanen's power dispersion theory assumes that modern societies are characterised by a wide dispersion of power resources, which in turn leads to a higher degree of democratisation, as no group would be able to displace competitors and maintain hegemony (Vanhanen, 2014).
- *Differentiation:* Niklas Luhmann emphasises the differentiation into subsystems/sectors outlined above as the most essential feature of modernity (and postmodernity). The various social spheres, such as law, politics, economics or religion, detach themselves from each other, functionalise themselves and create new spaces for communication and solutions. Unlike in traditional societies where, for example, religion, dominates everything, in differentiated societies the different functional areas have to be taken into account and weighed against each other (Luhmann, 2006).

What is striking about the discourse on modernisation is its Western-centrism. Do the principles presented here also apply to non-Western industrial societies? The discourse on modernisation has to date been concerned with the controversial question of the extent to which the transition to the developmental stage of a modern industrial society is linked to Western culture.

In view of the 'competition of the moderns' illustrating that modernisation does not necessarily require adopting Western culture, this question and the associated approaches[7] have become obsolete today. In this regard, the post-industrial societies of the West (above all large parts of Europe and the USA) are confronted with the emerging, new industrial and partly post-industrial societies of East Asia – above all the new world power China and the Tiger States.[8] The assumption that a modern industrial society always goes hand in hand with a democratic and pluralistic form of government, which was still taken for granted in the discourse on modernisation in the twentieth century, remains controversial today. WhileIt is true that some observers note that industrialisation in South Korea, Taiwan and, to a limited extent, Hong Kong is accompanied by a democratisation of political systems (Kirchberg, 2007; Castells, 2003), cross-culturally sensitive experts such as Martin Jaques (2012), Baron and Yin-Baron (2018), and Franka Lu (2019) dispute in their extensive analyses of China's rise whether modernity must necessarily go hand in hand with the communication and decision-making structures of an openly democratic society of Western standards. The case example of the city-state Singapore remains one of the most cited.

---

[7] In this context, the sociologist Max Weber attributed the rapid development of the West in the last centuries to the 'specific rationalism of occidental culture' (Weber, 1965, p. 20). Specifically, he postulated a special connection between Protestant ethics and the onset of industrialisation or capitalism. Weber denied non-Western cultures, for example Islamic and Confucian cultures, the ability to develop to the complexity level of modern industrial societies.

Another, typically Eurocentric, perspective was coined by the US political scientist Francis Fukuyama. He achieved fame in the early 1990s with his controversial thesis that followingafter the collapse of the Soviet Union the principles of liberalism in the form of market economy and liberalism would finally prevail everywhere in the world. With the victory of this model, the driving force of history, in the Hegelian sense, there would be a final synthesis in which there would no longer be any world political contradictions (1992). However, the 'end of history" which he propagated has not materialised. Today, Fukuyama argues that the *'end of history'* consists in the integration and assimilation of non-Western cultures into Western culture (Salzborn, 2014). However, in view of the competition of the moderns, this cannot be fully confirmed.

[8] The tiger economies currently include South Korea, Taiwan, Singapore and Hong Kong. Following the rapid economic rise of these economies, the so-called *panther states of* Thailand, Indonesia, Malaysia and the Philippines were accorded a similar development opportunity in the 1980s and 1990s. However, their upswing was so slowed down in the course of the Asian economic crisis in 1997 that none of the countries has yet been able to make the full leap to become an industrialised state.

Modernisation Without Democracy: The Case of Singapore
The former British colony became independent in 1965 after a brief federation with Malaysia. The city-state's prospects were gloomy at the time. The population was uneducated and fragmented into different ethnic groups; the city-state had neither mineral resources nor sufficient drinking water. The transition from a Third World country to one of the world's most prosperous industrialised states in just one generation is now largely credited to the leadership of Lee Kuan Yew, the city-state's first prime minister, who held that office from 1959 to 1990. He explicitly rejected liberal democracy. He tolerated no opposition and severely restricted freedom of the press. In return, he offered a high degree of political stability and economic development (Rist, 2017). Experts attribute Singapore's rapid and sustained development to the characteristic combination of at least three factors:

- *Rule of law:* Singapore is characterised by strict laws and a high level of surveillance, but also by a very low level of corruption. This leads to a relatively low crime rate and a high rule of law compared to the rest of the world.
- *Economic freedoms:* Singapore has a very successful market economy. The founding fathers were consistent in their relocation of multinational corporations and business-friendly laws. The pronounced entrepreneurial freedom attracted investors from all over the world and contributed significantly to Singapore's economic development (Lee, 2000).
- *Lifelong learning:* Singapore is characterised by a publicly communicated Confucian ethic that emphasises values such as discipline, diligence and education. This factor may have contributed to Singapore's strong learning culture. In all areas of the Pisa learning comparison, Singapore and most other tiger economies (including Japan and China) are worldwide leading and consistently score well above the OECD average (OECD; Spiewack, 2017). In addition, lifelong learning is systematically promoted among the population through programmes such as SkillsFutureCredit or the School of Continuing and Lifelong Education (Kamei, 2017).

Critical observers emphasise the elitism in Singapore and China that favours party-affiliated and certain population groups (in China it is the Han Chinese, in Singapore generally the Chinese ethnic group). This contradicts the above-mentioned heterogeneity factor for the emergence of collective intelligence. Another related criticism concerns the high degree to which development successes are tied to charismatic leaders. In Singapore it was Lee Kuan Yew, in China it was Deng Xiaoping and now it is Xi Jinping. This might indicate that success cannot be easily replicated and that sustainable development requires a lengthy process of institution-building (Rist, 2017; Ebbighausen, 2015).

The example of the city-state of Singapore, as well as the megacities of China and other emerging societies, highlight the key function of cities as hubs of development. Thus, regardless of whether a society is run democratically or autocratically, or by which national culture it is shaped, the modernisation of a society is likely to depend directly on the nature of the cities and how they expand. According to this line of argument, cities are not only transcultural expressions of modernisation, but modernisation itself could be specifically promoted by the formation of cities. In this context, the US economist Paul Romer caused a sensation with his proposal to establish so-called *'charter cities'* (often translated as 'special administrative zones' or 'free cities') in low-growth and structurally weak countries as a means of combating poverty. The concept implies that the government of a Third World country selects a non-settled piece of land that would be placed completely under the administration of a modern foreign government. The idea is to create a motor of growth in this artificially created special zone that attracts foreign investment and thus triggers development-promoting spill-over effects in the surrounding area. Romer often cites Hong Kong under British colonial rule as a success example and summarised his concept in a 2009 TED Talk with the provocative phrase 'Canada is developing a Hong Kong in Cuba' (Romer, 2009). A major incentive is supposed to be the legal certainty guaranteed by the external government in the charter cities, which automatically attracts investors and leads to impulses for economic and technological modernisation. From a critical perspective, this concept is assessed as neo-imperialist or neo-colonialist (Lenz & Rucklak, 2016). Romer counters that no one would be forced to move to the newly established city and that the land would be allocated voluntarily – hence, there could be no question of colonialism. Moreover, he rejects charter cities as a measure in humanitarian emergency areas, such as Haiti after the great earthquake in 2010. The fact that there would be no democratic elections in a charter city, as living conditions in the city would be dictated by politicians from abroad, is also seen as problematic. Romer counters that the residents would at least have the option to 'vote with their feet', i.e. the possibility of immigrating or emigrating. Furthermore, the financial feasibility of the concept is disputed – here the assessments diverge extremely (Himmelreich, 2010).

Regardless of the open outcome of this debate, the concept of the charter city highlights the increasingly central role of the city in the development discourse – this does not only apply to the development context of the modern industrial society, but especially to the post-industrial knowledge and information society.

5.2.2 The Threshold to Post-industrial Development Using the Example of the Smart City Debate

The post-industrial 'knowledge' and 'information society' is at the centre of an extended discourse on modernisation. According to Amitai Etzioni (2009) and Daniel Bell (1985), such a society is characterised by the priority that is given to information processing and so-called knowledge work as central factors in value creation. Here, in a society where many routine tasks will be taken over by algorithms in the future, people can only maintain their lead by learning skills, which include 'lifelong learning' and above all creativity and innovation (Spath, 2013).

The post-industrial information society is characterised by a form of production that is shaped by the *'second machine age'* (as it is described in English) or the 'fourth industrial revolution'[9] (as it is described in German). A key feature of this is the merging of technologies, in which the boundaries between the analogue (including the physical and biological) and the digital spheres are blurring. Today's mobile devices alone provide users with unprecedented capacities for information processing and storage, as well as almost unlimited access to knowledge.

When attempting to measure societal development in the context of the Second Machine Age, access to Information and Communication Technologies (ICT) and their degree of use has to be considered as crucial orientation factors. One of the most discussed concepts in this context is the digital divide in the population' access to these very technologies. Put simply, this concept states that the easier a member of the population has access to modern communication technologies, the better their chances for social and economic development (Marr & Zillien, 2010). Reducing digital divide would thus be related to improving the distribution of knowledge and wealth in a population. The ICT Index developed by the UN attempts to quantify this digital divide. The determining criteria are different levels of access (access, use, active appropriation) and six types of technology (fixed-line

---

[9] Industry 4.0 is a future project within the framework of the German government's high-tech strategy, which aims to advance the computerisation of traditional industries. The goal is the intelligent factory (smart factory), which is characterised by adaptability, resource efficiency and ergonomics, as well as the integration of customers and business partners in business and value creation processes. The term is intended to express the fourth industrial revolution. The first is characterised by mechanisation using water and steam power, the second by mass production using assembly lines and electrical power, and the third made use of electronics and information technology to automate production. Building on this, the digital revolution can be observed since the middle of the twentieth century, which characterises the fourth industrial revolution (FMER, 2019).

telephone, mobile phone, computer, digital television, internet, broadband at a certain speed) (Zillien & Haufs-Brusberg, 2014).

Other attempts to capture post-industrial development potential are currently the World Competitiveness Rating and the World Digital Competitiveness Rating, which are developed annually by the IMD in Lausanne. The former measures the competitiveness of societies based on the criteria of 'economic performance', 'government efficiency', 'business efficiency' and 'infrastructure'. In 2018, the US topped the ranking, followed by Hong Kong, Singapore, the Netherlands and Switzerland. The remaining places in the top 10 were predominantly occupied by Nordic companies (IMD, 2018a). In 2021, the US still tops the rankings for the fourth year running. China has risen 15 places in that time, from 30th to 15th (IMD, 2022a).

The second, the World Digital Competitiveness Rating, measures not so much the prevalence of ICT in the population, but rather the extent to which a society uses digital technologies. Using the factors of knowledge, technology and 'future readiness', it draws conclusions about the digital competitiveness of societies. In 2018, the USA led this ranking – in 2017 it was still in third place – followed by Singapore, Sweden, Denmark and Switzerland (IMD, 2018b). In 2021, Western societies are still leading, *topped with Switzerland (1st), Sweden (2nd), Denmark (3rd), the Netherlands (4th), and Singapore (5th) (IMD, 2022b).*

Following on from this, the EU's Technological Achievement Index (TAI) uses various indices (e.g. per capita consumption of electrical energy, share of hi-tech exports in total export volume) to determine the extent to which a society succeeds in developing and distributing high technologies and building up the necessary skills base among the population. Different to the other indices, there is no country ranking (Measuring-Progress.eu, n.d.).

Unlike the development of the first, second and third industrial revolutions, the process of the fourth revolution is exponential. The main reason currently cited is what is known as Moore's Law. This rule of thumb states that every 18–24 months the number of electronic circuits in the processor (transistors) doubles while costs fall. This dynamic is reflected in numerous current technical breakthroughs in other areas, which also appear to be mutually reinforcing. These include, for example, artificial intelligence (AI), robotics, the Internet of Things, autonomous vehicles, 3D printing, nano- and biotechnology, material sciences or quantum computing. The exponential dynamics make it impossible to estimate exactly when a technological singularity might occur. The following often-cited anecdote provides an impression of the erratic character of this dynamic:

> **The Chess Anecdote as an Example of Exponential Growth**
> A long time ago, the game of chess was invented in India. The Indian emperor was eager to reward the inventor of this game, as he had found great pleasure in it. As a reward he should make a wish and not be too modest about it. The latter said: 'My lord, give me one grain of rice for the first square of the chessboard, two grains for the second square, four grains for the third, and for each additional square twice as many grains as for the preceding one.' The emperor felt offended as he was not yet aware of the extent of the request. On the first fields, the number of grains of rice increased exponentially. On square 1 it was one grain of rice, on square 2 two, on square 3 four, on square 4 eight grains of rice, on square 5 sixteen, and so on. The number increased rapidly and kept increasing. On field 10 there were already 512 grains of rice and and all the previous fields taken together resulted in a total of 1023 grains of rice. From the tenth field onwards, the increase became even more drastic, so that at field 64 alone, the total number of rice grains reached was 9,223,372,036,864,775,808.
>
> This famous anecdote serves as a typical example to illustrate the extent of the current and future exponential development in the Second Machine Age (Brynjolfsson & McAfee, 2014).

Another cornerstone of the discourse in this chapter is the 'smart city'. It currently represents a 'solidified' form of a 'highly developed society'. The smart city harnesses technological innovations to better manage the global urbanisation trend[10] and associated economic, social and political challenges. Currently, there are various smart city projects. They differ in terms of their objectives, each addressing a different challenge. For example, they may focus on resource and energy efficiency or on adapting to demographic change and population growth. In the broadest sense the term also includes social innovations, albeit facilitated by the use of digital innovations, which contribute to a better and more sustainable life in the city. This includes, for example, concepts for citizen participation or sharing (share economy) (Vanolo, 2013; Libbe, 2014).

In a fully developed smart city, the entire urban environment is equipped with sensors that make all the collected data available in the cloud. This creates a permanent interaction between city dwellers and the technology surrounding them; they thereby become part of the technical infrastructure of a city. The potential of the smart city only unfolds within the framework of a 'cooperative network of relationships between citizens, city administration, business, science and politics' (Jaeckel & Bronnert, 2013, p. 16).

---

[10] According to forecasts, by 2050 more than two third of the world's population is expected to live in cities (Ritchie & Roser, 2019).

In Germany, the 'Leipzig Charter on Sustainable European Cities' adopted in 2007 created the basis for a sustainable European urban policy in the sense of the smart city. Across the EU, the Horizon 2020 programme (FMER, 2019) is one of the most important publicly funded initiatives with the aim of developing European cities into smart cities. One example of an EU-funded pilot project is 'Open Cities'. Here, from November 2010 to December 2013, various EU programmes supported cities in developing smart cities. The aim was to test methods of open innovation – such as open data, living labs, open sensor networks or crowdsourcing – on the basis of pilot projects as well as to network the stakeholders involved and to implement a better exchange of best practices (Mühlhans, 2018). Often, municipal political agents cooperate with universities or/and private research institutions in such initiatives. In addition to funding programmes initiated by politicians, the private sector often acts as an important initiator of smart city projects (see examples below).

In recent years, various areas of society have been discussed in connection with 'smartness'. Which dimensions are associated with smartness? In a country-comparative study by the Vienna University of Technology, the University of Ljubljana and the Delft University of Technology, the following dimensions are important to consider in dealing with smart cities: economy, population, governance/administration, mobility, environment and living (Giffinger et al., 2007). It can be observed that the projects differ in their focus due to their very different visions. In the following, I provide an overview of the relevant dimensions of smart cities:

- *Economy:* Smart economy refers to how economic productivity increases through the digital networking of a multitude of actors at local, regional and global level. Ideally, it is characterised by innovative business models and is closely linked to the knowledge society and the creative economy, which are promoted by the digital transformation. As a rule, smart city projects are carried out by heavyweights in the private sector, and it is not uncommon for cities to commit to one or more providers over a period of several years. IBM, for instance, is considered a pioneer when it comes to advising municipal stakeholders. In 2014, the company reported that it advised cities in the USA (4), Australia (2), Lithuania, Taiwan, Mexico, Japan, Belgium, China, South Korea, Kenya, South Africa, Ireland (one city respectively) on various smart city dimensions (Wilson, 2017). Similarly, the Siemens Group offers a wide range of consulting and services for cities worldwide (Hartmann, 2012). Other companies, such as the energy company Vattenfall or Cisco Systems have a more focused offering. Vattenfall specialises in building smart grids and regulating household energy consumption (Vattenfall). Cisco focuses more on areas of life that can be controlled by communication networks, especially traffic (Cisco).

- *Governance, politics and administration:* 'Smart governance' refers to the administrative dimension of smart cities. The focus is on citizen-oriented political decision-making in which the civil population is strongly involved in urban development processes. The aim of smart governance is therefore to make measures, planning and decision-making processes more transparent and participatory. In this context, forms of open government, e-participation and technologies play a major role (Chapman et al., 2006). Open government or e-government involves the optimisation of administrative processes through information and communication technology by making public services accessible online and thus turning citizens into 'customers' of government services. By contrast, e-participation emphasises the role of the citizen as a responsible partner in political decision-making (Kaiser, 2001). The henceforth resulting digital implementation of democracy is also referred to as e-democracy. E-democracy means the implementation of communication, information and decision-making processes within and between institutions of the legislature, citizens and the private sector as well as state institutions through the use of information and communication technologies (Meier, 2009). Typical examples include virtual town hall meetings on an Internet platform (USA) or the use of wikis and blogs to redesign the urban landscape (Melbourne, Australia) (Ramge, 2010).
- *Civil society:* Due to the permanent exchange of data, the relationship between the smart city and its population is regarded as interdependent. On the one hand, the life of the inhabitants is to be made easier and better through technical innovations and real-time information exchange; on the other hand, the residents, for their part, shape the city through their own initiative. This includes civil society initiatives that engage socially or creatively with urban space, and political decision-making processes related to smart governance, in which all social groups can influence the development of their city. Another particular form of smart civil society can be seen with regard to sharing culture, which can be both commercial and non-commercial. Sharing culture is about sharing devices and infrastructure for the purpose of ecological and economic benefits, for example car sharing or ride sharing. It is based on the commons theory developed by the political scientist Elinor Ostrom, according to which cities with their resources and spaces are understood as 'urban commons'. In this regard, the smart city would be a common in which all decisions were negotiated between all those who use the city (Helfrich, 2012). A distinction is made between different levels of decision-making – for instance between decisions at

the level of neighbourhood associations, at the district level, or at the city-wide level. Since it is more difficult to ensure equal participation at a larger scale, such as the city level (Harvey, 2013), Ostrom proposes a polycentric governance system that coordinates and merges decision-making processes at smaller scales (Carlisle & Gruby, 2017).

- *Sustainability:* Smart city visions also refer to many issues that have been formulated and addressed in the course of the sustainability discourse (Vanolo, 2013). This discourse has been increasingly present in urban development since the 1990s and is concerned with how to design development that meets the needs of the present without compromising the needs of later generations. The vision of the sustainable city, which is discussed in more detail below, emphasises above all the ecological, economic and social dimensions and includes, among other things, the following points in particular: Sustainable use of renewable resources and minimal use of non-renewable resources (ecological dimension); emphasis on a regional circular economy, minimal transport intensity and a sustainable economy (economic dimension); socio-cultural mixing of the city, co-responsibility and democratic participation of the population (socio-cultural dimension) (cf. the Brundtland Report, edited by Hauff, 1987). These and other sustainability topics are put into practice in the smart city. Sustainable smart city visions increasingly include ideas for regional commodity cycles in which things are produced where they are consumed. Examples of this in the energy sector are solar thermal energy so that hot water can be produced locally, photovoltaics for the production of solar electricity or geothermal energy for heating buildings. Current examples of sustainable energy supply and consumption are Vattenfall's 'Sustainable City' projects for Berlin, Hamburg and other major German cities (Vattenfall), or the 'Future Living Berlin' residential quarter in Berlin-Adlershof, constructed by the Japanese electronics company Panasonic. The latter is an ensemble of 69 residential units whose electricity and heat flow is to be automatically regulated in a circulation system (Lobe, 2018). Overall, it is still relatively undeveloped how the socio-cultural dimension of sustainability can be realised.
- *Mobility:* Smart mobility is characterised by improving the already existing infrastructure through the use of information and communication technologies to become more energy-efficient, cost-effective and with lower emissions. A typical example is the monitoring of traffic with video cameras, while at the same time citizens use mobile apps that provide them with accurate information on location and exact travel time, taking into account all data in real time (Giffinger et al., 2007).

5.2 The Developed Society in the Light of Four Discourses

In total, the smart city discourse is controversial. Advocates of the concept argue that new (information) technological achievements will supposedly make life easier and lead to greater efficiency and sustainability in the context of the urbanisation trend. The technology-driven permanent exchange of data between citizens and the city creates space for participatory urban development. The future will show to what extent smart cities actually enable institutional learning and resulting social innovations, such as e-democracy, and to what extent they prove to be functional answers to contemporary complex challenges.

From a critical point of view, it can be stated that smart city concepts are mostly based on technocratic assumptions that see technology as a panacea. Furthermore, critics complain that the smart city is nothing more than a fashionable term that, despite concepts for sustainability and climate change, also harbours potential dangers that have not yet been sufficiently considered. This concerns above all the possibilities of surveillance by sensors or the misuse of these technologies. In China, for example, there are plans to evaluate the social behaviour of citizens ('social scoring') (Trentmann, 2015), which could also serve as an inspiration for Western societies (Fathi, 2021b). Another point of criticism arises from the fact that, as far as the initiation of smart city projects is concerned, the private sector dominates. In his polemic work '*Against the Smart City*' (2013), Adam Greenfield criticises that the smart city is primarily a market 'in which technology companies can sell their products and services'. Here, the citizen appears as a consumer whose habits are observed and controlled by technical systems. There is no room for informal practices by citizens. The smart city is primarily a 'technocratic vision' in which the inhabitants are monitored. Greenfield's critique is based on an examination of the PR brochures and marketing materials of three exemplary projects – Songdo City in South Korea, Masdar in Abu Dhabi and PlanIT Valley in Portugal. They were all co-developed by large technology corporations, including Cisco, IBM, Siemens and Hitachi (Greenfield, 2013). Other critics, such as Jens Libbe of the German Institute of Urban Affairs (Difu), point out that the participation of citizens in concrete smart city concepts is apparently only a secondary consideration, while the actual focus is on technological aspects. This is reflected, among other things, in the fact that while numerous companies are represented in the EU's smart city advisory bodies, civil society initiatives are hardly represented (Libbe, 2014). Smart City is thus a concept that is predominantly driven by economic interests due to the dominance of the private sector in development and due to the principle of competition between cities.

5.2.3 Developmental Psychology Discourse

The discourse of developmental psychology focuses on the inner aspect of societal development. This perspective was popularised, among others, by Jean Gebser. Gebser is considered to be one of the first cultural science-oriented researchers of consciousness who developed a structural model of the history of human consciousness. Similar to the philosophers Hegel, Comte and Herbert Spencer, he assumed a progressive, evolutionary development of the history of human consciousness, in the course of which earlier structures of consciousness are recognised as 'errors' and replaced by new structures of consciousness that are more appropriate to the complexity (Gebser, 2015).

In addition to Gebser's model, there are a variety of other approaches today, each of which presents a different context for the development of consciousness. These include, for example, cognitive development (Jean Piaget and James Mark Baldwin), moral development (Lawrence Kohlberg, 1995), value development (Spiral Dynamics model according to Clare Grawes, further developed by Christopher Cowan and Don Beck (1996)), ego development (Jane Loevinger, 1976) or needs development (Abraham Maslow, 1954). From a transdisciplinary point of view, it is striking that all of these developmental psychological models (regardless of their differences in terms of the number of structures of consciousness) have one fundamental aspect in common: They all assume and describe a development of consciousness that proceeds from an egocentric via an ethnocentric to a world-centric perception of the self and the world. The developmental structures differ from one another in their degree of complexity. An egocentric view of the world, for instance, would only take one's own interests into account. Such a perspective would be far less complex than a world-centric view that includes not only one's own ego interest but various interests. More recent contributions increasingly refer to the Spiral Dynamics model to describe the different decision-making logics of individuals, organisations and societies. Therefore, a more detailed description of this model can be found in the box below:

> **Classification of Different Perspectives in a Complex Topic Using the Example of the Spiral Dynamics Model**
> The theory goes back to Don Beck and Chris Cowan and was based on the theories of Clare W. Graves. It was first presented in 1996 in the book of the same name. The theory was originally developed for a managerial audience, but has since found widespread application in other fields, such as the analysis of cultural development (Zollinger, 2005) in conflict regions (Stambolovic,

(continued)

5.2 The Developed Society in the Light of Four Discourses

continued
2002) or for political decision-making (Voros, 2006). Don Beck also advised on this basis in the Israel-Palestine conflict (Eldar, 2006).

Spiral Dynamics states a development spectrum of more than eight structures of consciousness, where structures 1–6 belong to the so-called 'first tier' and levels 7 and 8 to the 'second tier'. The model is open upwards, which means that more complex structures of consciousness would also be conceivable. The structures are also called *'value meme'*.[11] For the sake of simplicity, each structure/vMeme is assigned to a colour. Their inherent world views can be presented very briefly as follows (in the following Beck & Cowan, 1996):

First tier structures:
Beige: archaic, instinctive, survival-determined, self-acting, reflexological

- Since the Middle Palaeolithic (100,000 BC).
- Level of basic survival; food, water, warmth, sex, and safety are priorities. Habits and instincts are used for mere survival. Something like a distinguishable self has not yet awakened, and there is little effort to preserve one.

Purple: animistic, tribalistic, magical-animistic

- Since the Upper Palaeolithic (50,000 B.C.).
- Sacrifices to the ancestors and strict observance of customs subordinate the individual to the group. Magical spirits, good and evil haunt the earth, leaving blessings, curses, and enchantments that determine events. Ethnic tribes are formed. The spirits exist in the ancestors and hold the tribe together. Blood kinship and family establish political ties.

Red: egocentric exploitative violent gods

- Since 7000 BC.
- First appearance of an ego distinct from the tribe; powerful, impulsive, egocentric, heroic. Magical-mythical spirits, dragons, wild beasts and powerful people. Archetypal gods and goddesses, beings of power, forces to be dealt with, both good and evil.

---

[11] A meme is the cultural equivalent of the biological gene – it refers to a specific content of consciousness (e.g. a thought or, in this case, a complete world view) that can be passed on through communication.

continued

Blue: absolutist, obedient, mythical, orderly, determined, authoritarian

- Since 3000 BC.
- Life has meaning, direction, and purpose, the outcome of which is determined by an omnipotent Other or order. This just order enforces a code of conduct based on absolute and immutable principles of 'right' and 'wrong'. Violation of this code or rules has serious and perhaps 'eternal' repercussions. Following the code brings rewards to the believer.

Orange: diverse, efficient, scientific, strategic

- At the latest since 1700.
- Hypothetico-deductive, experimental, objective, mechanistic, operational – in other words, typically 'scientific' in the conventional sense. The world is a rational 'machine' with natural laws that can be recognised, mastered and used for one's own purposes. Strongly performance-oriented, and primarily for material gain. The laws of natural science dominate politics, economics and human society.

Green: relativistic, personalistic, communitarian, egalitarian

- At the latest since early twentieth century.
- Sense of community, human cohesion, ecological sensitivity, networks. The human spirit must be freed from greed, dogma and division; feelings and care take precedence over cold rationality; appreciation for the earth, for Gaia, for life. Against any hierarchy; making cross-connections and networking.

Second tier structures

These structures are called *'systemic-integrative'* (yellow) or *'holistic'* (turquoise). They have existed since the mid-twentieth century and, according to Beck/Cowan, are currently represented in about 1% of the world's population. The value system at this level is not only highly world-centric (this is also present in Green and partially in Orange), but also highly adaptable in terms of action, taking into account all levels of development (in contrast, Green tends to reject hierarchies) and promoting development.

Don Beck assumes that healthy and unhealthy (pathological) manifestations can emerge on each level of the developmental structure. On the one hand, each structure opens up more complex perspectives and associated achievements; on the other hand, it also opens up increasingly complex problem potentials. This characteristic is marked by the preposition 'mean'; e.g. in 'Mean Green Meme' (MGM) or 'Mean Orange Meme' (MOM). Crisis potentials arising primarily from the MOM would be, for example, predatory capitalism and the associated structural exploitation and environmental destruction in favour of an overemphasis on profit maximisation and rational values (Beck & Cowan, 1996). The representatives of developmental psychological models therefore always see the internal perspective on collective consciousness in close connection with the external perspective on economic and political structures.

5.2.4 Contributions from Happiness Research: Criteria of a 'Happy Society' and Social Conflict Potentials

Another, relatively new discourse that casts an internal and an external perspective on social development, focuses on 'happiness' or 'psychological well-being'. The assumption here is that conventional economic indicators such as GDP alone are not sufficient to measure the welfare and progress of a society. It is therefore necessary to look beyond the categories of national accounts ('Beyond GDP'). Against this background, an international meeting was held in Brussels in November 2007 with, among others, the OECD, the European Commission, the European Parliament, the Club of Rome and the WWF. At the beginning of January 2008, French President Nicolas Sarkozy commissioned Nobel Prize winners Amartya Sen and Joseph Stiglitz to develop an indicator to measure 'happiness' and the standard of living in the population (European Commission, 2022a). Against the background of this discussion, happiness research has become increasingly important. The insights gained from this discipline into the determinants of a 'happy society' therefore remain indispensable for our study.

Happiness research examines – as the term suggests – the determinants of 'happiness'. However, it does not do so in the sense of 'luck', but of 'well-being'. Philosophically, happiness or well-being is understood as the central meaning of life, since basically all other motivations, needs and goals relate only to one's own happiness or the happiness of others (and thus, again, one's own). This view was already held by Aristotle and has been taken up by many philosophers since. Several concepts of happiness can be distinguished. One widespread concept relates happiness to the best possible satisfaction of all needs, in the broadest sense of the experience of pleasure. Within the framework of hedonism, which was coined by Epicurus, happiness is primarily about experiencing well-being by living out pleasure and avoiding unpleasure (Bachmann, 2013). Another concept of happiness, eudaemonia, discussed by Aristotle among others, emphasises happiness as a state of inner balance (Kraut, 1989). More concretely and radically, this idea is found in Buddhist philosophy and psychology. Here, it is about overcoming suffering by letting go of all attachment to pleasure and displeasure and realising inner peace (Zotz, 2000).

How can happiness be measured? Methodologically, happiness research is geared towards deriving a person's state of happiness from their objective living conditions on the one hand and asking about subjective well-being on the other.[12] At present, a variety of indices can already be observed that measure not only 'well-being' but also 'trust' in a cross-national context. Research shows that trust is very closely related to happiness / well-being and, in turn, also to the propensity to or aversion to violent social conflict. The World Values Survey (Medrano, 2012) and the Gallup World Survey (Gallup, 2008) have become internationally established for measuring 'trust'. Among the best-known indices worldwide for measuring 'subjective well-being' are the extensive Dutch archive 'World Database of Happiness' (Veenhoven, 2010) and from the British area the 'National Accounts of Well-Being' (Michaelson et al., 2009) of the award-winning new economics foundation (nef) as well as the 'Satisfaction with Life Index' of the University of Leicester, which already dates back to 2006 (White, 2007). The Legatum Prosperity Index (2018) by the Dubai-based Legatum Institute is also significant and should be taken into account in this study.

Happiness research is currently reaching its limits in terms of intercultural comparability. At this point, comparative cultural studies demonstrate that definitions of 'happiness' and 'subjective well-being' do differ from culture to culture. Populations in Asia, for example, tend to describe themselves as significantly unhappier when asked questions about well-being that are more tailored to Western social contexts. This is particularly the case in countries like Korea and Japan, which are characterised by a relatively high level of socio-economic development compared to the rest of the world. From mystical perspective, there is in all wisdom traditions a cross-culturally identical concept of happiness in the sense of 'bliss' (in

---

[12] A typical question would be: 'All in all, on a scale of 0–10, how would you describe your state lately – would you say you are very happy (10 points) or unhappy (0 points)?' Experience has shown that despite very different subjective perceptions of happiness, survey results are to a certain extent comparable from person to person and especially from collective to collective (for more details, see Bruni & Porta, 2005).

5.2 The Developed Society in the Light of Four Discourses 155

Sanskrit, e.g., 'Ananda'). This concept of happiness is primarily accessed through the continuous practice of serenity-promoting mindfulness and meditation techniques. The ultimate goal of the practice is to overcome the origins of mental suffering – Buddhist psychology, for example, cites three 'mind poisons': craving, aversion, and delusion – and to find lasting peace and fulfilment (Zotz, 2000). In summary, it can be stated that in order to examine social happiness in a comparative cultural and religious context, there is still a need for further research. and the focus of the considerations in this book is on the context of Western societies.

Both variables, i.e. 'subjective well-being' and 'trust', provide information about the 'inner' potential for conflict and thus the risk factors that stand in the way of a developed society. In this context, the question of which welfare state conditions must be fulfilled on average in order to achieve the highest possible level of happiness and trust seems relevant. To answer this question adequately, an understanding of the basic psychological mechanisms of human happiness is essential. In simplified terms, two psychological mechanisms can be summarised:

- *Hedonistic adaptation:* Higher income and thus higher consumption levels, additional goods or services create satisfaction. However, we quickly get used to the standard we have achieved and life satisfaction does not increase significantly. The relationship between income or consumption level and happiness is therefore not linear – rather, there is a decreasing marginal utility. Economist Richard Layard calculated, 'If I earn one euro more this year, it will make me happier, but next year I will evaluate my income against a yardstick that is 40 cents higher. Consequently, 40% of 1 year's gains will be wiped out the following year' (Layard, 2005). This is also the main explanation for why, despite a large increase in material well-being across Europe (purchasing power doubled on average between 1950 and 2000), average life satisfaction has not increased. People have become accustomed to the new possibilities and have adjusted their expectations to technological progress (Binswanger, 2006; Layard, 2005).
- *We compare ourselves:* Studies show that the position on the income ladder is decisive for psychological well-being. It is not so much the absolute level of income that is of interest, but rather the social status in comparison to others, i.e., the relative income. Here, usually the comparison refers to people who have a higher income than oneself. Poorer fellow human beings are usually not considered. According to a study in the USA, the feeling of happiness decreases by one third more if everyone else also receives a 1% salary increase in salary than if the person in question was the only one to receive a 1% salary increase. Other studies show that people would even accept a decline in their quality of life if it meant a social advancement compared to others. Moreover, people

always compare their income with what they themselves are used to. For example, when asked what they need to live, rich people always name higher sums than poor people (Binswanger, 2006).

There are also exceptions to both mechanisms. With regard to hedonistic adaptation, there are also many nice things that never lose their appeal, such as friends, sex, and even marriage. As for our tendency to rivalry, studies show that this mechanism applies more to the relative amount of income, but less to the amount of free time (hence the widespread tendency to sacrifice free time for more income). Researchers suggest that there is an evolutionary purpose behind both mechanisms – they ensure one's own survival and the constant improvement of one's living situation. The

---

[13] Different disciplines such as emotion research or anthropology prove that biological and cultural evolution do not occur simultaneously. Since Homo Sapiens has not evolved biologically in the last 50,000 years, it is therefore still built for the struggle for survival in prehistoric times. All instincts, such as the instinct to fight and flight and, in the broadest sense, the instinct to 'be unhappy" as described above, ensured humanity's survival in the prehistoric struggle for survival. However, while biological evolution tends to proceed in cycles of at least 6-digit per year, cultural evolution has occurred in giant leaps over the last 50,000 years: from hunter-gatherer societies to horticultural societies, followed by arable societies, to industrial societies and today's increasingly globally interconnected post-industrial societies. Two conclusions emerge from this perspective: On the one hand, Dylan Evan's position could be confirmed to the effect that the biological instinct to physically struggle for survival and to be unhappy may itself have been the driving force behind humanity's comparatively rapid cultural evolution. On the other hand, these biological instincts, which were geared to the physical struggle for survival in the prehistoric context, have long since ceased to be tailored to the civilisational demands of today's highly developed affluent societies – this is particularly true for violent conflicts in highly developed affluent societies, for instance domestic violence, rampages or street crime. They lead to unnecessary violent conflict and social costs. Conversely, this could mean that even in the face of global interdependence, general well-being and happiness as well as the ability to cooperate are more significant than the instinct to survive and be unhappy. This position can be derived from the theses of the emotion researcher Daniel Goleman, who sees the development of emotional intelligence (and thus the capacity for empathy and the integration of rational intelligence and biological instincts) as the central challenge for a developed society (Goleman, 2001). The fundamental debate outlined here remains open. Nevertheless, several theses can be derived in advance, to which we will return in other chapters: First, it suggests that happiness and well-being, especially in the contemporary context, is a desirable individual and social goal. Secondly, however, humans seem to be designed for evolutionary reasons to constantly strive for happiness. This means, thirdly, that humans have always carried within them the potential for social conflict. The art of a sustainable (i.e. developed, sustainable and resilient) society is therefore, fourthly, to constantly anticipate and constructively manage this almost inevitable potential for conflict.

5.2 The Developed Society in the Light of Four Discourses

British developmental psychologist Dylan Evans takes up this agument and even claims that lasting happiness and contentment would mean the end of the human species. In order to survive evolution, he argues, it is vital for a society to be unhappy. This is debatable.[13] But in fact, 'well-being for all' cannot be fully shaped externally – studies show that well-being is highly biologically determined. According to the empirically saturated 'set point' theory, people always tend to automatically return to their personal sense of happiness, which always remains at a relatively stable level in the long term. Therefore, research concludes that environmental factors can only influence up to 50% of happiness in the long term. Happiness is therefore – according to the influential researcher David Lykken – already at least 50% genetically pre-determined (Lykken & Telegen, 1996).

All these psychological-anthropological mechanisms offer a sufficient explanation for why there is still a great potential for social conflict even in the highly developed welfare societies of the world. But there is also good news: Although studies show that there are psychological mechanisms that lead to lasting dissatisfaction, other findings in turn illustrate that social factors can have an impact on a person's well-being. Indeed, studies on economic growth show a seemingly ambivalent picture. On the one hand, a significant increase in socio-economic well-being since the mid-twentieth century has not led to a significant increase in happiness. On the other hand, studies show that people with higher incomes rate their psychological well-being higher than poorer people. Research has found that experienced well-being does not increase above an income of US$75,000/year (Killingsworth, 2021). Others even assume an income limit in the amount of US$25,000/year (Wilkinson & Pickett, 2010). This suggests that higher incomes still have the potential to improve people's everyday well-being and that many people in wealthy countries have not already reached a plateau. In this light, growth and increases in purchasing power can contribute to higher well-being to some extent (Killingsworth, 2021; Ball & Chernova, 2008).

In addition, there are other important happiness factors that are independent of income. Based on large-scale studies from the World Values Survey, researchers like Layard come up with a total of seven basic social factors that are indispensable for a happy society:

- Family relationships
- Financial situation
- Work
- Social environment
- Health
- Personal freedom and

- A philosophy of life that places the common good above self-interest and pressure to perform (Layard, 2005; Ball & Chernova, 2008).

A look at comparative studies shows that there are some differences in these factors between the highly developed societies of Europe and the Anglophone world (including, outside Europe, the USA, Canada, Australia and New Zealand). This in turn allow conclusions to be drawn about whether and in what respects certain types of society tend to be happier (and perhaps more developed in this respect) than others, and to what extent their welfare policies can learn from each other to develop even further.

- *Family stability:* Family stability is a fundamental factor linked to subjective well-being. Based on data from the World Values Survey (Medrano, 2012), studies for all OECD countries have consistently shown an increase in partner separations and divorces since the 1970s. In most cases, divorces and separations also involve parenting children, with the likelihood of separation or divorce decreasing the larger the family. The birth rate, which has been falling consistently since 1970, was highest in the 2010s in the USA and Europe, as well as in countries such as Iceland, New Zealand and Ireland, with an average family size of two children. The rate was lowest in Southern and Eastern European countries such as Hungary, Spain and Portugal with an average family size of one child. These trends are clearly due to the increasing changes in modern society. The by-products of this change are: the progressive dissolution of traditional gender roles, increasing financial independence for women (through greater involvement in the labour market), greater individualisation of values and the increasing acceptance of alternative life models.[14] Can we assume an increasing 'decline of the family'? The studies show contradictions. The quality of family life is increasing – more and more children say they feel comfortable with their parents, even if they are divorced. In addition, the family is becoming less and less dominated by violent relationships – studies consistently show a significant decline in domestic violence in all countries. What is worrying, on the other hand, is the increase in child poverty, which is also linked to the increase in single-parent households and poor adaptation of welfare policies (childcare, transfer payments, job placement). In this respect, Europe and the USA show a distribution in which the Anglophone countries, led by the USA

---

[14] The models vary, for example, from the highly emancipated 'super-family", in which men are also happy to put their careers on hold, to the 'patch-work family" or the 'single model" (i.e. the 'non-family").

and the Southern and Eastern European societies, perform worst on average. The five Nordic societies, led by Denmark, performed best in the OECD comparison, as did the continental European countries (including Germany, France and the Netherlands) with child poverty rates below the OECD average (OECD, 2011a).

- *Financial situation:* As indicated above, many empirical studies prove the factor 'income inequality' as a significant social conflict factor. According to the cross-country comparative studies by Fajnzylber et al. (2002) and Neapolitan (1999), there is a strong link between criminal violence and income inequality worldwide. Other studies suggest that societies with less income inequality have higher levels of interpersonal trust (OECD, 2011b). The currently most discussed study on the negative social effects of income inequality was conducted by Richard Wilkinson and Kate Pickett (2010). In their extensive comparison of 23 of the world's richest societies, both conclude that social inequality is linked to a number of social hotspots. These include: social disintegration, mental illness, health deficiencies and declining life expectancy, analphabetism, increase in violence and drug use, prison overcrowding, lack of social mobility, lack of opportunity for social advancement, loss of future, and life-ethical apathy. They are all, according to the diagnostic thesis, effects of inequality and could, according to Wilkinson and Pickett, be combated by appropriate redistributive measures, e.g. by raising the average income. According to their study, Japan and the Nordic countries, as well as the US state of New Hampshire, have the highest levels of equality, while the USA as a whole, as well as England and Portugal, are affected by the greatest inequality. The figure below provides a rough overview (Fig. 5.1).

In addition to relative income, absolute income also plays a role that should not be underestimated, although people tend to get used to the consumption standard they have reached all too quickly. As indicated above, an increase in income by a factor of two, for example, would not lead to an increase in subjective well-being to the same extent. In other words: Well-being actually increases with higher income, and without a specific 'saturation point' (Hagerty & Veenhoven, 2003). At the same time, the increase in well-being decreases significantly as soon as a subsistence level is reached. This confirms to some extent complex basic needs models such as Maslow's, according to which more than just the satisfaction of material needs is

---

[15] This was measured by the average responses to questions about whether respondents had volunteered, the extent to which they had volunteered and whether they had helped a stranger in the last month (OECD, 2011a).

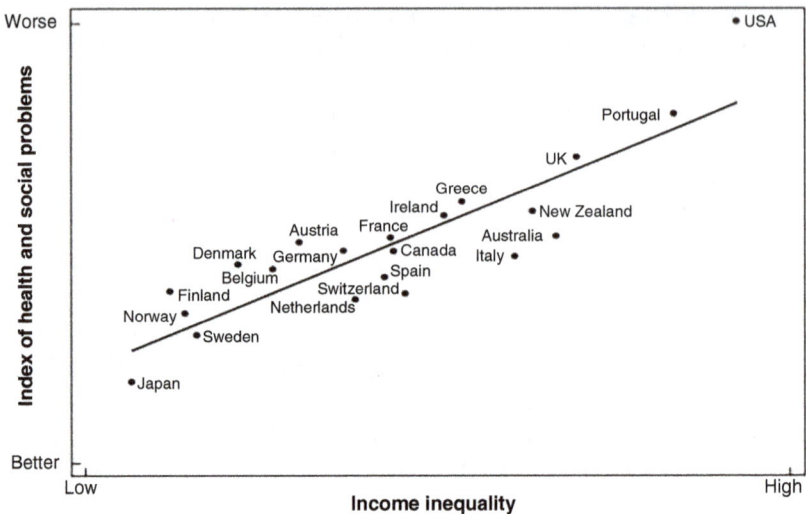

Fig. 5.1 Income inequality in relation to health and social problems according to Wilkinson and Pickett (2010)

crucial for life satisfaction. An OECD study confirms a significant correlation between absolute average income and interpersonal trust in a society (OECD, 2011a). According to the same study, there is even a correlation with pro-social behaviour.[15] Highest levels of pro-social behaviour were measured in the Anglophone countries, which also have the highest average income, whereas in the OECD comparison the Southern and Eastern European societies scored lowest on both criteria (Fig. 5.2).

- *Work:* Similar to relative income, the employment situation has a very strong influence on subjective well-being. As Voltaire wrote, 'Work keeps away three great evils. Boredom, vice, and misery.' Many recent studies confirm this thesis and prove a direct link between unemployment and the risk of poor mental and physical health, higher stress and greater dissatisfaction (Halvorsen, 2001). Other studies even show a link between the 'commodification' of work, i.e. the market dependence of the providers of work (self-employed, bogus self-employed, employees, unpaid workers and welfare recipients) and anti-social behaviour, including homicide.[16] Conversely, studies suggest that feeling in

---

[16] The theory derived from this is called the Decommodification of Labor Theory (DLT).

5.2 The Developed Society in the Light of Four Discourses

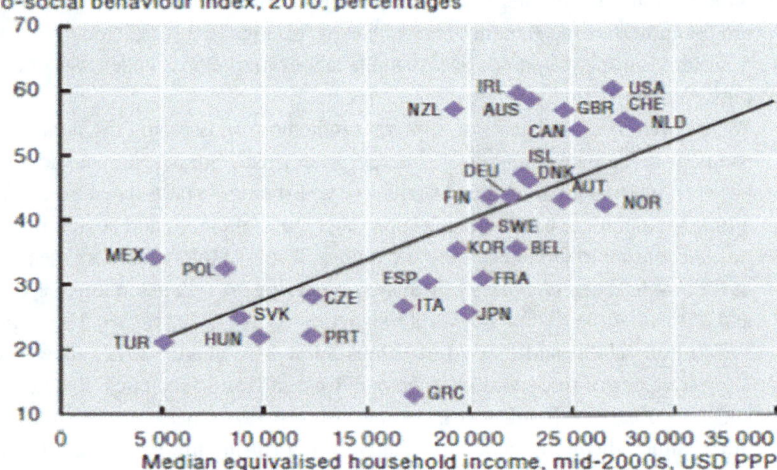

Fig. 5.2 Relationship between pro-social behaviour and absolute average income. (OECD, 2011a, p. 95)

control of one's life is central to one's subjective well-being and self-esteem (Radcliff & Pacek, 2008). Accordingly, for a 'well-being-promoting' labour market policy, it can be derived that a 'passive' labour market policy based only on transfer payments does not promise nearly as much well-being as 'active' measures based on needs-adapted placement in suitable jobs. This again confirms complex basic needs theories, according to which people are not only interested in the satisfaction of purely material needs (which can be covered by social benefits), but also in the satisfaction of intellectual and spiritual needs of meaning and the feeling of being 'part of society'. In the OECD comparisons of the 2010s, the Southern European countries, above all Spain, Portugal and Greece, had the highest unemployment rates. In 2009, Spain's unemployment rate was six times higher than that of Norway (OECD, 2011b). This may be due to the global economic crisis of 2008 and the COVID-19 crisis, which hit the Southern European countries particularly hard. In general, in all Western societies, young people and migrants are on average more affected by unemployment than other segments of the population. Experience has shown that they represent a particularly explosive potential for conflict (OECD, 2011b) compared to other segments such as 'old people' (generational conflict) or 'women' (gender conflict). According to the Migrant Integration Policy Index (MIPEX) studies, in which labour market integration is of particular importance, the top ten societies are composed of the Benelux region, North America, the Nordic countries and Southern Europe. Migrant's access to the labour market in Sweden was rated as outstanding (MIPEX).

- *Social environment:* The most important indicator for assessing the social environment is the measurement of 'interpersonal trust'. Studies show a clear link between interpersonal trust and subjective well-being. In part, they also confirm a link to tolerance, namely being more accepting of ethnic and sexual minorities. According to OECD studies, the highest levels of tolerance were measured in the Anglophone countries of Canada, Australia, New Zealand and the USA, and also to some extent in Nordic societies (OECD, 2011a, b). The Gallup World Poll (2008) and the World Values Survey (Medrano, 2012) showed the highest values of measured interpersonal trust in Nordic societies, Switzerland and New Zealand. Nordic societies are also top performers in trust in institutions, although this is generally relatively high in highly developed Western societies (Ortiz-Ospina & Roser, 2016).
- *Health*: Health has been shown to be directly related to subjective well-being. Several trends in the development of physical and mental health can be observed for Western societies. As far as physical health is concerned, there is a trend towards improvement, which is also expressed in a steady increase in life expectancy and which in turn brings with it a 'new' perception of illness. This is because higher life expectancy is also accompanied by an increase in chronic diseases such as diabetes, dementia, etc. Another accompanying trend is the increasing gap between socio-economic groups. In the Czech Republic, for instance, a 65-year-old man with a high level of education lives on average 7 years longer than a man of the same age with a low level of education. Mental health[17] is currently the biggest challenge for OECD countries as the number of mental illnesses has increased over the last two decades, despite a slight decline in suicide rates and regardless of rising income. Studies show that this trend is largely related to two societal developments: on the one hand, general job insecurity and the pressure to perform and compete at work are increasing in connection with globalisation (OECD, 2012); on the other hand, in the wake of family breakdown processes, occupational mobility and more frequent separations, there is a 'reduction in sustainable social relationships, and this in

---

[17] Mental health is defined by the World Health Organisation as 'a state of well-being in which the individual realises his or her abilities, can cope with the normal stresses of life, can work productively and fruitfully, and is able to make a contribution to his or her community" (OECD, 2008).

both qualitative and quantitative terms' (Galuska et al., 2010). Again, all OECD societies are more or less equally affected by this. Depression alone is becoming an increasing economic burden for welfare states, as it is responsible for an increasing number of protracted and costly sickness absences from work (OECD, 2012). One of the most prominent studies currently in the health field, directly related to the concept of 'evolved societies', was presented by the Canadian Institute for Advanced Research (CIFAR). The Institute concludes that well-developed social relations in civil society are the most crucial factor in ensuring effective and efficient health care. According to this study, strong civil society cohesion (including intact families) 'from below' (bottom-up) would prove even more relevant than politically imposed universal access to public health care 'from above' (top-down). In summary, it can be concluded that there is a wide ramified relationship between health, well-being and social cohesion. A new discussion is emerging on the relationship between health or health promotion and economic growth. According to Erik Händeler, the greatest untapped resources of the national economy lie in the health of its citizens. The futurologist emphasises that a health care system with a stronger focus on health promotion and prevention could also lead to the elimination of the currently rising economic and social costs caused by the increasing number of people losing their health[18] (Händeler, 2005).

- *Location attractiveness and innovation:* Further studies point to an indirect link between psychological well-being, location attractiveness and a society's ability to compete and innovate. In this regard, it doesn't surprise that the Second Machine Age is increasingly characterised by the 'War of Talents'. At the same time, research confirms a connection between a society's ability to innovate and its ability to attract a so-called 'creative class' – which is regarded as a central driver of value creation, especially in the post-industrial knowledge society. Above all, studies show a link between innovation, personal freedom and the economic prosperity of a location. One of the best-known economic theories in this context was coined by Richard Florida. His famous model of the 'three Ts' stands for the key indicators 'technology', 'talent' and 'tolerance'. 'Technology' stands for concentration of knowledge-intensive economy in general or high-tech sectors in a region in particular. 'Talent' is determined

---

[18] Confirming this, the above-mentioned open letter on the psychosocial situation in Germany diagnoses, among other things, that approx. 30% of the population suffer from a diagnosable mental disorder within 1 year and that the share of mental illnesses in early retirement is continuously increasing. At the same time, mental illnesses and behavioural problems among children and adolescents are also steadily increasing (Galuska et al., 2010).

by the number of employees in creative professions in the region – the indicator thus determines the creative potential. 'Tolerance' stands for the openness of a society that attracts a wide range of different personalities and fosters a high exchange of new ideas. This theory is confirmed by the results of the Global Innovation Index 2018. This ranking compares 126 countries/regions, with Switzerland, the Netherlands, Sweden, Singapore, the UK, the USA, Finland, Germany and Ireland taking the top places (INSEAD et al., 2018). The emerging concept of the so-called creative industries is considered to be the economic sector with the greatest innovative power. It comprises a total of 11 core industries: Design industry, book market, music industry, art market, film industry, broadcasting industry, performing arts market, architecture market, software/games industry, advertising market and press market. Among them, the design economy had the largest export volume with a share of 61%. According to UNCTAD's 'Creative Economy Outlook 2002–2015', during this period the world market for goods and services has developed from US$208 billion to US$506 billion. Despite the severe financial crisis in 2008, the average growth rate of the creative industries was about 7%. The study emphasised the growing importance of Asian emerging markets and ranked China, Hong Kong (China), India, Singapore, Taiwan, Turkey, Thailand, Malaysia, Mexico and the Philippines among the top global performers. Among First World nations, the study ranked the USA, France, Italy, England, Germany, the Netherlands, Poland, Belgium and Japan among the top ten exporters of creative goods. It concluded that ASEAN countries and Europe in particular were among the largest regional exporters. In 2015, the volume of the EU was about US$ 171 billion (in 2002 it was US$ 85 billion); in East and Southeast Asia it was even US$ 228 billion (of which China with a share of 14%) (UNCTAD, 2018). Modern or post-modern societies that want to maintain and increase their competitiveness, especially in terms of their ability to create economic value, will have to promote this increasingly important sector more systematically. This will be mainly achieved by increasing the attractiveness for investment and the 'War of Talents' as a location, which in turn is closely linked to overall subjective well-being.

So which societies might be the happiest? Although the comparative studies measuring subjective well-being sometimes use very different criteria, there might be a trend for the decade 2010. Several studies, such as the Legatum Prosperity Index (2018), the OECD Better Life Index (2011) and the Satisfaction with Life Index 2006 (White, 2007) consistently see Nordic societies as well as Australia, Switzerland, Canada, the Netherlands and New Zealand, among the top 10% happiest societies in the world. In parallel, studies show a high degree of consistency with other measured subjective values, such as impressions of 'positive' and 'negative experiences',[19] which is largely uniform within the country groups 'Nordic societies', 'Continental European societies' and 'Anglophone societies'. Nordic societies had the highest level of 'positive experiences' and the lowest level of 'negative experiences'. The Anglophone societies showed high values for both positive and negative experiences, while the Continental European societies were always in the middle.

5.3 Measurement Criteria and Principles of the Developed Society

5.3.1 Measurement Criteria

From the previous considerations, at least the following interrelated dimensions and measurement criteria can be summarised (Table 5.1):

This list (which is by no means exhaustive) allows for a multidimensional, cross-country comparison of the realisation of the key principles of the developed society. As already illustrated by the example of the concept of resilience, this approach also proves to be limited. Neither are all criteria directly comparable with each other, nor is their respective weighting clear. Which, for example, weighs more heavily: the ICT Development Index or the Human Development Index? Nor does the list as such make the interrelationships between the criteria clear.

5.3.2 Principles

From the perspective of the simplify tradition, the question arises to what extent all the above-mentioned criteria can be broken down to a few generalisable principles, especially since many of these criteria are directly or indirectly interrelated. Thus, studies such as those listed above demonstrate that there is a correlation between psychological well-being, location attractiveness and innovation capacity in the Second Machine Age. In sum, similar to the multi-resilient society, at least four general orientation principles can be derived for a developed society:

Principle 1: A developed society promotes competence development of its citizens
Principle 2: A developed society fosters collective intelligence

---

[19] The 'experience indices' ask about the respondent's well-being on the day before the survey. Among other things, they ask about stress, physical pain, worries or depression (negative experiences) or about the degree of restfulness, whether one laughs often, is treated with respect, etc. (positive experiences) (OECD, 2011a, p. 82).

Table 5.1 Selected criteria for assessing developed societies

Selected development Dimensions	Selected individual criteria
Psychological	Psychological well-being Skills development/education (share of highly qualified) Developmental psychological complexity level (proportion of orange meme and higher)
Social	Public health (life expectancy in good health) Social cohesion (including volunteering) Crime/anti-social behaviour Subjective well-being Human development (e.g. HDI) Trust (interpersonal and institutional)
Economy	Indebtedness Growth Inoccupation Income distribution (relative income) Competitiveness (e.g. competitiveness index) Share of creative industries/innovation Purchasing power Absolute income Patent intensity/start-up intensity Investment rate of the industry Proportion of employees in third and fourth sector[a]
Technological progress	Urbanisation level Dissemination of information/communication technologies (ICT Development Index) Technology achievement index Digital competitiveness

[a]The last three criteria of this compilation have been extended by some criteria from the 'Future Atlas 2010' of the German consulting firm Prognos AG – this is a 'set of indicators for regions of the future' (Prognos, 2010)

Principle 3: A developed society is characterised by a hot learning culture
Principle 4: A developed society is characterised by a high level of well-being

Regarding *Principle 1*: The modern industrial society – and even more so the post-industrial information and knowledge society – are essentially based on the competences of their citizens. On the one hand, these include technological and professional know-how for the development and operation of universal technologies. At the complexity level of post-industrial knowledge societies, these technologies include information and communication technologies (ICT), but also other highly competed disruptive technology areas that will gain importance in the future, such as for example green tech, nanotechnology, biotechnology, AI. Analogously, from

5.3 Measurement Criteria and Principles of the Developed Society 167

the developmental psychology perspective, the goal is to have as many developmental lines of a critical mass of the population as possible reach at least Orange Mem and above. The foundation for this is laid by educational policies that aim to develop the awareness and skills of the population accordingly. From the perspective of the complexify tradition, such an educational policy would have to cover as many lines of development of the population as possible, i.e. not only cognitive development, but also others, such as social and personal competencies. Personal competencies can also mean resilience, social competencies are important for the development and implementation of collective intelligence.

Regarding *Principle 2*: Any collective approach to complex problems requires collective intelligence. In a wide variety of contexts, complex problem-solving and necessary innovation requires collaboration of specialists from different backgrounds – be it, for example, in interdisciplinary research associations and expert conferences, in interdepartmental high-performance teams, in multi-stakeholder dialogues or in interdisciplinary think tanks. In all these and other contexts, the complex problems of our time are addressed, highly complex situations are analysed, and innovations are developed. The quality of problem solving depends on the fulfilment of several factors that determine collective intelligence (under Sect. 4.4. these criteria were summarised with the following terms: diversity of opinion, independence of opinion, decentralisation, aggregation, trust in the collective group). Above all, it is an intellectual and communicative achievement in which the different perspectives of the manifold knowledge agents are brought together.

Regarding *Principle 3*: From a cultural perspective, the model of the developed society can in many respects be assigned to the ideal type of a hot culture defined by Claude Lévi-Strauss. Key motifs are progress and the adaptation of the environment to culture. The Swiss ethnologist and psychoanalyst Mario Erdheim expanded the model by recognising that in all societies there are both 'cooling' and 'heating' institutions that influence the respective culture in one direction or the other. For instance, according to him, the church, the military, and the school are likely to represent cold institutions in an otherwise hot, modern society (Erdheim, 1988). This seems plausible at first glance, but is put into perspective in view of the central importance of education for the development of the competences and consciousness of citizens necessary in a developed society (Principle 1). Consequently, the education system would not only have to teach proven, traditional approaches, but also promote competences that go hand in hand with the ability to open up new, unknown ways and development paths.

Regarding *Principle 4*: The *raison d'être* of any society is to create the necessary framework conditions in which its members can develop their potential and satisfy their (basic) needs. In a developed society, the framework conditions should be designed in such a way that they lead to a high level of welfare. High welfare means a high level of meeting the (basic) needs of the population as a prevention of social conflicts, since all conflict is based on frustrated needs. In addition, a location with such framework conditions appears attractive for investment and for the best minds. From this perspective, the indicator 'happiness' or 'psychological well-being' points to a high level of welfare, an attractive location and a high potential for preventing social conflicts. Since this principle is also central to multi-resilience and sustainability, and thus to any future-preparedness of societies, it has to be described in more detail in Chap. 8.

If the measurement criteria listed above in Sect. 5.3.1 prove to be very concrete but difficult to compare with each other and confusing in their diversity, the principles approach could be criticised as being very general. Ultimately, both approaches are to be understood as complementary to each other and as necessary preliminary considerations for the comparison of the three guiding principles made in Chap. 7.

5.4 Conclusion

To summarise: The guiding principle of the developed society is a subject of debate in a wide range of disciplines and discourses. It includes multiple facets, such as the 'happy society', the 'psychologically developed society', the 'economically efficient society', the 'healthy society' or the 'conflict-preventing welfare society'. All these different facets imply several index-based criteria which, in combination, provide an appropriately multi-dimensional view of the developed society. However, further research needs to explore these criteria further in terms of their reciprocal dynamics and discuss them critically in terms of cross-societal comparisons. These reflections could contribute to unifying and more firmly establishing the discourse on the developed society in a transdisciplinary, possibly even transcultural, context.

All in all, a developed society is characterised by its ability to cope with its environment and secure the (basic) needs of its population in the best possible way. Both, the aspect of exponential efficiency increases in numerous technological areas (Moore's Law), and the 'War of Talents' in today's knowledge societies are likely to dominate the further discourse. In most of the above-mentioned aspects, Western societies continue to lead the world, although the trend of an intensifying 'competition of the moderns' is already emerging today, and the field of technological innovation is particularly highly contested.

Within the Western community of states, the Anglophone and Nordic societies may have a leading position regarding most of the above-mentioned criteria. However, they are not free of crises themselves, e.g. as a reaction to the pressure of world markets and the increase in mental illness. This is probably partly due to the fact that, according to current studies, approximately about 50% of human happiness can be directly influenced by socio-political factors. The concept of the developed society remains important for the overarching discourse on societal future preparedness.

The Sustainable Society   6

'Sustainability' (German: *Nachhaltigkeit*) can be understood as a concept for the use of resources, in which a lasting satisfaction of needs for future generations can be ensured by preserving the natural regenerative capacity of the systems involved (including living beings and ecosystems) (World Ocean Review, 2015). Etymologically, the term 'sustainability' is derived from the Latin *sustinere* (*tenere*, to hold; *sub*, under), meaning in terms of *sustain* 'to maintain', 'to support', 'to uphold' or 'to endure' (Online Etymology Dictionary n.d.). This means that the systems involved can 'permanently withstand' a certain level of resource use without suffering damage. Historically, the concept goes back to the Saxon tax accountant and mining administrator Hans Carl von Carlowitz (1645–1714), who wrote the first comprehensive treatise about forestry and is considered to be the father of sustainable yield forestry. The concept involved the basic idea of using natural resources such as forests mindfully, so that the supply never runs out (Environment & Society Portal, n.d.; World Ocean Review, 2015). In 1972, the term was used in the Club of Rome report *'The Limits to Growth'* in the broader sense of a 'state of global equilibrium' and thus achieved international attention (Meadows et al., 1972). Since the UN World Commission on Environment and Development in the 1980s, 'sustainability' has been understood primarily in the sense of 'sustainable development'. The Commission was chaired by Gro Harlem Brundtland, then Prime Minister of Norway and leader of the Social Democratic Labour Party. Under her leadership, the United Nations Commission on Environment and Development published a report in 1987 entitled 'Our Common Future', which shaped the concept of sustainability as we know it today. It states:

> Sustainable development is development that meets the needs of the present without compromising the ability of future generations to meet their own needs (Hauff, 1987, p. 46).

There is still a broad consensus in today's sustainability debate about this 'Brundtland definition', which is still quite general. A more concrete definition of sustainability has not yet been able to establish itself.

6.1 Key Dimensions of the Sustainable Society in the Current Discourse

The three-pillar model, also known as the 'magic triangle of sustainability', closely follows the Brundtland definition and distinguishes between an ethical/social/human, an economic and an ecological pillar of sustainability. Two types of this model dominate today's discussion: The classical pillar model is nowadays mostly presented in the form of three overlapping circles. This emphasises that all three areas are to be regarded as equal and equally important. The resulting statement is that sustainability can only be achieved, if all three areas are considered simultaneously. The other model type is called the priority model and is presented in the form of nested concentric circles. It states that individual areas are to be considered in their relationship and dependence on each other. The resulting key statement is that the ecological dimension limits social and economic action. With regard to the other two dimensions, there are two interpretations. Observers such as Nico Paech see society as the limiting framework for the economy (Paech, 2012), other observers rather see the economy as the framework for the social dimension (Fig. 6.1).

In practice, a political dimension is sometimes included as a fourth pillar, which includes, for example, gender equality, human rights, democracy and education (Stockmann, 1996). In German politics, special emphasis is placed on education policy (Federal Ministry of Education and Research [FMER]).

In today's prevailing social discourse, the *social pillar* of sustainability emphasises the aspect of combating poverty and securing basic needs by ensuring equitable access to opportunities and distribution of resources (Bauer, 2008). Firstly, the context of social sustainability extends to the internal social sphere, i.e. similar to the model of the developed society outlined above, which aims to secure the broadest possible basic needs of the population and to prevent social conflicts. In the sense of sustainability, prevention should also include future generations. Secondly, social sustainability also refers to the global context, especially with

6.1 Key Dimensions of the Sustainable Society in the Current Discourse

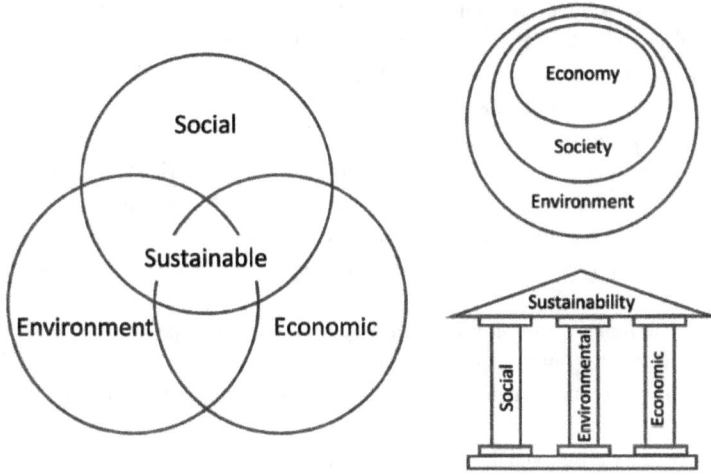

Fig. 6.1 Three pillars of sustainability. (Purvis et al., 2019)

regard to the worldwide fight against extreme poverty worldwide, e.g. through sustainable international development cooperation and fair trade relations.[1] The latter is mainly discussed at the intersection with the economic sustainability pillar – currently dominated by the 'fair trade' discourse[2] and criticism of

---

[1] Extreme or absolute poverty is defined by the World Bank as poverty characterised by an income of about one US dollar per day. Another characteristic of extreme poverty is the phenomenon of the so-called 'poverty trap'. According to this, affected states are so poor that they cannot develop at all, which makes them dependent from external support. Only from the next higher development level of so-called moderate poverty is there the potential for independent development (Sachs, 2005). Globally, about 1.2 billion people fell into this category in 2010 (UN, 2010). In 2015, the UN estimated 836 million people (UN, 2015).

[2] The discussion refers to pros and cons of controlled trade, in which the producers are paid at least a minimum price for the traded products set by fair trade organisations, which is set above the respective world market price.

agricultural subsidies, especially in the USA and Europe.[3] These factors can be seen as a major structural cause of economic migration.

At the *intersection with the ecological sustainability pillar,* at least two approaches to social sustainability can be distinguished: From an anthropocentric perspective, social sustainability is understood as necessary for the maintenance of the human livelihood, while in the biocentric (also ecocentric) perspective social sustainability serves only to preserve nature. In the ecocentric approach, nature is therefore often ascribed an intrinsic value, whereas in the anthropocentric perspective nature is only a means to secure human existence (Littig & Grießler, 2004).

The *economic sustainability pillar* currently encompasses at least three dimensions of discourse. The first refers to the economic stability of national economies: Various national and international policy measures are being discussed against the background of preventing global economic and financial crises. At the national and international levels, these include increasing security reserves and increasing currency stability (e.g. backed by gold reserves), reducing debt levels, market-based regulation, and increasing the ability of the economic system as a whole to avoid speculative bubbles (Luks, 2002). The second discourse critically refers to the global financial system, particularly the process of economic and money creation and the question of how to prevent future bank and debt crises more efficiently. Several interrelated measures and topics are critically discussed in this context. Without claiming to be exhaustive, a sketchy list:

---

[3] What is being discussed here is the dismantling of agricultural subsidies by the Western industrialised countries, which means that small farmers from poorer regions in the Third World like for example in India, Latin America or Africa are unable to compete with these subsidised cheap products without corresponding protective tariffs. The result is that these small farmers are squeezed out of their domestic market, gradually stalling agriculture as a key driver of development. A major study published in September 2021 by FAO (Food and Agriculture Organization of the United Nations), UNEP (United Nations Environment Programme) and UNDP (United Nations Development Programme) came among others to the conclusion that 87% of the agricultural subsidies distort competition and harm the environment or small businesses. 470 billion US $ of the total 540 billion US $ flowing annually would have to be used differently in order to be sustainable and fair (FAO, 2021).

6.1 Key Dimensions of the Sustainable Society in the Current Discourse 173

- the infinite growth compulsion for growth resulting from the current money creation system,[4]
- stabilising the financial system through measures like a gold backing,[5]
- increasing the risk-bearing capacity of the banking system by increasing the capital base[6]

---

[4] Money creation refers to the increase in the money supply and thus the creation of new money. When credit is granted by a commercial or central bank, the amount booked to the borrower (minus the minimum reserve) is literally 'recreated'. When the loans are repaid, the credit money is 'destroyed' again. The impetus for money creation comes from the demand for credit by private banks and non-banks, as well as by central and commercial banks. Commercial banks borrow from the central bank when their reserve requirement ratio is insufficient to lend themselves. The sustainability debate now starts from the assumption that the economic growth compulsion of the capitalistic system arises from the money creation cycle, which is driven by the need to repay an ever-increasing amount of credit money. At the centre of the debate is primarily the following mechanism: banks require borrowers to pay interest on the credit money they create. However, even after the loan is granted, the banks do not dispose of the money repaid *and* the interest paid. Although the interest on loans collected by credit institutions flows back into the economic cycle, the increase in the quantity of credit money can never be completely destroyed. Against this background, the central banks, which are at the top end of the money creation hierarchy, are at the centre of criticism. The critics stress that central banks charge interest on newly created money which means that they are at the same time the highest instance of money creation. Ultimately, more debt would be created than there is money available. This results in an urge and a compulsion to grow, which ultimately only leads to the debt merry-go-round spinning faster (cf. among others Wolff, 2017; Homm, 2016; Krall, 2017; Rickards, 2012).

[5] The background to this discussion is about the increase in hyperinflations since the spread of government-issued currency that is not backed by a physical commodity (fiat money) since the beginning of the twentieth century in general and since the collapse of the Bretton-Woods gold-backed international exchange rate system in 1973 in particular. Currently, the main debate among economists is about gold backing of the monetary base, whether in the form of a fiduciary system or proportional backing. The latter provides that a certain proportion of the banknote money supply is covered by gold. The fiduciary system, on the other hand, provides that a certain proportion of the banknote money supply is not covered, but that the money supply in excess of this is 100% covered, in order to limit the expansion of the central bank money supply. However, at least two arguments are put forward against gold backing: First, the amount of available gold would not be sufficient for this purpose, although this would be a question of the degree of coverage and the gold price to be set. Secondly, it is doubted that the expansion of the money supply would keep pace with the expansion of the economy, which, if not, would entail the risk of deflation (cf. among others Wolff, 2017; Homm, 2016; Krall, 2017; Rickards, 2012).

[6] Proposals in this direction are already being made in the context of the Basel III regulatory drafts. However, it remains questionable whether it is sufficient in view of the thin capital base of many banks (Wolff, 2017).

- or more profoundly, the enforcement of the principle of liability in the financial markets and narrow banking (one of the most extreme forms of which would be an increase in the minimum reserve through the so-called 100% money)[7] and
- the debate on the introduction of the financial transaction tax (e.g. Tobin tax).[8]

At the *intersection with the ecological pillar* is the so-called 'ecological economy' or 'sustainable management' (Luks, 2002). Based on the economic assumption that the concept of capital can be transferred to nature, a distinction is made between 'weak' and 'strong sustainability'. From the perspective of weak

---

[7] The background is the discussion about the lack of market discipline in the financial sector and the devastating effect that systemic banking crises have on the real economy. This effect stems from three essential economic functions that banks currently perform and that cannot be fulfilled if a bank fails: they act (1) as intermediaries between capital supply and demand (financial intermediaries); (2) as payment service providers; and (3) creators of a predominant part of the money supply through credit creation. Of particular concern here is the level of reserve requirements – currently about 10% of the world's money in circulation is collateralised. Against this background, approaches such as the separation banking system and deposit insurance are discussed. The overarching concept of 'narrow banking' is understood as a requirement for banks to ensure that the bank can meet its payment obligations under all circumstances. The main focus is to reduce the risk of bank illiquidity and to create greater independence from the banking system. In practice, banks today are no longer subject to such rigid regulations as in narrow banking, but – also on the occasion of the current economic crisis – to intensive market regulation in order to avoid banking crises. In Germany, the so-called Liquidity Regulation governs the extent of liquidity to be held (cf. among others Wolff, 2017; Homm, 2016; Krall, 2017; Rickards, 2012).

[8] The financial transaction tax is a very low tax (approx. 0.05%) on all international currency transactions. The aim is to curb currency fluctuations and generate money for international development, international poverty reduction and a global environmental policy. The best-known models in the international sustainability discourse are the 'Tobin Tax' proposed by James Tobin in 1972, which led to the founding of the globalisation-critical organisation 'attac' in 1997 (Kaiser et al., 2006), and the 'Robin Hood Tax', which has been supported by a coalition of over 50 charitable institutions and organisations (including Greenpeace, Oxfam and UNICEF) since 2010. The debate on the concrete introduction of financial transaction taxes has gained momentum in Europe in the wake of the debt and currency crisis (Sachs, 2010). In 2012, France introduced a limited financial transaction tax. In addition, the European Union (EU) has included the introduction of an EU-wide financial transaction tax in its draft for the EU's multiannual financial framework for the period 2014–2020 (European Parliament, 2012). The main criticism of the introduction of the financial transaction tax is that it could lead to an increase in tax avoidance strategies, i.e. to a greater shift of trading transactions to trading venues where this tax is not levied. The financial transaction tax would therefore have to be introduced worldwide if possible and at the same time go hand in hand with an abolition of tax havens – but this, according to the critics, is not realistic at present (Palley, 2000).

6.1 Key Dimensions of the Sustainable Society in the Current Discourse

sustainability, resource scarcity only becomes a problem when no suitable substitutes can be found. In contrast, ecological economics, in its call for strong sustainability, stresses that certain natural resources cannot be substituted and that therefore not a relative, but an absolute scarcity prevails. This position criticises the current economic approach for attaching too much importance to the present as opposed to the future (present preference) – this contradicts the goal of intergenerational justice (Luks, 2002).

Ecological sustainability is most strongly oriented towards the original idea of not overexploiting nature. Thus, ecological sustainability would be a way of life that makes demands on the natural foundations of life, but only to an extent that they can regenerate accordingly. Compared to the other pillars, the ecological context enjoys priority in the overall sustainability discourse. Here, the assumption is that the protection of natural living conditions is also the basic prerequisite for economic and social stability. The thematic focus extends to at least six areas, which are discussed primarily within a domestic social framework, but also against the background of a world environmental policy: Ozone policy, climate policy, waste policy, biodiversity, water policy and renewable energy supply. The ozone, climate and energy issues are considered to be the internationally most explosive ones. For various reasons, ozone policy has already made the most progress at the international level (Simonis, 2004). In contrast, the issues of 'global warming' and 'energy supply' are still poorly regulated at the international level, despite the great need for political action. They therefore currently occupy a special place in the ecological pillar of the social sustainability debate.

From a meta perspective, the sustainability discourse primarily relates to the controversial question of whether and to what extent the sustainability focus should be placed on all three pillars as equally as possible or rather with a focus on the ecological pillar. This debate is primarily ignited by the distinction between 'weak' and 'strong sustainability' already outlined above. 'Weak sustainability' – understood as the notion that environmental, economic and social resources can be balanced against each other – is criticised by most proponents of the ecological pillar. They advocate the concept of 'strong sustainability', according to which natural capital can only be replaced by human or physical capital to a very limited extent or not at all. This approach also corresponds, for example, to the environmental space concept, the well-known 'ecological footprint' or the 'guard rail model'. The latter means that ecological parameters are developed in such a way as to ensure stable living conditions on earth in the long term – only within this 'development corridor' is there room for the implementation of economic and social goals (Littig & Grießler, 2004; Luks, 2002). Against this background, critics argue that the three-pillar model is a weak model of sustainable development because it calls for the mutual integration of economic, ecological and social concerns. Despite all the criticism, however, only the three-pillar model has been able to gain acceptance in the sustainability discourse to date.

For all the diversity in the various sustainability dimensions and discourses, the basic ethical idea of intergenerational and global justice can be identified as an essential common denominator. The question of the justification of this ethic has received little attention in the discourse so far.

6.2 Ethical-Religious Influences on the Concept of Sustainability

Compared with the other two guiding concepts, the sustainability context is the most normative one. The normative-ethical reference of the sustainability discussion is reflected in at least two perspectives: On the one hand, the idea of sustainability is grounded in the human rights concept stemming from the tradition of humanism and natural law. On the other hand, the ethical reference in the sustainability discourse has a religious-metaphysical component.

With respect to the first, the concept of human rights assumes that all human beings are endowed with equal rights simply by virtue of being human. These are considered universal, indivisible and inalienable. This idea is closely related to the idea of natural law and humanism developed in the Age of Enlightenment. To date, human rights are systematised in various national and international agreements and institutions, which cover the three pillars of sustainability to varying degrees. The ethical-human pillar, and to some extent the economic pillar, is mainly covered by the so-called 'first generation' (civil and political rights)[9] and the 'second generation' of human rights (economic, social and cultural performance rights).[10] Only the comparatively new 'third generation' of human rights explicitly refers to

---

[9] Civil and political human rights include the liberal rights of defence and democratic rights of participation (human dignity, equality before the law, freedom of assembly and expression, right to life, liberty and security, etc.) and are reflected above all in the constitutions of Western states and in the liberal-legal theory of fundamental rights.

[10] These rights include entitlements and participation rights and are granted by the state in the form of positive benefits. These include labour rights, social security, food, health and education.

6.2 Ethical-Religious Influences on the Concept of Sustainability

the ecological dimension (collective rights of peoples, including the right to development).[11]

From another perspective, the concept of sustainability also has a religious-metaphysical justification dimension, which has had considerable influence on culture, politics and the economy since the mid-twentieth century. Unlike the humanistic-natural human rights discourse outlined above, the ecological sustainability pillar appears to have a stronger emphasis in the religious-metaphysical discourse. The influence of religion on the sustainability discourse manifests itself in several temporally staggered and intersecting tendencies.

The first one resulted from a post-materialist change in values that began in Western industrial societies in the 1970s and continues until today. This change in values was fostered by two factors: On the one hand, with the end of the post-war period Western societies experienced a consistently increasing prosperity, which made it possible to demand 'renunciation on a high level'. On the other hand, since the 1970s, which were influenced by the *'Limits to growth'* study (Meadows et al., 1972) and the oil price crisis, awareness of ecological challenges has significantly increased. Additionally, the post-materialistic value patterns touched on the questions of meaning surrounding the position of human beings in the cosmos:

> Human beings should once again become aware that they are integrated into the fragile mesh of the earth's ecosystem and show respect for the ecological balance understood as a moral norm (Vogt & Ostheimer, 2006, p. 14).

Secondly, work on ecological issues in the context of scientific research established a distinction from the modern positivist approach. The alternative scientific paradigm views nature as an interconnected system that requires appropriate thinking in terms of relationships and interdependencies and seeks to sensitise science to questions of value. In parallel, the ecology movement opened up to esoteric movements (especially the New Age) and became their leading mediator, so that a religious-spiritual transfiguration of these scientific approaches could and can be observed. Perhaps the best-known example is represented by the 'Gaia hypothesis' developed by natural scientists Lynn Margulis and James Lovelock in the mid-1960s. The name is derived from Gaia, the 'Great Mother', the goddess of

---

[11] These rights, which are still comparatively weakly institutionalised, relate primarily to the international context and are understood as collective solidarity rights of the Western industrialised countries vis-à-vis the Third World/the Global South. They include, among other things, the right of people to self-determination and the associated right to development, the right to a clean environment, to peace, to communication, and to a fair share of the treasures of nature and culture.

the Earth of Greek mythology. The approach states that the Earth and its entire biosphere – the totality of all organisms – can be viewed as one living being. According to this, a living being is an open and entropy-producing system that can reactively adapt to its environment in such a way that it is able to keep its entropy dynamically below its maximum entropy through entropy export. A central characteristic of living beings is also reproduction (Lovelock, 1991). The Gaia hypothesis has now found many adherents, most of whom portray the Earth as an 'ensouled' organism that – like an Earth goddess – punishes and rewards. However, the founders of the hypothesis have always distanced themselves from such an interpretation of their hypothesis (Lovelock, 1992).

The second trend is reflected in the growing influence of the classical world religions in sustainability discourses. As early as the end of the 1980s, the Worldwide Fund Conference for Nature (WWF) took place in Assisi, the first conference of its kind, in which representatives of the world's five largest faiths discussed sustainability strategies and environmental protection. Since then, there have been and still are a wide variety of religious initiatives and partnerships with international politics on the topic of (ecological) sustainability, such as the 'World Council of Churches Climate Change Programme' e.g. in 1988, the 'Global Forum of Spiritual and Parliamentary Leaders' e.g. in 1988, 1990, 1992, 1993, 'Religion, Science and Environment Symposia' e.g. in 1994, 1997, 1999, 2002, or the 'Millennium World Peace Summit of Religious and Spiritual leaders' e.g. in 2000, to name but a few (Bright et al., 2004). At this point, it can be concluded that international politics interested in implementing sustainability strategies (e.g. UN, WWF) have learned relatively quickly to take up the role of religion as an important framework for action for most people in the world. It was only in this spirit that religion and culture could be consciously integrated into political sustainability initiatives. This is reflected, for example, in the UNESCO programme *'Culture and Religion for a sustainable Future'* (UNESCO).

In the field of economics, against the backdrop of increasing criticism of 'predatory capitalism' and growing environmental awareness, a more active role of religion in justifying business and economic ethics can also be observed, but only in isolated cases.[12] References to Christian (especially Catholic) social teaching can often be observed. More or less widespread examples from other cultures, on the other hand, can be found, for instance, in transfers of the Tao Te King to

---

[12] As contradictory as it sounds, not only sustainable economic activity, but also predatory capitalism (especially of the US kind) is partly based on religion. In this case, the Calvinist-Protestant ethic, as the sociologist Max Weber noted, might have also have had a considerable influence (Weber, 1965).

6.2 Ethical-Religious Influences on the Concept of Sustainability

sustainable corporate management (Wielens, 2004), in the guiding idea of 'karma capitalism' coined by the US business guru Vijay Govindarajan (Govindarajan), or in the comprehensive neo-Hindu social concept of the so-called 'Progressive Utilization Theory' (PROUT).

The examples of sustainability principles found in religious teachings are numerous – some of them overlap, others set their priorities differently. The Abrahamic religions (Judaism, Christianity, Islam) for example emphasise that man is 'God's governor on earth' and thus also the 'custodian and steward of God's possession' – nature. In contrast, Buddhism and Hinduism emphasise in the context of their cause-and-effect doctrine of 'karma' that every physical and mental action (karma literally means 'act') has a direct or indirect influence on fellow human beings, the environment and future generations.

In addition to these different emphases, teachings supporting the principle of ecological, economic and social sustainability can be found in all world religions. In the context of *ecological-economic sustainability*, the following is an exemplary collection of teachings that reject wasting material resources and thus share the same basic idea of a moderate lifestyle:

- Christianity: 'For God did not send his Son into the world to judge the world, but that the world through him might be saved.' (Bishops of Germany Austria Switzerland et al., 2017: *The Bible*, 1 John, 3.17)
- Islam: 'You Children of Adam! Put on your ornaments at every place of worship, and eat and drink! And be not wasteful (in this)! Allah loves not those who do not moderate.' (Asad, 2011: *Qur'an*, Sura 7.31)
- Hinduism: 'In case of obtaining anything in excess, one should not hoard it. One should abstain from acquisitiveness' (*Tulsi & Mahāprajña*, 2001: *Acarangasutra*, 2.114–19).
- Buddhism: 'A deed (kamma), monks, done out of greed, arising out of greed, caused by greed, arising out of greed – such a deed will come to maturity wherever the person concerned is reborn; and wherever that deed comes to maturity, that is where the fruit of that deed will come to one, whether in this life, in the next, or in a later life' (Neumann, 1922: *Pali Canon*, Chap. 4, Nidāna Sutta).
- Taoism: 'He who is content is rich.' (Laotse: *Tao Te King*, Chap. 33)
- Confucianism: 'The Master said, "Too much is just as [wrong] as too little."' (Kungfutse: *Conversations*, 11:15)
- Judaism: 'Why do you weigh money for that which is not bread, and your earnings for that which does not satisfy?' (Bishops of Germany Austria Switzerland et al., 2017: *The Bible*: Isaiah 55:2)

In the context of *ethical-social sustainability,* a tradition of charity can be found in almost every religious ethic. The best-known principle corresponds to the so-called 'Golden Rule': 'Treat others as you would have them treat you' or 'do unto others as you would have them do unto you.' The following is a sample selection:

- Hinduism: 'One should never do something to others that one would regard as an injury to one's own self. In brief, this is dharma. Anything else is succumbing to desire.' (Dharma, 2020: *Mahabharata:*13.114.8)
- Confucianism: 'This is "mutual consideration" (shu). What one should not do to me, I do not want to do to other people.' (Kungfutse: *Conversations,* 15:24)
- Buddhism: 'What is an unlovely and unpleasant thing for me is an unlovely and unpleasant thing for the other person. What is an unlovely and disagreeable thing to me, how could I charge it to another?' (Neumann, 1922: *Pali Canon,* Samyutta Nikaya)
- Judaism: 'You shall not take vengeance or bear a grudge against your kinsfolk. Love your neighbour as yourself: I am the LORD.' (Bishops of Germany Austria Switzerland et al., 2017: *The Bible,* Leviticus 19:18)
- Christianity: 'Whatever you expect of others, do likewise to them.' (Bishops of Germany Austria Switzerland et al., 2017: *The Bible*: Luke 6:31)
- Islam: 'None of you is a believer unless he wishes for his brother what he wishes for himself.' (al-Nawawi, 2007: *Book of 40 Hadith*: Hadith 13)

Ethical-social sustainability is not only based on the idea of solidarity, but also on the development paradigm that we have already presented above against the background of the 'developed society'. The development paradigm is systematised in the third generation of human rights as the 'right to development' and is considered both an individual and collective human right (General Assembly of the UN, 1986). From a religious metaphysical perspective, the ethical justification of a 'right to development' goes far back to the doctrine of emanation in ancient Greek philosophy. The doctrine was systematised by Neoplatonism and had an enormous influence especially on the mystical traditions of the Abrahamic religions (Jewish Kabbalah, Christian mysticism, and Islamic Sufism), German idealism (especially Hegel), and all the Masonic, Rosicrucian, and Theosophical traditions of Europe. The term 'emanation' comes from the Latin *emanare* meaning 'to flow from' or 'to pour forth or out of'. It denotes the way in which all things are derived from the first reality or the ONE through stages of differentiation and degradation to ever lesser degrees and ever more differentiation of the first reality. Accordingly, the differentiated is merely the unfolding of potential that is contained and involuted. From this perspective, everything in the world is an expression of the same universal origin and fulfils its purpose of existence in the expression of its inherent or involuted potentials through 'evolution' or 'unfolding' (Störig, 2002). Understood in this way, the doctrine of emanation has a normative core that is relevant to international development policy and also in the context of sustainable development, even if neither explicitly refers to it. For instance, development policy expert Franz Nuscheler defines 'development' as the active 'evolving [evolution] through the unfolding of one's capacities [involution] (Nuscheler, 2006, p. 226).'

In summary, the sustainability discourse has a pronounced ethical value reference, both in the sense of humanistic tradition of human rights and above all in the religious-metaphysical sense. The latter has an underestimated influence on the sustainability discourses in various societal areas, such as science, politics, economics, and culture.

6.3 Measurement Criteria and Principles of Societal Sustainability

6.3.1 Measurement Criteria

Today, there are a large number of sustainability indices worldwide – most of them revolve around the ethical-ecological assessment of companies.[13] In the context of societal sustainability, the Sustainable Society Index (SSI) is probably one of the world's leading ones. It gives an impression of the criteria for assessing social sustainability and the current global progress towards worldwide sustainability.

Based on the Brundtlandt definition and the fundamental distinction between a human, economic and ecological dimension, the SSI assesses between 154 and 213 countries/territories with the help of 24 individual indicators. The following figure provides an overview that could be self-critically reflected upon and deepened in the context of further research, but is suitable here as a working basis (Table 6.1).

It is striking that the indicators for 'human well-being' and partly also for 'economic well-being' are also found in their entirety in the model of the developed

---

[13] The most prominent of these are the Dow Jones Sustainability Index (DJSI), which was created in 1999, and the Frankfurt-Hohenheim Guidelines, which were developed in 1997 under the leadership of the economist Gerhard Scherhorn and the theologian Jürgen Hoffmann. The latter forms the assessment basis of oekom Research AG, one of the world's leading (and Europe's leading) rating agencies in the sustainable investment segment for the assessment of companies, supranational institutions and, in some cases, countries, based in Munich. Overall, however, there is no globally shared catalogue of criteria for assessing sustainability.

Table 6.1 Three pillars and 24 criteria for assessing social sustainability using the example of the SSI (2017)

Human/social dimension	Subjective well-being Human development (e.g. HDI) Public health Gender equality Good governance Income distribution Population growth
Economic dimension	Material consumption Savings Gross domestic product Employ Indebtedness
Ecological dimension	Quality (soil, air, water) Renewable energies Greenhouse gas emissions Energy consumption Renewable water resources Available forest areas Biodiversity Organic farming

society – only in the case of the indicators from happiness research, which are rather less taken into account in the sustainability discourse (but can certainly be associated with it in case of doubt), is there a deviation. Since large parts of the social and economic indicators overlap, it is certainly possible to find many societies that are equally at the top of the most developed and most sustainable societies (Table 6.2).

Among the developed societies, it is noticeable that the sustainability ranking is almost universally led by the European societies – and here in particular the Scandinavian societies – while the Anglophone societies also performed comparatively less well. This is probably due to the tendency to pay less attention to environmental sustainability in the Anglophone countries, but also to the more short-term principle of the free-market economy that prevails there. In contrast, European countries tend to be characterised by the principle of a coordinated, mostly 'social' market economy, which has a relatively long-term economic focus (I will come back to that in a later chapter).

The SSI assesses sustainability using a point system in the spectrum from 0 (no sustainability) to 10 (full sustainability). For all countries studied, a trend increase

6.3 Measurement Criteria and Principles of Societal Sustainability

Table 6.2 Top five sustainability rankings from 2006 to 2016. (SSI 2017)

Rankings	2006	2008	2010	2012	2014	2016
Human development						
Finland	1	1	1	1	1	1
Germany	9	10	2	5	4	2
Netherlands	7	5	5	6	3	3
Iceland	4	15	14	2	2	4
Norway	5	4	4	4	5	5
Ecological well-being						
Burundi	30	24	12	3	1	1
Togo	15	5	2	44	23	2
Lesotho	1	4	4	2	3	3
Central African Republic	8	7	9	4	6	4
Uganda	12	13	14	7	7	5
Economic prosperity						
Norway	6	4	2	1	1	1
Switzerland	1	1	1	2	2	2
Estonia	10	3	5	13	5	3
Sweden	3	6	3	4	3	4
Czech Republic	11	9	6	8	7	5

was noted in 2016 compared to previous years, but still about 40% below the overall level required for the world. Comparing the three pillars of sustainability, the study concludes that progress is greatest in the area of human well-being, whereas the environmental and especially the economic pillars lag significantly behind. Figure 6.2 provides an overview – the outdoor area represents the full score (full sustainability) and the centre a score of 0 (no sustainability):

In the context of a sustainable society, too, the same advantages and disadvantages of the approach of measuring sustainability by means of measurement criteria arise as described elsewhere. It can be summarised as complementary to the principles outlined below.

6.3.2 Principles

Analogous to the guiding principles of the developed and resilient society, at least four orientation principles can be derived from the perspective of the Simplify tradition. The extent to which they overlap, complement and contradict those of the resilient and sustainable society will be examined in more detail elsewhere.

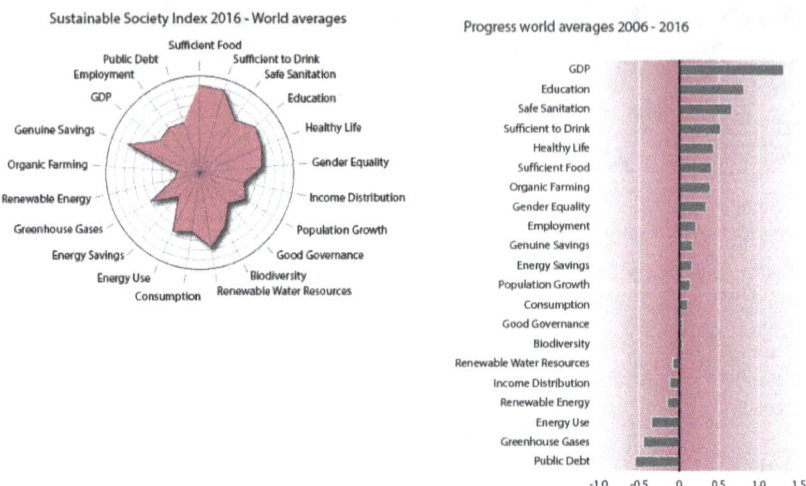

Fig. 6.2 Sustainable Society Index, 2016 and 2006–2016 trends for the world. (SSI, 2017)

- Principle 1: The sustainable society promotes the realisation of the Inner Development Goals (IDGs)
- Principle 2: The sustainable society is tendentially based on a cold learning culture
- Principle 3: The sustainable society develops collective intelligence
- Principle 4: The sustainable society develops collective wisdom

Regarding *Principle 1*: Perhaps the most important prerequisite for a sustainable society is correspondingly mindful and competent citizens. This includes not only raising awareness of the manifold social, economic and particularly ecological consequences of individual actions in space (consequences for the entire planet) and time (consequences for future generations); sustainable awareness-building also implies corresponding development of values. In this context, sustainability would have to be a central component of education policy. Depending on the type of society, knowledge transfer can have different content foci, depending on the already existing resources, challenges, and framework conditions. Traditional societies, for instance, could have a solid knowledge base in agriculture and how to maintain a subsistence economy, but might have a lack of knowledge in technologically efficient and effective means of waste management, food and energy production. By contrast, citizens of relatively more urban modern industrial societies or postmodern information societies, are often estranged from nature.

6.3 Measurement Criteria and Principles of Societal Sustainability

Accordingly, education policy could also focus on bringing citizens closer to nature again. Regardless of the society type, education policy should incorporate interrelational or systems thinking, and other forms of cross-disciplinary thinking that are important to anticipate complex cause-and-effect relations. All of these and further criteria are incorporated in the framework of the so-called 'Inner Development Goals' (IDGs). The IDGs is a not for profit and open source initiative to develop 'transformational skills for sustainable development (IDGs, 2021). The framework includes a set of 23 skills and qualities which can be categorised as follows (Table 6.3):

The IDGs can help accelerating the work towards the UN's Global Goals or Sustainable Development Goals (SDGs). The SDGs or Global Goals is a collection of 17 interlinked global goals designed to be a 'blueprint to achieve a better and more sustainable future for all' (General Assembly of the UN, 2017) and is intended to be achieved by 2030.

Regarding *Principle 2*: From a collective-psychological perspective, too, the implementation of the guiding principle of the sustainable society corresponds more to the ideal type of a cold culture. The knowledge and learning focus are less on the question of how the environment can be controlled and adapted to one's own culture, but conversely on how the environment can be preserved as best as possible and how one's own culture can be brought into harmony with the environment. This, however, requires a differentiated approach. In Lévi-Straus's original model,

Table 6.3 The IDG framework at a glance. (IDGs, 2021, p. 12)

Being – relationship to self	Thinking – cognitive skills	Relating – caring for others and the world	Collaborating – social skills	Acting – driving change
Inner compass	Critical thinking	Appreciation	Communication skills	Courage
Integrity and Authenticity	Complexity awareness	Connectedness		Creativity
		Humility	Co-creation skills	Optimism
Openness and Learning mindset	Perspective skills	Empathy and Compassion		Perseverance
Self-awareness	Sense-making		Inclusive mindset and Intercultural competence	
Presence	Long-term orientation and visioning		Trust	
			Mobilisation skills	

the distinction between cold and hot culture corresponds to the demarcation of traditional (pre-modern) from modern value structures (Treichel & Mayer, 2011). In fact, awareness of sustainability is conceivable in both traditional and (post-)modern types of society, but is likely to vary in the concrete content of educational policy, as already outlined in the context of Principle 1. What is certain is that the collective focus of values is more likely to be on preservation.

Principle 3: Similar to the (multi-)resilient and the developed society, the sustainable society requires a high degree of collective intelligence. Awareness of sustainability requires anticipating unforeseeable side-effects of collective action and developing preventive solutions to problems. For this reason, sustainability research has always been transdisciplinary. It presupposes a high degree of knowledge integration and requires close collaboration between knowledge agents from all sectors of society – civil society, academia, the private sector and politics. Think tanks, multi-stakeholder initiatives, cross-disciplinary symposia, research associations and/or conferences – in all these and other constellations experts from different fields of expertise and practice come together to manage complex problems.

Principle 4: While collective intelligence asks what is possible in terms of new developments and innovative problem solutions, collective wisdom supports reflection on what is the deeper motivation behind any decision and development path. Wisdom is understood here as a profound understanding of the interrelationships in nature, life and society, as well as the ability to make the most coherent and sensible decision in the face of challenges. It enables one to take a step back, to look critically at development (as well as the possible side-effects that accompany it) by reflecting on what the actual goal of development is. Thus, while collective intelligence is necessary to create and to follow a development path, collective wisdom supports questioning and choosing between different possible development paths, always keeping in mind what really matters. In the context of wisdom, one might ask, for example, whether moving towards a technologically highly complex affluent society leads to more well-being and happiness than achieving a post-growth society?

Two things become clear from the considerations so far: firstly, the concept of sustainability goes beyond the context of national society – it always encompasses the entire planet or 'world society'. Secondly, it is noticeable that sustainability, although it is an all-encompassing concept, usually focuses on the ecological-economic aspect. That means, it is usually directly concerned with the carrying capacity of the earth, which is overtaxed by today's prevailing economic practices, which also affects the economic and social spheres.

6.4 Three Concepts in the Growth-Critical Discourse

Similar to the concepts of resilience and development, the discourse on sustainability is heterogeneous and includes several other concepts. The sustainability concepts all have in common that they share a thoroughly global claim. They do, however, differ in how radical their critique of economic growth is. Despite their differences, the question of a national and supranational social order that prevents the danger of particularly ecological and social crises is always at the centre. Mainly inspired by the research of the Club of Rome, the basic motive of the sustainability debate is to avert global negative scenarios that will occur 'if we continue as before'. Among the possible future scenarios of global development, Franz Josef Radermacher's reflections, including the concept of a so-called 'Global Marshall Plan', are currently one of the best known worldwide. He distinguishes between the worst-case scenario of a 'collapse', the likeliest case scenario of a 'neo-feudalisation/brazilianisation' and the best-case scenario of a 'balance/sustainability':

- Negative scenario (worst case): According to Radermacher, an ecological collapse will occur if humankind continues to operate in the traditional sense, i.e. if technical know-how and resources are not used in a mindful way. He estimates the probability of this scenario occurring at around 15%.
- Most likely scenario (likeliest case): Radermacher describes 'Brazilianisation' as the rule of a few rich people over an impoverished, uneducated and powerless mass, who are 'condemned' to a life with minimal resource consumption – an 'eco-dictatorship'. This could avoid ecological collapse, but the result would be a global gap of about 98% poor and 2% rich – a gap that has already occurred in Brazil. Radermacher assesses the probability of occurrence at about 50%.
- Positive scenario (best case): In the best possible scenario, humanity understands how the first two scenarios can be averted and reacts in a timely and appropriate manner. A reasonable regulation of market activity, social balance and good governance as well as sustainable development would be possible for up to 10 billion people. Radermacher rates the probability of occurrence at about 35% (Radermacher & Beyers, 2011).

In the context of these scenarios, the dominant discourses revolve around the question of how economic systems can be designed in the future that, on the one hand, enable the greatest possible safeguarding of the population's needs in relation to the social sphere and, on the other hand, provide the best possible relief for the ecological sphere. The following concepts and corresponding discourses differ in their critical attitude towards economic growth and in their radicalism with regard to the question of restructuring and regulating the market economy system. At least three more or less competing concepts and corresponding discourses and visions can currently be cited, which differ from one another in their radicalism with regard to growth and the need for transformation of the market economy system:

1. The *'green growth'* concept is relatively uncritical of the idea of growth and does not call for a radical transformation of the economic order. It is usually linked to a conventional concept of the market economy, usually the social market economy.
2. The 'post-social *market economy*'[14] largely retains the concept of green growth, but envisages a further development of the market economy order. The best-known visions include the 'eco-social market economy' and the 'ethical-ecological market economy'.
3. The *'post-growth economy'* is characterised by a strong critique of growth and envisages a comprehensive restructuring of the market economy. As a rule, it is linked to an economic form of sufficiency and/or subsistence. This discourse could also include other visions and concepts that vary further in their degree of radicalism, such as the 'Donut economy' or the 'zero-growth-economy'.

6.4.1 Concept 1: Green Growth

Green growth (sometimes also called 'sustainable' or 'qualitative growth') assumes that economic growth can be decoupled from material and energy flows through technological and system innovations. Consequently, the concept envisages a society that continues to maintain economic growth and its material comfort, consumption and capitalist economic order. The assumption here is that the negative consequences of economic activity will be compensated by technological innovations, ensuring high ecological efficiency and maybe also the possibility to reverse ecological damages.

Currently, the sector of 'green' and 'smart technologies' is gaining momentum – not only because of its ecological necessity, but above all because of its economic potential. Sustainable urban development (including energy-efficient buildings and intelligent infrastructure projects) is the most promising and money-intensive area

---

[14] This term is a neologism of mine.

6.4 Three Concepts in the Growth-Critical Discourse

of business. It is no wonder that, given the ongoing urbanisation, more than half of the world's population already lives in urban agglomerations – and the trend is rising. Cities account for around two-thirds of global energy consumption, 70% of greenhouse gases and 60% of water consumption (Ritchie & Roser, 2019). Since cities with sustainable infrastructures are not only ecologically necessary, but also provide a higher quality of life, they increase their competitiveness. There are already hundreds of sustainable urban development projects worldwide, that combine comfortable living and environmental sustainability, supported by the latest technology. In terms of 'high standard of living' with 'low ecological footprint' and 'lowest CO_2 emissions', the comparatively greenest and most environmentally friendly cities worldwide are still concentrated in the European region, including for instance Copenhagen, Stockholm, Madrid, Amsterdam. Aside from this, Vancouver and Wellington are mentioned for the Anglophone region and Singapore for the Asian region. However, in Singapore there is a massive increase in interest in sustainable urban development, particularly in the economically booming mega-cities of East Asia.[15]

Green growth is best realised in the smart city that aims for sustainable development. Here, there is a clear intersection between the guiding principle of the developed and the sustainable society outlined above. In addition to the projects mentioned above, the 'ecological prefabricated city' developed by electronics giant Panasonic is currently considered ground-breaking. Like a car, it is supposed to be turnkey and can be ordered with all options from the manufacturer (Koelling, 2011). The first model town, 'Fujisawa Sustainable Smart Town', is considered to be climate-friendly, fully networked and capable of covering its own electricity consumption. It is expected to provide space for 3000 people from 2013 (Fujisawa Sustainable Smart Town). The world's most prominent smart city project, 'Masdar City', is located in the Persian Gulf, in Abu Dhabi. US, Japanese and above all European companies are involved in the construction project. As the 'world's first eco-city', Masdar City stands above all for an independent regenerative energy supply and a completely CO_2-neutral, car-free and even waste-free city life. The investment volume is around US$25 billion. After completion, which is scheduled for 2030, the city is expected to accommodate about 40,000 inhabitants. Currently,

---

[15] There are various indices measuring the 'greenest cities' in the world, such as sustainable jungle's 'Top 15 Greenest Cities', Resonance Consultancy's 2020 World's Greenest Cities Index, Schroders European Sustainable Cities Index, Arcadis Citizen Centric 2018 Sustainable Cities Index, The Global Destination Sustainability Movement (considering a city's sustainability from a tourism perspective), Sustainable Development Solutions Network' 2019 US Cities Sustainable Development Report, Economist Intelligence Unit's 2021 Liveability Ranking.

about 500 people live there. The International Renewable Energy Agency (IRENA) has its headquarters there (Masdar City).

In addition to these urban development concepts, 'Desertec' should be mentioned as another major project. This concept envisaged generating green electricity at the world's most energy-rich locations – the sun-rich deserts in the Middle East and North Africa – and transmitting it to centres of consumption by means of high-voltage direct current (HVDC) transmission. On 14 October 2014, the downsizing of the initiative was announced. The 17 shareholders decided to dissolve the planning company and to move the headquarters from Munich to Dubai in order to ensure optimal promotion of the initiated projects through physical proximity. A small part of the company was converted into a consulting firm supported by three shareholders (Desertec). Four times larger than Desertec and currently one of the largest sustainable energy projects in the world is China's aforementioned giga park initiative with a capacity of 450 gigawatts which is about twice the capacity of all German green power plants combined. Similar to Desertec, China's giga parks are to be built in desert regions, especially in the Gobi desert (Schultz, 2022). These and other projects prove that new potentials for economic growth arise precisely from ecological necessities. The extent to which this 'green growth' can actually be considered sustainable is disputed by representatives of the other two concepts, especially the 'post-growth economy'. This will be discussed in more detail below.

Another characteristic feature of green growth is that it does not question the established market economy order. Currently, the realisation of green growth concepts can mostly be observed on the basis of a *free market* or *social market economy order.*

The concept of a social market economy is understood very differently around the world. Essentially, the underlying core idea can be interpreted as combining the advantages of a free market economy (especially a high supply of goods and efficiency) with the welfare state (which is intended to compensate for possible negative effects of market processes). Across Europe, the concept of the social market economy is widely accepted across political party lines and is explicitly stated as an objective in the EU Constitution, including in the context of social and environmental sustainability. Thus, in connection with the European Single Market, Article 3 of the Treaty of the European Union states that the EU shall act

> towards the sustainable development of Europe based on balanced economic growth and price stability, a highly competitive social market economy, aiming at full employment and social progress, and a high level of protection and improvement of

the quality of the environment. It promotes scientific and technical progress. (TEU, 2008, p. 17)

This formulation was incorporated into the Treaty of Lisbon, which came into force in 2009. In the Anglophone countries, such as the USA, Australia or New Zealand, on the other hand, the concept of the liberal market economy prevails. This type of capitalism is characterised by pure control of all economic processes via the free market. Whether one of the 'varieties of capitalism' (Hall & Soskice, 2001) will prevail over the other, is unlikely at present. From the perspective of comparative capitalism studies, both (ideal) types are characterised by different types of innovation and thus different levels of competitiveness. While the Anglophone liberal market economy has a short-term return mentality, its institutional context fosters sectors characterised by radical, disruptive innovations, such as the development of software solutions, semiconductors, biotechnology or nanotechnology.[16] All these sectors are essential for the development of renewable energies and smart technologies, as well as the clean technology sector. On the other hand, the coordinated market economy –which the social market economy is also a part of – is likely to have specific competitive advantages in relatively long-lasting, incremental innovations, such as those observed in the chemical industries

---

[16] The *liberal market economy* type ideally includes all Anglophone countries, such as the USA, Britain, Australia, New Zealand. The activities of this economic system are primarily controlled by the market and the competition – the result is a low protection against dismissal, but a high flexibility on the labour market, broadly applicable educational qualifications for employees and a high degree of economic freedom in general. Companies finance themselves mainly through the stock market (shareholder value), which leads to a short-term return mentality. Moreover, the institutional context encourages firms to pursue radical innovation, as they can lay off employees at short notice if projects fail, but can also poach and hire personnel for new projects at short notice (Hall & Soskice, 2001).

or automotive sector.[17] These mainly concern long-lasting consumer goods and are also of high ecological relevance, due to their resource-saving character. Both types of innovation will continue to be of great importance in the future.

6.4.2 Concept 2: Post-social Market Economy

The concept, which can be roughly summarised under the collective term 'post-social market economy', is not fundamentally sceptical of the basic idea of 'green growth'. However, its proponents emphasise the need for a more far-reaching transformation of the market economy system in order to flank green growth in the sense of the principle of sustainable development. The aim is above all an economic order that expands the social market economy to include further political categories, namely at least environmental protection, ethical-humanistic values and sustainable economic activity. To date, there are a number of names and models for such an economic order, which can be subsumed under the umbrella notion of a post-social market economy. Two inherent concepts are, among others, the 'humane market economy' and the 'eco-social market economy'.

The *humane market economy* represents an economic model oriented towards people.[18] The main focus is on (1) education, training and further education, (2) the creation of fair conditions of competition, (3) sustainable economic activity with resource conservation and (4) qualitative growth and an economic approach

---

[17] The group of the so-called *'coordinated market economies'* includes all Continental European societies, such as Germany and France, as well as the Scandinavian societies. This economic order is not exclusively regulated by the market, but also by additional institutions and consensus-building bodies. For instance, while in a liberal market economy a company is free to agree on the wage with an employee, in a coordinated market economy this is done through collective bargaining between employers' associations and trade unions. The consensus reached then applies equally to all competitors in the national market and thus has an overarching effect. As a result of regulation, there is a comparatively higher level of protection against dismissal and a hiring relationship that is geared towards long-term industry- and sector-specific training and employment. Accordingly, firms pursue product strategies that rely on the willingness of employees to share information and to cooperate. The innovation character is 'incremental' – that means, innovations develop step by step on the basis of existing competences and improvements of existing products and processes, especially with regard to product quality. The sectors concerned are durable consumer goods, such as automobiles. Companies therefore have access to long-term credit, i.e. 'patient' capital. The profitability mentality is more long-term and characterised by 'stakeholder value' (Hall & Soskice, 2001).

[18] A standard work on this subject: Jäger (2008).

6.4 Three Concepts in the Growth-Critical Discourse

oriented towards the true needs of people. The economic model was developed in the early 1980s at the Albert-Ludwig's-University Freiburg. Since 2008, facets of this concept have been publicly discussed and debated by representatives from business (e.g. Claus Hipp), academia (e.g. Viktor J. Vanberg), journalism (e.g. Hans-Ulrich Jörges) and politics (e.g. Christian Lindner) – but so far with little influence and international dissemination (Jäger, 2008).

By contrast, the *eco-social market economy* is by far the best known and most widespread among the concepts of the 'post-social market economy'.[19] It is based on the pillars of (1) an efficient, innovative market economy, (2) social justice and (3) ecological responsibility. The concept was developed in the 1970s and 1980s based on the research of Swiss economist Hans Christoph Binswanger. The term was and still is particularly coined in the Austrian political landscape, where it finds supporters not only among the Greens but above all also among conservative and Christian Democratic politicians. The Austrian People's Party (ÖVP), including Josef Riegler, ex-Vice Chancellor, and Franz Fischler, for a time EU Commissioner for Agriculture and Rural Development, clearly refer to the eco-social market economy. In the Federal Republic of Germany, the demand for an eco-social market economy as a guiding principle of global economic policy is represented in particular by the above-mentioned scientist Franz Josef Radermacher. In addition, Klaus Töpfer (CDU), Friedbert Pflüger (CDU), Heiner Geißler (CDU) and Ernst Ulrich von Weizsäcker are prominent politicians who support this guiding principle. The concept can also be found to some extent in some European Green parties. In particular, the Global Marshall Plan Initiative founded by the US politician Al Gore,[20] whose Austrian coordinator is Josef Riegler, has made a worldwide eco-social market economy its concern (Radermacher & Beyers, 2011).

---

[19] A standard work: Radermacher and Beyers (2011).

[20] The Global Marshall Plan Initiative is essentially based on five interlinked building blocks that are already included in the outcomes of past UN summits, as well as being part of EU policy and the demands of various institutions and non-governmental organisations:

1. The further development and implementation of the UN Millennium Development Goals (for more details see next footnote).
2. Achieving the 0.7% target and providing the additional resources needed (US$100–150 billion per year).
3. Fair taxation of global value creation processes: Levy on global financial transactions, trading of CO_2 emission rights in the context of climate justice and a kerosene tax.
4. Fair, global partnership and effective use of funds.

Establish a regulatory framework for the global economy that is compatible with sustainability (Radermacher & Beyers, 2011).

Essentially, then, at least *two target areas* can be named that representatives of the post-social market economy discourse want to realise. The envisaged path is that of 'institutional change' by making use of public, predominantly civil society campaigns to persuade political decision-makers to enact appropriate legislation:

1. Economic growth – including green growth – is to be flanked even more profoundly by *ecological, social and ethical regulations*. The aim is thus to achieve a balance between economic, ethical-social and ecological objectives by enforcing environmental protection and social and ethical guidelines. A variety of tax instruments are envisaged, usually consisting of statutory bans and prohibitions combined with market-based incentives. Among these, incentive taxes or stricter environmental liability are among the most widely discussed steering instruments. Their aim is to reintegrate external costs into the polluter's individual cost accounting ('internalisation of externalised costs'). In this way, environmental protection should become economically cheaper than environmental pollution. Concrete demands are therefore the inclusion of a strict polluter pays principle in competition law as well as the introduction of an eco-tax and ecologically oriented laws. Pioneering in this context is the push developed by the Frankfurt-based project group 'Ethical-Ecological Rating' (in German: Forschungsgruppe Ethisch-Ökologisches Rating) for a legal restriction of property rights through a 'property obligation'. Juristically, the approach, which was developed primarily by the economist Gerhard Scherhorn, who died in 2018, refers to the German constitution, in which the preservation of the commons is laid down by Article 14.2 ('Property obliges') and Article 20a ('Protection of the natural foundations of life'). In the opinion of the project group, it would be necessary to expand Article 903 of the German Civil Code (BGB): Here, the owner's right to do with the thing 'as he pleases' would have to be limited not only by the rights of third parties, but also by the duty to preserve the common goods (Forschungsgruppe Ethisch-Ökologisches Rating).
2. Secondly, it is also about a *'new global economic miracle'*, which envisages to sustainably develop untapped potential – especially in the form of human resources. The focus here is primarily on resources that are still largely untapped due to global poverty. From this perspective, sustainable development can also be understood as 'immaterial growth'. The concept of 'sustainable development' has been an integral part of the implementation of international policy at least

6.4 Three Concepts in the Growth-Critical Discourse

since the establishment of the Millennium Development Goals[21] and since the World Summit on Sustainable Development in Johannesburg in 2002. The Sustainable Development Goals (SDGs) succeeded the MDGs in 2016.[22] The established Global Marshall Plan Initiative is currently the most influential and best-known movement seeking to implement and promote the Sustainable Development Goals and sustainability strategies. In order to transform the global economy, the initiative envisages various regulatory measures, such as investment, coordinated market openings and co-financing in many areas. These are to be provided in return for the alignment of ecological and social standards. In the course of the global fight against poverty, strong new economic impulses are to be released for the affected regions, but also for the entire global economy. At the same time, linking new growth with clear ecological standards should promote a more environmentally friendly way of doing business (Radermacher & Beyers, 2011)

6.4.3 Concept 3: Post-growth Economy

The most radical undercurrent of the social sustainability debate in every respect is determined by various representatives of the critique of globalisation and growth, which can be roughly summarised under the guiding principle of the *'post-growth economy'*. According to the premises of this perspective, permanent economic growth – also 'green' or 'qualitative growth' – can neither be ecologically mitigated nor sustained. At best, 'clean technologies' bring about a reduction of ecological wear and tear, which is measured in terms of waste, emissions, materials, land,

---

[21] The United Nations Millennium Development Goals are eight development goals for the year 2015, which were formulated in 2001. They include (1) Eradicate extreme poverty and hunger, (2) Provide primary education for all, (3) Achieve gender equality, (4) Reduce child mortality, (5) Improve maternal health care, (6) Combat HIV/AIDS, malaria and other major diseases, (7) Achieve environmental sustainability, (8) Build a global partnership for development.

[22] The 17 SDGs are: (1) No Poverty, (2) Zero Hunger, (3) Good Health and Well-being, (4) Quality Education, (5) Gender Equality, (6) Clean Water and Sanitation, (7) Affordable and Clean Energy, (8) Decent Work and Economic Growth, (9) Industry, Innovation and Infrastructure, (10) Reduced Inequality, (11) Sustainable Cities and Communities, (12) Responsible Consumption and Production, (13) Climate Action, (14) Life Below Water, (15) Life On Land, (16) Peace, Justice, and Strong Institutions, (17) Partnerships for the Goals (General Assembly of the UN, 2017).

water and biodiversity – but it cannot be reduced to zero in this way. Relief is therefore only possible step by step, but this is not sufficient to improve the overall ecological balance of the system. The adequate answer can therefore only lie in a conscious departure from the 'affluent society' and in an economic model that focuses on deceleration and decluttering. The guiding principle of the post-growth economy therefore considers the concepts of green growth and the post-social market economy outlined above as insufficient preventive measures against the threat of ecological collapse.

The area of the post-growth economy is also subdivided into further concepts and corresponding discourses, such as 'de-growth' (or: 'zero-growth') and 'post-growth'. Typically, post-growth and de-growth do not specify the answer to the limits-to-growth challenge, but instead develop emerging solutions that are appropriate with regards to place, time, resource and cultural factors (Treehugger, n.d.). Although both post-growth and de-growth differ in the radicality of their demands, both terms are often used synonymously. Currently, Kate Raworth's model of a post-growth 'Doughnut Economy' and Niko Paech's 'zero-growth' approach (although he often used to call it 'post-growth') belong to the most popular ones that shape the discourse.

Based on the Oxford University economist Kate Raworth's 2017 bestselling book '*Doughnut Economics: Seven Ways to Think Like a 21st-Century Economist*' (Raworth, 2018), the city of Amsterdam officially decided to be the first city worldwide to adopt and to implement a variation of the post-growth concept (Purdy, 2020). Raworth's 'Doughnut Model' attempts to balance people's needs without harming the environment. It is illustrated with two 'rings' reflecting the basic needs of the population (inner ring of needs) and the ecological needs of the planet (outer ring of needs).

The inner ring represents the minimum framework for a good life, derived from the United Nations' Sustainable Development Goals (SDG's) long agreed upon by world leaders of all political stripes. These needs range from food and clean water to a certain level of housing, sanitation, energy, education, healthcare, gender equality, income and political voice. Anyone who would not attain these minimum standards would live in the doughnut's hole. The outer ring represents the ecological limits drawn by earth-system scientists. It highlights the limits of the planet's resources beyond which humanity should not go. These limits are intended to prevent damage to the climate, soils, oceans, the ozone layer, freshwater and biodiversity. Between the two rings lies the dough on which people's needs and that of the planet are met (Raworth, 2018). Thus, the 'Doughnut' is a model to consider 'the core needs of all but within the means of the planet' (Boffey, 2018) in the economic activity of a society.

6.4 Three Concepts in the Growth-Critical Discourse

A more radical approach of post-growth, which could also be regarded as 'zero-growth' or 'de-growth', is emphasised by Germany's most popular representative of 'pluralist economics' Niko Paech, Professor of Pluralist Economics at the University of Siegen. He questions economic growth as a whole, including green growth, and argues for a rethinking of prosperity. Paech emphasises the need to reduce global consumption and production in order to achieve a socially, economically and ecologically sustainable society based on well-being as an indicator of prosperity instead of GDP. Paech advocates deglobalisation, as it would reduce the cost advantages of the unlimited division of labour and thus lead to higher prices but at the same time would lead to more crisis stability and new jobs, albeit less in the academic than in the craft sector. According to Paech, living in a post-growth economy would contribute to a more stress-free and responsible life at the price of less consumption and travel opportunities, which might perhaps not be a bad deal at all (Paech, 2012).

Despite the differing discourses, at least two aspects can be found that are common to them and that need to be promoted. First, the establishment of a relocalised, i.e. largely decentralised economic order is sought in order to become independent of national and international policies. Second, the aim is to establish moderate supply and value-added structures and a cultural return to the essentials. The defining key concepts here are 'sufficiency' and 'subsistence' (self-sufficiency). The basic idea here is not a general renunciation of consumption per se, but rather a reduction of industrial production with a simultaneous upgrading of non-commercial supply, e.g. through self-work, manual skills, community gardens, barter rings, networks of neighbourly help, give-away markets, communal use of equipment (e.g. drills, cars) and regional cycles based on interest-free circulation-secured complementary currencies.

Since the mid-2000s, there has been a growing number of grassroots initiatives that have developed cultural techniques of community-based management, decentralised energy production and self-sufficiency. The best known of these is the Transition Town movement, which originated in the Anglo-Saxon world and is spreading mainly in the Western societies. It was initiated in 2006 by permaculturalist Rob Hopkins and students from Kinsale Further Education College in Ireland, among others. The aim is firstly to enable the transition to a relocalised and post-fossil economic order 'from below', and secondly to respond from a community level to a future of dwindling raw materials and fuels and to create agricultural and social systems that are similarly resilient to ecosystems (Transition Network). Transition Town initiatives are therefore not only a remarkable example of a 'sustainable', but also a 'resilient' concept of society (as will be discussed in more detail below). Advocates of the post-growth economy, such as the economist Niko

Table 6.4 Three concepts in the discourse on sustainable societies at a glance

Dimensions	Guiding concepts		
	Green growth	Post-social (eco-social) market economy	Post-growth economics
Growth orientation	Green growth	Green growth; Intangible growth	Zero-growth or post-growth
Economic order	Free or social market economy	Post-social (eco-social) market economy	Subsidiary economic order Sufficiency and subsistence
Agent level (implementation)	Politics (national governments) Private sector	Politics (supranational organisations [e.g. UN])	Civil Society
Determinant variable for future development	Technological innovation	Institutional innovation	Cultural innovation

Paech, see a large-scale spread of these grassroots initiatives as the best way to achieve the guiding vision of 'post-economic growth' (Paech, 2012).

At a glance, the three concepts outlined above that dominate the social sustainability discourse can be summarised as follows (Table 6.4):

It can be assumed that there are other, less well-known, hybrid approaches which fit 'somewhere in between' or maybe integrate the three concepts presented above. Contributions like these could be of importance insofar as they invite a broader and deeper concept of sustainability that, for example, takes into account sustainability areas that are currently relatively neglected, such as the economic or technological one. Hybrid approaches could also help to build bridges between the concepts described above.

6.5 Conclusion

The guiding principle of a sustainable society is characterised above all by consideration of the consequences of social, political and above all economic action. The sustainability discourse focuses on measures to prevent crises in the ecological, economic and ethical/human/social dimensions. Mostly, the dimension of 'ecologically sustainable economic activity' in the sense of a concept of 'strong sustainability' is given the most attention. Compared to the discourses on resilience and development, it can be observed that the sustainability discourse is more strongly normative, including ethic-humanistic and religious.

6.5 Conclusion

In an international comparison, European societies clearly dominate the claim to realise sustainable cities and societies – also in comparison to Anglophone societies. Moreover, it is obvious that the sustainability discourse goes beyond the social context. In its temporal dimension, the concept is designed for the longest possible term (intergenerational), and spatially the focus is always global (international). As the discourse is relatively well established, several concrete concepts can already be found, which differ mainly in terms of the radicality of their demands for a transformation of the existing economic order. Most dominant are the concepts of green growth, post-social market economy (including in particular eco-social market economy) and post-growth economy (including Donut economics and zero-growth). Currently, all three concepts are being advanced in parallel by agents at different levels – in other words, at the economic and national policy level (green growth), the supranational policy level (post-social market economy) and the civil society level (post-growth economy).

Whether and to what extent one of the three schools of thought will prevail over the others in the foreseeable future remains an open question. It is also conceivable that one or more visions will prevail that integrate the basic ideas of the currently dominant three concepts. The core contribution of such conceivable hybrid social models is to integrate and expand the sustainability discourse, also with regard to neglected areas, since the focus of the sustainability discourse is currently primarily on avoiding ecological collapse and social crises. In contrast, comparatively little attention is paid to the danger of economic collapse or other areas like the technological one. In view of the diversity of current crises, societal future preparedness would have to be multi-dimensional.

Development Versus Sustainability Versus Resilience: Commonalities, Intersections and Contradictions

7

Resilience, development and sustainability are key concepts in the current discourse on societal future preparedness in the twenty-first century. What do they have in common, in how far do they contradict and complement each other?

7.1 Main Features of the Three Main Concepts

All of these three major discourses imply more or less concrete concepts for a 'future-prepared' society, yet in different ways and with different emphasis:

The discourse and concept of *developed society* is the oldest one. It has existed for as long as there have been settled communities. This may be explained by the fact that the development of a society is essentially linked to the most existential of all questions – that of inner cohesion allowing it to co-develop. This cohesion is ensured above all by the fact that society must enable its population to meet its (basic) needs and develop its own creative potential. The issue of development contains many different facets, such as the 'happy society' or the 'economically effective' and/or 'economically efficient' or the 'conflict-preventing welfare' society which may all be drivers of productivity and wealth generation. Given exponential progress in key development technologies for the future, such as in the areas of AI or microbiome research, it can be observed that the context of technological development is likely to dominate further discourse of development in to. To date, western societies are (still) considered developmental leaders in global comparison, although observers claim that the twenty-first century will be an 'Asian century' (Khanna, 2019). The major contribution of the developed society concept is that it drives technological and economic progress and achievements necessary to enable socio-economic satisfaction and social cohesion and thus prevent social conflict potential. However, development also unavoidably drives the above-mentioned development paradox – in other words, more and more achievements increase complexity, and more complexity brings more and more complex undesir side effects.

The concept of *sustainable society* extends the view of economic and social development to include the aspect of inter-generational equity. At the intersection of the ecological and economic pillars, representatives of this concept are discussing whether 'clean' or 'green' growth is possible by means of technological and social innovation. Overall, at least three schools of thought can be distinguished within the sustainability discourse, which differ in terms of the degree of transformation of the current paradigm of socio-economic growth in relation to a finite planet. These are the schools of 'green growth', 'post-social market economy' and 'post-growth economy' (Paech, 2012). One – albeit often little-noticed – characteristic of the sustainability discourse is its comparatively high normative orientation; another special feature is its broad focus, which traditionally goes beyond the societal context and encompasses the entire global environmental context. The core focus of the concept of a sustainable society consists above all in its inbuilt holistic crisis-preventive measures, which covers all ethical-human-social, ecological and economic pillars. The concept of sustainable society overlaps in the social-ethical field with the claim of the developed society to secure the needs of the citizens. However, the sustainability discourse has a much sharper eye on the ecological impact of economic growth and therefore criticises the growth paradigm of the concept of a developed society. In international comparison, Nordic and Continental European societies are currently leading the discourse on sustainable societies.

The concept of *resilient society* is the comparatively newest one. It represents, to a certain extent, the continuation of the sustainability debate without replacing the model of a sustainable society, but aiming at taking it to a new level. If the motive of the sustainability debate existed in the question 'How can we avoid humane, economic and ecological crises through inclusive and holistic foresighted actions (both in politics, the social sphere and economics)?', the resilience perspective is analogously characterised by the reflection: 'How can societies be made more robust in the face of humane, economic and ecological crises, which are in any case inevitable and sooner or later will hit and be endured and mastered, in the ideal case triggering progress instead of regression?'. Compared to a sustainable society, the resilient society is relatively free of ideology and the respective strands of ethics. Although it is also a matter of preventing social, ecological and economic crises in the best possible way, the basic motive of resilience is of purely pragmatic nature – it is simply a matter of 'survival'. As outlined above, the discourse about the resilient society is per se already very complex. On the one hand, it refers to all levels of a system (individuals, teams, organisations and societies); on the other hand, it goes beyond the three pillars of sustainability and addresses any possible types of crises (e.g. also technological crises not covered by the three pillars). Although we can outline at least five dimensions of resilience that together form the overarching advanced concept of multi-resilience and can in principle be applied to any country or system, it is not yet possible to determine which country and which political system in the world might be the most (multi-)resilient. The reason for this is that resilience is not easy to measure. The effectiveness of resilience capacities can only be clearly assessed in the face of concrete and direct crises. Here, resilience shows to be highly context-dependent. During the Corona crisis, for example, we could observe several very different countries with widely different strategies, all defended by politicians as particularly resilient and effective. An often-cited 'success example' is Sweden, which had a very open pandemic policy without any restrictions, but recommendations to keep only the risk groups isolated on a voluntary basis (Savage, 2020). Other examples are South Korea (Kim, 2020) or Taiwan (Griffiths, 2020) that similarly implemented a pandemic policy without lockdown, but with broad testing and highly elaborated contact-tracing management. At the other end of the spectrum, for example, was New Zealand (Gulley, 2020), where relatively strict confinement rules combined with one of the world's most effective home-schooling approaches significantly reduced the spread of the pandemic (Hasel, 2019).

The discourse around the resilient society is particularly complex; first, because it addresses all system levels (individuals, teams, organisations and societies), and second, because resilient society goes beyond the three pillars of sustainability addressing all possible types of crises.

The following table represents a first attempt at comparing the core characteristics of all three guiding principles using the classic PESTE criteria (political, economic, social, technological, ecological) (Table 7.1).

7.2 General Commonalities, Complementarities and Contradictions Between the Three Main Concepts

The most obvious commonality between all three guiding concepts lies in their future orientation. They reflect different facets of an overarching discourse on societal future preparedness. There are clear overlaps with regard to the relevance

Table 7.1 Overview of the core features of all three societal guiding concepts

Five societal dimensions	Three guiding concepts		
	Sustainable society	Developed society	Resilient society
Politics	Sustainability policy	Competition of the moderns (also geostrategic competition)	Crisis (preparedness) policy
Economics	Three discourses: Social market economy Post-social market economy Post-economic growth	Focus on economic growth Great importance of post-industrial value creation (e.g. creative economy) Beyond GDP. Mental well-being as a measure of societal development	Shaping a crisis-proof economic order, e.g. through new civil society models of self-sufficiency (this is also related to post-economic growth)
Ecology	Ecology as a central object of sustainability policy	Ecology as an inhibiting factor of economic competitiveness or as a motor of new post-industrial forms of value creation? ('green growth')?	Ecological crises comparatively most likely and severe. Strong focus in the resilience discourse
Technology	To what extent can technological innovations prevent ecological crises? (post-social market economy vs. post-growth) Preventing new man-made crises (relatively little considered)	Technology as an expression/motor of societal development	Technological innovation as a resilience factor (e.g. earthquake-resistant architecture), but also as a development driver of new man-made crises (e.g. cyberterrorism)
Social	Welfare policy for prevention of social crises, within society and worldwide (social balance)	Meeting the (basic) needs of the population through a developed welfare policy	Social innovations Social cohesion

of the dimensions of economy, ecology, technology and social affairs. Perhaps the greatest common denominator is to be found in the social sphere. It refers to the internal cohesion of a society and thus represents the most fundamental component of a developed, sustainable and resilient society.

7.2 General Commonalities, Complementarities and Contradictions...

Another overlap between all three concepts is the central role of innovative capacity. It is seen as a core collective competence of a high-performing developed society, a forward-thinking sustainable society and a prepared and adaptive (multi-) resilient society. In the sustainability and development discourse, technological innovation capacity is mainly discussed in the context of green growth. This is justified by the fact that green growth is not only seen as ecologically relevant, but also as a lucrative economic sector and thus an important factor for developing welfare (even if green growth is controversial within the sustainability discourse). Social innovations are mainly discussed in the context of the principle of self-sufficiency and are rather part of the sustainability and resilience discourse. Particularly striking at this point is the cross-discursive consideration of civil society approaches to regional and community self-sufficiency, such as the Transition Towns initiatives. The approach towards more community orientation and decentralisation is also at least mentioned in the guiding vision of the developed society.[1] In the development discourse, however, social innovations are usually discussed in conjunction with technological innovations, e.g. e-democracy[2] would be a social innovation of the smart city.

In addition to overlaps, mutual complementarities can also be identified, which are derived from the different core concerns of all three guiding concepts: The developed society focuses on the exploitation of its own growth potential and thus also on securing welfare and satisfaction of needs of the population. The sustainable society tries to secure these needs for future generations and to politically prevent unintended side effects in the social, economic and ecological spheres that result from individual, corporate and societal actions. Finally, the resilient society prepares itself against the unintended side effects that cannot be prevented by sustainability policy.

From an abstract systemic perspective, all three guiding concepts are characterised by the fact that they provide a conceptual response to complexity management in an unpredictable environment. Taken together, all three guiding concepts create a multi-dimensional basis for this. In order to elaborate this, we

---

[1] (Exemplary for Germany is the advance of the theologian Lothar Schneider in his book published already in the early 1980s: '*Subsidiäre Gesellschaft - Entwickelte Gesellschaft. Implikative und analoge Aspekte eines Sozialprinzips*' (Subsidiary Society - Developed Society. Implicative and analogous aspects of a social principle) (Schneider, 1990).

[2] E-democracy (a combination of the words electronic and democracy), also known as Internet democracy or digital democracy incorporates information and communications technology to promote democracy and is a form of government in which all adult citizens are presumed to participate equally in the proposal, development and creation of laws (Pautz, 2010).

must first take a closer look at the points of friction between the guiding concepts, which exist primarily between the development concept on the one hand and the sustainability and resilience concept on the other. No comparable frictions can be identified between the resilience and sustainability concepts, presumably because both concepts have always been directly related to each other and therefore appear to complement each other above all (Dissen et al., 2009).

7.2.1 Development Versus Sustainability

The most obvious point of friction is between the demand for ecological sustainability on the one hand and the demand of developed societies for economic growth on the other. Currently, this contradiction is being addressed in two contexts in particular. In the international context, it is the clash of interests between the economically catching-up countries of the Second and Third World (above all China) that is being played out at the expense of compliance with ecological and social standards repeatedly demanded by the developed countries of the West (especially Europe). Countries like the emerging superpower China might take a special role here, as this country is the developer of the worldwide biggest renewable energy projects (see Schultz, 2018, 2022), but also – with more than ten billion tonnes of carbon dioxide, which is twice as much as the emission of the USA – by far the largest CO_2 emitter in the world (Schultz, 2022). In the domestic context, representatives of the sustainability discussion draw particular attention to the devastating effects of urbanisation, which, however, according to representatives of the development discussion, turns out to be an opportunity to improve human living conditions, as cities are engines of growth and centres of productivity. Both contexts – i.e. the international and the intra-societal – are interlinked, since especially the megacities in developing and emerging countries are less and less able to cope with population growth. An uncontrolled, i.e. non-sustainable, development leading to an aggravation of social and ecological crises is the consequence.[3]

Inherent to these and other discourses is a fundamental conflict between the concepts of the developed and the sustainable society, which results from a very different assessment of growth and technological innovation. The growth-optimistic

---

[3] Currently, about 280 million people live in the twenty largest metropolises. And the trend is rising. The negative consequences include scarce housing, congested roads, increasing waste production, air and water pollution, an increasing risk of flooding (due to a loss of natural flood regulation power), and even food supply crises (Hansjürgens & Heinrichs, 2007).

representatives of the developed society concept emphasise the idea that the scarcity paradox is dissolving, which can be seen as a fact of economic history.[4] In other words, progress and innovation can overcome the zero-sum constellation between the not-having of poor people and the having of the rich people ('scarcity paradox'), since general economic growth promises an increase of welfare for all people. Accordingly, it is not necessarily a matter of equitable distribution in the context of a zero-sum game, but of increasing the size of the 'pie for all'. Overcoming scarcity – be it socially in the context of rich and poor or ecologically in the context of planetary resources – could, from the growth optimist's perspective, be managed primarily through economic and technological innovation. The growth-optimist position of the developed society thus does not exclusively stand for a weak sustainability, i.e. a balancing of the social and economic pillar against the ecological pillar, but for the assumption of the possibility of green growth.

Closely related to this is the technology debate in the broadest sense, which deals with the question of whether technological innovations can ensure the survival of humanity from an ecological collapse of the Earth. Among the issues discussed here are the possibility of renaturalisation and thus the restoration of planetary resources and the possibility of colonising outer space through technological innovations (including nanotechnology, advanced AI technology etc.). The last point in particular will become increasingly relevant in the future, and interestingly enough, it is being driven forward especially by those societies that are not characterised by a high development focus. Already today, a race is emerging between the space-faring nations, in particular between the USA, Russia and China – so far, the only countries with the capability of manned space flight.[5] Additionally, tech giant companies like SpaceX (Tesla) or Blue Origin (amazon)

---

[4] Accordingly, industrialisation in the eighteenth century made it possible to overcome the biological limits of food and wood production by means of machine power and thereby break out of a universal poverty that had hitherto been more or less equally pronounced in all parts of the world. Put simply, one could also conclude that today's deep divide between rich and poor countries is primarily the result of the fact that the wealth of the Western states has grown faster than that of the other countries since 1820. According to the economic historian Maddison, the per capita income of the European population in 1820 corresponds to about 90% of that in present-day Africa (Sachs, 2005).

[5] The space-faring states and groups of states currently include the USA, Russia, China, India, Argentina, Brazil, Europe (ESA, Iran, Israel, Japan, North Korea, South Korea, whereby the first three states are regarded as world leaders). The USA is planning a manned flight to Mars by 2025. China, on its way to becoming the second largest space nation, is currently pursuing the goal of a manned moon landing in addition to building its own space station, Tiangong 1. Russian spaceflight is (still) considered the world leader in unmanned spaceflight and in the operation of space stations in Earth orbit (Campbell, 2015).

are highly involved in the ongoing 'space race' (Huddleston, 2015). In particular, the increasing involvement of the private sector in these multi-billion dollar programmes is not driven exclusively by scientific and prestige interests, but above all by pragmatic interests, with the long-term goal of exploiting resources and maybe colonising space in the inner solar system before the end of this century.

These two essential topics – the technology issue and the growth issue – are critically addressed in various debates. Is a fruitful, dialectical development of the discussion possible that contributes to future preparedness?

- Firstly, it can be argued from a technology-sceptical perspective that with the increasing spread of technology in people's everyday lives, new dependencies and challenges will arise, which in principle would already have to be critically reflected in the future sustainability discourse. One facet of this topic area is the still little-established debate about the popular movement of transhumanism,[6] which in turn today determines the normative discussion and parts of the political discussion about the future of man: the critical and still little-tapped question would be: what is left of being human if he/she merges more and more with technology? Another facet would be the question of whether man, should he/she manage to colonise space in time, would not repeat the same sustainability ethical mistakes, only on a larger scale. Even the most significant technological innovations in the future could not replace awareness of sustainability (meaning: the ability to consider the future consequences of one's actions), but would have to go hand in hand with it. From another perspective, the question of space mining and colonisation could not only solve the problem of resource scarcity on the planet, but also pave the way for a change in consciousness that would involve all of humanity across national and cultural boundaries.
- Secondly, and following on from this, there is a fundamental debate to be had about the extent to which the central desirable motives for human life that are recognised across philosophies – in other words, bliss, well-being and the absence of suffering – can be achieved exclusively, or even at all, through technological innovation. Not to be neglected are post-materialistic, above all decelerating ways to general well-being, which precisely do not aim at

---

[6] *Transhumanism* is a techno-optimistic, globally expansive ideology that dominates the discussion about the future of humanity today. In essence, it is about enhancing humans 'beyond' their physical limits through the application of new and future technologies. Technological applications include, among others, lifetime extension via genetic engineering, cyborg technologies, brain-computer interfaces, uploading of human consciousness into computers, the development of superintelligence or the further development of cryonics (Benedikter & Fathi, 2013).

controlling the environment, such as overcoming suffering in one's own psyche through meditation practice with the result of being at peace with all that is.
- Third, the strong-sustainability discourse in particular contradicts the technology- and growth-optimistic assumptions of the developed society. In particular, growth critics emphasise that economic growth is not only ecologically but also economically and socially problematic for ecological and social reasons – which are expressed in the scarcity paradox mentioned above. Economic growth and with it the general 'enlargement of the pie' is seen above all as a wishful thinking which, since the eighteenth century, as a guarantee of progress, should make the great social game of haves and have-nots manageable. Ultimately, however, this suggestion distracts from the fact that resources are being used in the present for which future generations will only have to pay in the future. Economic growth that is only foreseeable in the short term and that excessively depletes the life resources of future generations could well impair social stability and lead to generational conflicts already in the present. This assumption is also advocated by the systems theorist Niklas Luhmann, among others (Luhmann, 1988).

The discussion about the contradiction between the demand for sustainability and the demand for growth therefore remains open at several points. The discussion of 'subsidiarity', 'community orientation' and, in the broadest sense, 'self-sufficiency', are also strongly emphasised in the resilience discourse. As these aspects are taken into account in all three guiding concepts (albeit with a stronger emphasis in the resilience and sustainability discourse) and could provide another starting point for the coexistence of different models of life and society that are largely independent of each other in terms of supply. The mega-city, which is economically strong and gradually tending towards green growth, and the Transition Town-style community economy therefore do not represent contradictions to each other, at least in terms of implementation.

In sum, different models of economic activity and cohabitation are emerging and implemented. The extent to which one of the models will prevail will depend on various factors, such as technological breakthroughs and value developments. A technological breakthrough or even a singularity would favour a growth model in the sense of the developed society model. By contrast, the scenario of an energy and resource crisis that can no longer be offset by technological progress would tend to lead to an economic model inspired by the post-growth economy. In both cases, however, it can be assumed that the models will continue to retain their fundamental relevance for societal future preparedness. The development paradigm enables innovation in a world with increasingly complex problems and – as will be shown in more detail below – seeks to pave the way to unlimited prosperity through a technological singularity. The flip side, however, is that every development involves potential problems and even crises. This requires a corresponding awareness of sustainability in order to anticipate unintended side effects.

7.2.2 Resilience Versus Development

Another typical point of friction, this time between the resilience and development concepts, arises from two issues of conflict: first, the contradiction between economic efficiency and the need for redundancy, and second, the need for problem-solving through technological innovation vs. new vulnerabilities through technological interdependencies.

The first conflict does not only concern societies, but also organisations. Accordingly, resilience requires, among other things, redundancies to be set up, i.e. backup systems that remain unused for a long time and are only used when necessary. This is at odds with the postulate of economic efficiency, which in recent decades has sought to streamline public administration and save unused resources. Ultimately, according to Walter Perron of the Freiburg Centre for Security and Society, this contradiction will be resolved by asking

> what price a society is willing to pay to establish security. It is already apparent that while many innovative developments in the field of security technology promise a considerable gain in security, their widespread deployment is not considered affordable, at least at present (Perron, 2011, p. 110).

The second conflict concerns the contentious issue of the benefits of the technologisation of society, which can be illustrated by the example of the smart city. In the smart city, understood as a 'solidified' form of a 'highly developed society', technological innovation represents a central variable. It enables, at least in theory, social innovations such as e-democracy and resulting opportunities for participatory urban development. The resilience concept has an ambivalent relationship towards the technologisation trend. On the one hand, it can be argued that technologisation, especially in the course of the spread of new information and communication technologies, certainly contributes to the resilience of a society by strengthening its network, especially through the capacity for self-learning, and that it contributes to a democratisation of knowledge. On the other hand, it should be critically noted that technologisation also leads to new dependencies and vulnerabilities. For instance, a power failure, a software error or a cyber-attack in a smart city could already lead to the most devastating crises. A resilient society rather emphasises the dimension of the analogue (non-digital) compared to the smart city. When comparing the resilient society and the smart city (as a manifestation of a highly developed society), two further differences stand out, although these are less important in the overarching debate. First, it is striking that both concepts fundamentally share the challenge of institutionalising equal cross-sectoral collaboration between the private sector, policymakers, civil society, and academia. Second, a different orientation in complexity management can be identified. The resilient society focuses its attention on anticipating complex, unpredictable external and internal crises and aligning resources to enhance responsiveness. The focus of the developed society is to create framework conditions to effectively and efficiently unfold inherent human and technological potential. As a particular example, the smart city attempts to ensure that urban development is permanently adapted to the behaviour of citizens.

7.2.3 Resilience Versus Sustainability

The concept of resilience is comparatively value-neutral and without normative implications. A criminal organisation or a terrorist cell can serve as an inspiring object of study with regard to its best practices just as well as a disease virus. It is therefore all the more surprising that resilience is sometimes discussed as the 'new sustainability' and is increasingly confronted with ethical-normative expectations. This gives rise to contradictions with the concept of sustainability, since the interest in promoting societal resilience may well lead to unethical consequences. Geopolitics in particular makes it clear that a society's strategy to ensure its survival can be pursued at the expense of the survivability of other societies, and even of the global social system. As will be shown in more detail below, this is exemplified by the special case of the USA, whose resilience is based in particular on its global position as the world's only superpower. Overcoming this contradiction could make national-society agents more aware of the needs of the global system and thus flank resilience strategies of states with sustainability values in favour of promoting global resilience.

7.2.4 Mutual Contradictions Between All Three Guiding Concepts, Using the Superpower USA as a Particular Example

The example of the USA includes further contradictions or at least negative correlations between the three guiding concepts. According to this, the USA, which can (still) be regarded as the world's only superpower, is characterised on the one hand by a policy that is highly unsustainable in ecological terms (environmental pollution, high dependence on fossil fuels) and above all in economic terms (extreme consumer orientation, highest level of foreign debt in the world). Parallel to this, the dilapidated infrastructure in many cities, the relatively high homicide rate (compared to other Western societies), or its vulnerability towards natural disasters such as 'Katrina' or 'Sandy' (to name but a few) might indicate a lack of resilience potential. From another perspective, the USA prove to be highly developed on many levels – especially due to its global leadership in science, social and technological innovations, and its economic and cultural dominance in the world. Moreover, despite its economically precarious situation, the USA could be described as economically resilient due to the worldwide recognition of the US dollar as the reserve currency. From this perspective, the economic resilience of the US is based not on its self-sufficiency – an essential criterion by which the resilience of a society is usually judged – but on the great interdependence between the world's last superpower and the rest of the world. 'When the US sneezes, the rest of the world gets a cold' – so goes an old stock market adage. Seen in this light, the USA would be 'too big to fail' – in other words, in order to maintain the stability of the prevailing world economic structure, the USA would be propped up by the community in case of doubt, even at its own expense. As a consequence, the resilience of the USA as described above could even be interpreted as a 'door opener' for a continuation of its very unsustainable national-egocentric policy. This can be seen, among other things, in the fact that social peace is maintained through an economic and consumption policy 'on credit' (consequence: debt bubble) and the avoidance of ecological standards and social redistribution in favour of high economic dynamism (consequence: further vulnerability to social crises and natural disasters). This highly simplified example suggests how resilience and development, when the concepts are applied only to the national context, can lead to short-term successes and to a lack of sustainability and even greater vulnerability of the global system. Likewise, it is apparent that vulnerability on the one hand (here: exposure to natural disasters and inadequate protection of critical infrastructures) can lead to new forms of resilience (here: the trend towards increased prepper activities organised by civil society). The discourse therefore remains very ambivalent between the three guiding concepts. The overarching debate on societal future preparedness in the twenty-first century requires integration of the core aspects of all three guiding concepts to overcome the points of friction.

The essential criteria and interrelationships of all three guiding concepts can be summarised roughly as follows (Table 7.2).

7.3 The Development Paradox

The development paradox describes the crisis-like dynamics inherent in the concept of development and the need to prevent unintended man-made side effects. In this context, the British historian Ian Morris states that the development of any society always produces precisely those forces that hinder its further growth. Morris calls this phenomenon the 'development paradox', in which every achievement and every solution to a problem creates new problems (Morris, 2011).

Analogously, it is derived from the perspective of developmental psychology that as a society develops, the complexity of its needs structure and the problems associated with it also increase and evolve. A pre-industrial society, such as for example large parts of Afghanistan, might be relatively highly affected by material-existential challenges related to securing basic needs such as food, care, shelter and security. In contrast, an affluent society such as Norway is affected by 'higher' needs – the population faces primarily existentialist challenges (not to confuse with 'existential challenges'), such as the search for meaning, occupational burnout, depression, and pressure to perform and to consume. These different vulnerability profiles make it difficult to give a generalisable answer as to which society is more sustainable. What is certain is that each level of complexity produces new challenges which – and this is crucial to bear in mind – are man-made. Social scientists such as Giddens, Luhmann and Beck separate man-made problems from unintended side-effects that societies have to face. Giddens uses the image of the 'Jagannath chariot', which comes from Hindu philosophy. Jagannath is a Hindi word that roughly means 'Lord of the World' and denotes a title of the Hindu god Krishna. The Jagannath Chariot is a huge vehicle that used to be driven through the streets of India once a year with an image of the god Krishna. It is said that some of his followers even threw themselves under this chariot in worship (Giddens, 1995). Applied to the present, this image symbolises, according to Giddens, an

> unrestrainable and enormously powerful machine that we as humans can collectively control to a certain extent, but which at the same time urgently threatens to elude our control and could smash itself to pieces (Giddens, 1995, p. 173).

Table 7.2 Development, sustainability and resilience as interacting core aspects of societal future preparedness

Contents	Societal concept		
	Developed society	Sustainable society	Resilient society
Focus	Securing and increasing well-being/welfare Economic growth and, where appropriate, efficiency	Focus: Preventing the consequences of individual/societal action Generational justice	Preparing oneself against the consequences of (man-made) problems Adaptability
Concepts included	The modern/post-industrial society The happy society The developmentally psychologically evolved society	Innovation-based ('green') growth Post-social market economy (e.g. eco-social market economy) Post-growth economics (e.g. sufficiency and subsistence [Transition Towns])	Crisis management (maintaining the system's ability to avert problems and to function) Crisis transformation (capacity for continuous change, e.g. through Transition Towns)
Dimensions (focal points)	Social: well-being; health; social balance Economy: growth orientation; location competition; Technology: Technological innovation; creative industries	Social: Social balance (society); protection against exploitation (global) Economics: Reserve management; critical examination of growth and the global economic system. Ecology: Priority of environmental protection (strong sustainability)	Social: Social balance; social cohesion Economy: 'Robust' economy; self-sufficiency Ecology: Anticipate ecological exposure Infrastructure: Priority to critical infrastructure protection (including virtual) Precaution: Ensuring independence from the environment (survivalism) Culture: Collective psychological resilience
Main contradiction	Development versus sustainability: economic growth interests and social welfare interests (weak sustainability) versus environmental protection (strong sustainability) Development versus resilience: technological innovation versus technological dependencies (and possibly new crises)		

7.3 The Development Paradox

In this context, the sociologist Ulrich Beck coined the term 'risk society'. The term describes the fact that in advanced modernity the social production of wealth is systematically accompanied by the social production of risks. According to this, it is not only a matter of distribution problems and conflicts of the society of scarcity, which are to be solved by economic growth, but also of

> problems and conflicts that arise from the production, definition and distribution of risks produced by science and technology (Beck, 1986, p. 25).

With regard to the three guiding concepts, it could be assumed that all dangers emanate from the development concept and that this guiding concept only needs to be supplemented by the sustainability and resilience concept. However, this assertion, which seems quite conclusive at first glance, falls short of the mark. A closer look reveals that all achievements, including those from the sustainability and resilience context, can lead to new vulnerabilities. The resilience researchers Zolli/Healy even claim that per se every new resilience strategy would be accompanied by new vulnerabilities. For instance, the Internet provides extensive protection against the loss of information as a result of physical events (such as a natural disaster) because it is stored virtually and cannot be destroyed, as is the case with a physical library, for example. Likewise, this very problem-solving opens up new vulnerabilities to new virtual threats, such as cyber viruses and cyberattacks (Zolli & Healy, 2013).

Even achievements that are more likely to be attributed to the field of sustainability harbour the potential for serious unintended side-effects. This applies particularly to new technologies that are developed to prevent the occurrence of a certain ecological crisis, such as for instance the so-called geoengineering. The term *geoengineering* refers to large-scale interventions by technical means in (bio-)geochemical cycles of the Earth. Their main aim is to stop man-made global warming and to reduce the CO_2 concentration in the atmosphere. One major approach is to influence solar radiation (Solar Radiation Management (SRM)), which involves increasing the reflection of sunlight by aerosol application methods in order to counteract a global rise in temperature. The second approach involves reducing the concentration of CO_2 in the atmosphere (Carbon Dioxide Removal (CDR)) through direct methods such as air filtration, CO_2 sequestration (CCS), as well as indirect methods such as fertilising the oceans with iron or phosphorus (Caldeira & Bala, 2016). While these measures could help stop global warming, they would come with numerous other social and environmental risks (Baum & Barrett, 2017; Global Challenges Foundation, 2017). In 2008, climate scientist Alan Robock published a 20-point list of potential dangers in the use of geoengineering. He states that at least 13 of the 20 points represent side effects and dangers for the climate system and the environment (Robock, 2008):

- Regional temperature changes
- Changes in precipitation patterns
- Damage to the ozone layer (with aerosol geoengineering)
- No reduction of the CO_2 content of the atmosphere (with SRM methods)
- No prevention of ocean acidification
- Negative effects on flora and fauna
- Intensification of acid rain (in case of application of sulphur dioxide)
- Effects on natural (cirrus) cloud cover
- Sky bleaching
- Lower power output for solar systems
- Strong temperature rise when project has to be stopped
- Human or technical failure
- Unknown, unpredictable effects
- Negative impact on willingness to reduce CO_2 emissions
- Abuse for military purposes
- Danger in the case of commercial control of techniques
- Contradiction with the ENMOD Convention[7]
- Possibly extremely high costs (exception: aerosol geoengineering)
- The need for supranational control
- No decision-making framework in place
- Incompatible conflicts of interest of individual states (Who determines the global temperature?)
- Considerable potential for conflict (political, ethical, moral)

Progressive development not only carries the potential of unintended risks that must be anticipated for the purpose of system preservation, but is often accompanied by extensive destruction and transformation of the system that is meant 'to be preserved'. This discussion primarily encompasses current and future engagement with transhumanism and posthumanism. Both terms refer to worldviews that place the further development of humans through new and future technologies, which will fundamentally change the human mind and social order, at the centre of their considerations and goals. While transhumanism refers to the aspect of human enhancement via new technologies, the term posthumanism suggests that a new intelligent species could evolve, maybe in terms of an artificial austerity, to replace

---

[7] The *ENMOD Convention* (Convention on the Prohibition of Military or Any Other Hostile Use of **En**vironmental **Mod**ification Techniques) translates as *'Environmental Warfare Convention'* and is an international treaty drawn up by the UN Disarmament Commission on the Prohibition of Military or Any Other Hostile Use of Environmental Modification Techniques (ICRC).

7.3 The Development Paradox

humans in the same way that humans replaced primates (Loh, 2020). Since the turn of the millennium, science fiction literature (especially under the influence of the sub-genre post-cyberpunk and with a view to the allegedly approaching technological singularity) has increasingly dealt with this topic.

The following is a short exemplary paraphrase from the '*Post-Human Omnibus*' novel series by Dave Simpson illustrating the ambivalent interrelation between technological development, man-made (existential) crises and the need to anticipate them in terms of sustainability and (multi-)resilience. In his second book (out of in total currently six books), the author describes a technologically rapidly evolving society in which all post-human members of society are physically nearly immortal and invulnerable by means of nanotechnology. By today's standards, they are highly intelligent, their consciousnesses multiply interconnected and in constant contact with a super intelligent AI that monitors all processes. When an upload fails that was supposed to enhance all humans IQs up to 200, almost all humans connected to the system are wiped out with one blow. It turns out that the omnipresent nanotechnology had developed its own consciousness and hacked the super-intelligent AI. The protagonist, who by chance was not affected by the upload and was one of the only five posthuman survivors, eventually manages to regain control of the AI and uses reprogrammed nanobots to undo all the damage. A revival of the post-humans also succeeds because their consciousnesses were uploaded to a cloud and could be retransferred to newly created bodies (Simpson, 2013). This story exemplifies the ambiguity and many open questions related to highly developed, potentially transhuman or posthuman future societies. First, the notions of trans- and posthumanism imply the question of what humanity is meant to be (and what not). Second, posthuman society in relation to this story implies the question of whether the development into post-humans would be the result of poor resilience in humanity (since humans as we know them today no longer exist in this society) or rather a symptom of evolutionary readaptation (according to which humanity would have evolved, which would be a sign of transformative resilience). Third, the story illustrates the ambivalence between high amenities and high resilience on the one hand (e.g. almost infinite prosperity and development opportunities and no physical threats prevail anymore) and on the other hand existentially dangerous new vulnerabilities (due to the omnipresence of AI and nanotechnology). Even if these considerations may seem improbable from today's perspective, they make the core challenge and demand for societal future preparedness plausible: to develop while at the same time anticipating and preventing complex (existential) risks and bundles of risks.

Part III

What Does Societal Future Preparedness Mean in the Twenty-First Century? An Outlook

Future Preparedness in the Twenty-First Century: Contours of an Integrative Concept

8

It can be summarised from the previous reflections that at least three guiding principles have to be considered, which overlap in large parts, but also complement and contradict each other in some points. To what extent can the guiding principles of a resilient, developed and sustainable society be brought together to form an integrative concept for societal future preparedness?

8.1 Criteria and Principles for Societal Future Preparedness – An Overview

A comparison of the respective dimensions, measurement criteria and principles of all three guiding principles illustrates a variety of interrelationships, some of which are ambivalent. The following figure provides a summary list of all dimensions and associated measurement criteria of the three guiding concepts. Again, it should be noted that there is no claim to completeness and that the associated indices have the limitations already mentioned above. The main purpose of this comparison is to illustrate the many commonalities and complementary points between the models (Table 8.1).

Some dimensions are likely to have comparatively prominent importance in the respective guiding principles – as defined and contextualised in the context of this book. For example, the technological dimension is particularly relevant in the context of the developed society, 'ecology' in the sustainable society, and 'preparedness' and 'infrastructure' in the resilient society. The most important implication of this presentation is that all three guiding principles make

Table 8.1 Dimensions and measurement criteria of the three guiding principles (selection)

Criteria dimensions	Developed society	Sustainable society	Resilient society
Psychic	Psychological well-being Skills development/education (proportion of highly qualified) Developmental psychological complexity level (proportion of orange meme and higher)		Psychological resilience (resilience quotient) Meditation practice
Social	Health (life expectancy in good health) Social cohesion (including volunteering) Unemployment[b] Crime/anti-social behaviour Subjective well-being[c], life satisfaction Trust (interpersonal and institutional) Human development (e.g. HDI)	Psychological well-being, life satisfaction Income distribution Population growth Unemployment Income distribution (indicator of conflict potential) Human development (e.g. HDI)	Health Number of individuals at risk (including young early school leavers, child poverty etc.) Unemployment Social cohesion (including volunteering) Trust (interpersonal and institutional)
Economy	Indebtedness Growth Income distribution (relative income) Competitiveness (e.g. Competitiveness Index) Share of creative industries/innovation Purchasing power Absolute income Patent intensity/start-up intensity Investment rate of the industry Proportion of tertiary employees[a]	Material consumption Gross domestic product Savings rate/debt Gold and other reserves to ensure currency stability	Diversity of the economy (also any imports) Savings rate Debt Gold reserves

(continued)

8.1 Criteria and Principles for Societal Future Preparedness – An Overview

Table 8.1 (continued)

Criteria dimensions	Developed society	Sustainable society	Resilient society
Technological progress	Urbanisation level Dissemination of information/communication technologies (ICT Development Index) Technology Achievement Index (TAI) Digital Competitiveness		
Ecology		Quality (soil, air, water) Renewable energies Greenhouse gas emissions Energy consumption Renewable water resources Available forest areas Biodiversity Organic farming	Quality (soil, air, water) Renewable energies Renewable water resources Available forests Organic farming Land use/available forest land per capita Self-sufficiency through agriculture/organic farming
Infrastructure			Number of days that central utility power can be restored after a power/system outage Vulnerability of critical infrastructure and digital/automated control systems to disruption Degree of Cybersecurity (protection against hacker attacks and other disruptions from the Internet) Expansion of local public and passenger transport (dependence on individual transport) Dependence on supra-regional mobility

(continued)

Table 8.1 (continued)

Criteria dimensions	Developed society	Sustainable society	Resilient society
Preparedness			Bunkers: Available number and accessibility of bunkers per household Precaution: number of days on average that a household can survive without outside supply

[a]The last three criteria of this compilation have been extended by some criteria from the 'Future Atlas 2010' of the German consulting firm Prognos AG – this is a 'set of indicators for regions of the future' (Prognos., 2010)
[b]As an indicator of conflict potential
[c]As an indicator of satisfaction of needs

indispensable contributions that cannot be replaced by one another and that must be given appropriate consideration in the overarching context of future preparedness.

As shown above, there are also contradictions between the guiding principles at some points. As far as contradictions are concerned, the different perspectives of the concepts of the sustainability and the development concept on economic growth and (currently neglected) on technological progress may appear to be the most serious. In summary, societal future preparedness means not only taking into account the different contributions of the respective concepts, but also integrating them. Thinking back, the respective guiding principles include the following orientation principles:

(Multi-)Resilient Society

- Principle 1: The multi-resilient society is fostered by resilient citizens.
- Principle 2: The multi-resilient society is comfortable with unknowns.
- Principle 3: The multi-resilient society is based on decoupling and knowledge integration of its subsystems.
- Principle 4: The multi-resilient society makes collectively intelligent decisions.
- Principle 5: The multi-resilient society is based on a strong learning culture.

Developed society

- Principle 1: A developed society promotes competence development of its citizens.
- Principle 2: A developed society fosters collective intelligence.
- Principle 3: A developed society is characterised by a hot learning culture.
- Principle 4: A developed society is characterised by a high level of Well-being.

8.1 Criteria and Principles for Societal Future Preparedness – An Overview

Developed society Sustainable society

- Principle 1: The sustainable society promotes the realisation of the IDGs.
- Principle 2: The sustainable society is tendentially based on a cold learning culture.
- Principle 3: The sustainable society develops collective intelligence.
- Principle 4: The sustainable society develops collective wisdom.

Following the logic of *complexify*, the commonalities, complementarities and differences between all three guiding concepts result in the following overview (Table 8.2):

Following the logic of *simplify*, the common denominators, complementarities and contradictions between all three concepts can be integrated into at least eight guiding principles.

- Principle 1: Development of universal personal key competences.
- Principle 2: Complexity-adequate problem-solving taking into account knowns and unknowns.
- Principle 3: Decoupling and knowledge integration of the subsystems.
- Principle 4: Collective intelligence.
- Principle 5: Learning culture.
- Principle 6: Securing (basic) needs and preventing social conflicts.
- Principle 7: Development and preservation.
- Principle 8: Collective wisdom.

The first five principles have already been presented in detail in Chap. 4 and are briefly summarised below with regard to the different facets of the three guiding concepts. The last three principles require a more detailed description in the subchapters below.

Regarding *Principle 1 ('Development of universal personal key competences')*, societal future preparedness requires multi-layered competency development of citizens. In the context of the resilient society, this means the promotion of individual resilience, which can ideally be implemented via the curriculum of schools and in adult education. The context of the developed society also emphasises the need for lifelong learning in the face of increasing and constantly changing knowledge. Lifelong learning is considered a key competence of the post-industrial knowledge society. Finally, competency development and knowledge transfer in the context of sustainability also include awareness of the manifold intergenerational and international consequences of one's own actions. This can also include

Table 8.2 Essential orientation principles of all three societal guiding concepts

Concept Principle	Development	Resilience	Sustainability
Individual competence development	Learning to learn in the knowledge society	Promotion of individual resilience	IDGs (including sustainability awareness; values)
Dealing with knowns and unknowns		Focus: Innovative practices	
Decoupling and knowledge networking		Autarkic (subsistent) units capable of making decisions in crisis situations; simultaneous communicative networking of all units	
Collective intelligence	Focus: Development of technological innovations	Focus: High responsiveness and innovation in crises; development of appropriate social and technological innovations; time-efficient decision making	Focus: Development of sustainable innovations (social and technological)
Learning culture	Emphasis: Hot culture; continuous improvement; control of nature	Learning from mistakes; high cohesion; balance between adaptation and preservation of core values	Focus: Cold culture; harmony with or preservation of nature; anticipating mistakes
Securing (basic) needs and preventing social conflicts	Ensuring welfare		
Development and mastery of universal technologies	Promote technological and economic development		Technological and economic development not at any price
Collective wisdom			Anticipation of man-made rebound effects; critical perspective on achievements of collective intelligence (e.g.) technological innovations

8.1 Criteria and Principles for Societal Future Preparedness – An Overview

corresponding imparting of values. As early as the 1990s, the US journalist and emotion researcher Daniel Goleman pleaded in his popular book 'EQ – *Emotional Intelligence*' (2001) for an education system that promotes this eponymous competence in the population. According to him and other scholars, emotional intelligence involves multiple personal and social skills that are of societal consequence. These include:

- *Emotional self-awareness:* this also includes the perception of one's own gut feeling, i.e. intuition (Damásio, 1994);
- *Emotional self-regulation:* this means the ability to regulate and release one's own feelings of stress;
- *Self-motivation and reward deferral skills:* the ability to not let one's mental performance be affected by stressful emotions such as anger, fear, worry, and sadness;
- *Empathy is* widely understood as the ability to perceive the feelings of others. It presupposes emotional self-awareness (self-empathy) and forms the starting point for morality and altruism;
- *Social competence:* The foundation of this competence, which is also called 'interpersonal intelligence', is the ability to deal with the feelings of other people. It includes other individual competencies such as teamwork and leadership and contributes to effective communication.

Emotional intelligence appears here as a 'universal competence', i.e. a basic skill that can be applied to many different problem areas, which therefore has a great social impact and thus indirectly contributes to a resilient, developed and sustainable society. Emotional intelligence leads to a more constructive handling of one's own emotions and therefore promotes one's own ability to cope with stress. It contributes to more effective teams and to more empathetic cooperation. Emotional education would also significantly prevent the following social problems in adolescence (Goleman, 2001):

- School shootings;
- Juvenile delinquency;
- Bullying;
- School performance failure and dropouts;
- Teenage pregnancies;
- Drug addiction;
- Mental disorders, including eating disorders and depression;
- Suicide.

Another universal competence that is increasingly being discussed is *Futures Literacy*, which can be seen as a meta-approach that integrates various tools, methods and models that enable the management of the future by 'working actively with it'. An important axiom is hereby that 'anticipation' is the basic operation of any future reflection. The long-time head of UNESCO Futures Literacy and one of its inventors, main theorists and mentors, Riel Miller, defines Futures Literacy as:

> the capacity of an organism to incorporate the later-than-now into its functioning in ways that are relevant. (Miller, 2018).

Building on that, Futures Literacy in his view means the:

> ability to imagine outside predefined paradigms, or to sense and make sense of phenomena that may not belong to pre-existing models. Imagined futures that do not arise from efforts to address what is currently deemed probable or desirable have no place in mainstream thinking. As a result, given the power that images of the future have over what we perceive and do, most novel phenomena remain invisible, bereft of meaning, because they are excluded from our images of the future. (Miller, 2018).

In other words, Futures Literacy is the ability to overcome this 'present bias' (which perhaps may even include some kinds of 'presencing bias') in order to:

> combin[e] a broad palette of planning horizons and methods with what might appear to be a surreal or absurdist imagination, detached from deterministic purposes and methods. (Miller, 2018).

Futures Literacy is already incorporated in the curriculum for Master's students at the Hanze University of Applied Sciences in Groningen, Netherlands, under the leadership of Loes Damhof, UNESCO Chair in Futures Literacy. In this context, students and teachers not only learn the capability of working about possible and probable futures, but also on how to design and facilitate them in order to apply their anticipated values and benefits in their studies and work (Larsen et al., 2020).

These universal skills cover most of the above-mentioned IDGs framework to help accelerating the work towards the UN's SDGs (IDGS, 2021).

In summary, societal future preparedness implies a multi-facet education policy which does not one-sidedly focus on developing professional competences. Rather, it strongly considers the development of universally applicable key features, such as personal resilience or emotional intelligence, futures literacy and in the broadest sense the IDGs.

Principle 2 ('Complexity-adequate problem-solving considering knowns and unknowns') considers and integrates different practices to adequately cope with

8.1 Criteria and Principles for Societal Future Preparedness – An Overview

manifold problem types with varying degrees of complexity and unknowns. These include good practices (simple problems), best practices (complicated problems), emerging practices (complex problems) and innovative practices (chaotic problems). Basically, most practices are considered in all three guiding concepts. For instance, more and more organisations in post-industrial knowledge societies require competencies in adopting and implementing agile project management practices. These are characterised by developing products through step-by-step (incremental) processes based on experimental trial and error and rapid adjustments for further development. This approach is found primarily in software development, but increasingly in other industries as well. In the context of sustainability, this approach can be found above all in the context of social innovations by testing new forms of coexistence. In this sense, Transition Towns represent typical social innovations that are tested in real life experiments. The most explicit approach to chaotic phenomena is the resilience concept, which focuses on developing mindfulness and the ability to cope with unpredictable problem situations. As with Principle 1, it is also noticeable with Principle 2 that the three guiding concepts do not contradict each other and that synergies largely arise here.

Principle 3 ('Decoupling and knowledge integration of the subsystems') is of particular importance for the resilient society and contains two aspects which are integrated and theoretically justified by the Viable Systems Model. Accordingly, societal systems should include, first, decentralisation, decoupling and, in the broadest sense, autonomy of the subsystems, and second, knowledge distribution between all subsystems, so that collective intelligence emerges at the higher system levels. The concept of the developed society has no direct reference to this, but an indirect one, which is derived from the systemic modernity theory. Thus, from Luhmann's systemic viewpoint, the complexity level of modernity is inevitably characterised by its functional differentiation into several subsystems – e.g. politics, civil society, private economy, media. Differentiation in this sense presupposes a decentralised approach and differs from the governance approach of pre-modern societies in which all competences in the state are coordinated by a central authority (Luhmann, 1993). Consistent with the resilience concept, sustainability considers decentralisation more in terms of local self-sufficiency in order to avoid an unnecessary exploitation of the planetary resources. Taking into account the main ideas of all respective guiding concepts, societal future preparedness should therefore include subsidiarity which ensures self-sustaining independence of subsystems, effective local decision-making, but also overall transparency as well as knowledge distribution and knowledge integration.

Principle 4 ('Collective intelligence') refers to the enabling factors for the development of collective intelligence and effective decision-making, building on the aforementioned principles. Both competencies emerge in different manifestations in the context of societal future preparedness: be it an interdisciplinary research team that attempts to gain new insights into several future scenarios of society, an agile performance team developing technological or social innovations, or a think tank developing policy recommendations. In all of these and other contexts, practices are being developed or weighty decisions are being made to solve a wide range of societal challenges. Collective intelligence and effective decision-making are inevitable core competencies of societies with high inherent complexity, and are also essential features of resilient systems for coping with crises. Collective intelligence is also relevant in the context of sustainability, but here rather in the development of sustainable technologies (here there are synergies with the guiding principle of development) and social innovations in the sense of new forms of coexistence. With this principle, too, there are no significant contradictions, but synergies between the three guiding principles.

Principle 5 ('Learning culture') raises awareness of the aspects of societal future preparedness that go hand in hand with the collective view of the world and the culture we live in. In this respect, all three guiding principles set different emphases, whereby a contrast arises between the guiding principles of the developed and the sustainable society. The developed society contains essential features of 'hot cultures' that focus on control of nature and constant self-improvement. By contrast, the sustainable society reflects typical values of 'cold cultures', which focus more on harmony with or preservation of nature. Both concepts seem to contradict each other and reflect the above-mentioned disagreement about the meaning and sustainable design of technological and economic development. Within society, this dissent is reflected in the competition between opposing institutions. Here, the heating tendencies of the market economy and globalisation are opposed to the cooling-regulating institutions of religion (this would be particularly the case in traditional societies) and/or civil society (this would be more the case in modern and post-industrial societies with an ecological value orientation). If they peak, cooling institutions and agents would influence the political sector to enforce legally binding regulatory norms. From an inter-societal perspective, the contrast can be seen in the extent to which societies influence each other culturally (the related technical term is 'transculturation'). For instance, modern societies with hot culture seem to have a generally more dominant influence on traditional societies with cold culture than vice versa (Groh, 2004). From an intra- and inter-societal perspective, the resilience concept seems to bridge the gap between the two value systems. Intra-societally, it includes the idea of development in the form of active self-improvement through further development of technological solutions (typical for hot cultures) and social innovations (typical for cold cultures).

Moreover, resilience culture is characterised by learning from the dynamics of the environment, and also by enduring ambiguity in a way that it enables constant adaptation and the integration of new information, while at the same time preserving its integrity and a firm core of values. Thomas Bauer calls this 'tolerance of ambiguity' (in German: 'Ambiguitätstoleranz') and sees it as the key competence of a society in dealing with today's complexities. In this sense, he counts the rise of right-wing populism in the 2010s as the exact opposite strategy of complexity management: here, complexity is reduced to a simplistic either-or perspective (Bauer, 2018). The common denominator between the different cultural manifestations of all three guiding principles can possibly be summarised under the term 'learning culture'. Analogous to Principle 1, societal future preparedness thus requires the collective ability to acquire knowledge – firstly to influence the environment (development), secondly to preserve the environment (sustainability), and thirdly to achieve balanced self-change (resilience) (Table 8.3).

All in all, these five principles prove to be indispensable for the development of societal future preparedness. The following three principles, which result in particular from the contributions of the developed and the sustainable society, are presented in detail in the next subchapters.

8.2 Principle 6: Securing (Basic) Needs and Preventing Potential Social Conflict

If we assume that the raison d'être of a society is to secure the (basic) needs of its population and thus its internal cohesion, the present principle proves to be particularly fundamental for societal future preparedness. A look at the various current and past social crises of this century proves to be very revealing. Despite an increase in prosperity in most regions of the world and a simultaneous decrease in extreme poverty, social conflicts appear to be intensifying worldwide. They affect not only the poorer regions of the Third World, e.g. Latin America and Africa, and since 2011 especially the Middle East region in the wake of the Arab protests, but also the welfare societies of the OECD. In 2011 and 2012, Spain, England, France and the USA was particularly affected and the Occupy Wall Street movement emerged. More recently, with the culmination of the refugee crisis around 2015 and the socio-economic impact of the Corona crisis in 2020 and 2021, almost all European societies have been affected by social conflict. A particularly strong escalation was observed in France in the course of the so-called 'yellowjacket protests' in 2018/2019 (Michael, 2019).

Table 8.3 Summary of the first five guiding principles for societal future preparedness

Concept Principles	Development	Resilience	Sustainability
1: Individual competence development	Emotional intelligence, learning to learn in the knowledge society	Personal (psychological and physical) resilience	Sustainability awareness and values
2: Dealing with knowns and unknowns	Distinction between good practices, best practices and emerging practices. Focus on development and sustainability: Emerging practices (experimental approaches). Focus on resilience: Innovative practices (mindful and immediate response in crises)		
3: Decoupling and knowledge integration	Implicit reference to decentralisation.	Autarkic (subsistent) units capable of making decisions in crisis situations; simultaneous communicative networking of all units	Emphasis on self-sufficiency and self-sufficiency (subsistence). e.g. transition towns
4: Collective intelligence	Focus: Development of technological innovations	Focus: High responsiveness and innovation in crises; development of appropriate social and technological innovations; time-efficient decision making	Focus: Development of sustainable innovations (technological and particularly social)
5: Learning culture	Emphasis: hot culture; continuous improvement; control of nature	Learning from mistakes; high cohesion; balance between adaptation and preservation of core values	Focus: cold culture; harmony with or preservation of nature; anticipating mistakes

As early as 2012, Dan Smith, the Secretary General of International Alert, suggested that peacebuilding, which normally focuses on war zones in the Third World, should be given a new priority for application to the inner-EU region (Smith, 2012). This consideration is surprising, given that Western societies have developed welfare systems and were therefore not affected by the factors that led to the social protests that had been taking place simultaneously across the Middle East and beyond since 2011. Factors that led to the Arab protests included poverty, human rights violations, limited freedom, limited democratic participation, corruption, and political repression (Fathi & Karolewski, 2015) – factors that do not apply to Western affluent societies in identical form. Nevertheless, the highly escalated social riots prior to the Corona crisis, such as the yellowjacket protests in France, show that despite the relatively high level of prosperity, some essential needs in the population do not seem to be satisfied. Social protests in both poor and affluent societies express a pervasive atmosphere of discontent.

> where people's sense of social belonging and engagement in the common good is challenged [...] by economics as job opportunities and the belief in a better future diminish before our eyes (Smith, 2012).

How is it that in a society like France there are currently more massive social protests than in a society that is relatively calmer in this respect, but much poorer, such as Senegal? A preliminary partial explanation could be provided by the psychological mechanisms of happiness described above in Sect. 5.2.4. According to this, happiness that results from material prosperity is relativised by two psychological principles. On the one hand people get used to what they already possess (hedonistic adaptation) and on the other hand people constantly compare themselves with people of the next higher (not lower) social status level. This may explain why social protests worldwide have not decreased in proportion to the increase in actual wealth in the last decades. As outlined above, the absolute wealth, the household purchasing power and access to knowledge of most populations has in fact significantly increased and is higher than ever before. At the same time, however, many people's discontent seems to be shaped by the impression that relative inequality, i.e. the gap to the rich, has increased. And under the influence of digital communication media, the impression of not being able to afford as much as others seems to be increasing, too. In addition, referring to the needs pyramid by Maslow (1954), the structure of needs in affluent societies is far more complex than in poorer societies. In societies like France it is by no means only a matter of existential (basic) needs, but at the same time also of higher needs, such as self-realisation, self-discovery, meaning and, above all, the prevention of social decline. As outlined above, it may come as no surprise that affluent societies in East (e.g. Japan, South Korea) and West (e.g. Europe, USA) have been affected by an increase in mental stress disorders, depression, and suicides for years. Accordingly, the social situation in affluent societies appears to be by far more complex than in poorer societies.

It can be assumed within the framework of this principle that welfare is an essential criterion for preventing social crises and securing the internal cohesion of a society. Although no society in global comparison is currently free of social conflict potentials, it is worth taking a look at the different welfare regimes in the relatively highly developed OECD world in terms of their characteristics, advantages and disadvantages. It could be assumed, for example, that the liberal market economies of Anglophone societies with relatively unambitious social policies contain more conflict potential than the coordinated market economies in continental Europe, which have a relatively higher degree of regulation and thus more social equality. However, people in societies with lower levels of regulation seem to benefit from greater freedom. Do either of the two key values – freedom and equality – weigh more heavily in terms of social conflict prevention? What can be deduced from this for societal future preparedness?

8.2.1 Overview of Welfare Regimes

Comparative welfare research is an interdisciplinary discipline that integrates social, economic and political models. To this day, this discipline is dominated by Gøsta Esping-Andersen's model coined in the 1990s – the *'three worlds of welfare capitalism'*. In his model he distinguishes ideally between several 'welfare regimes'. This term is not to be confused with 'welfare states' – rather, 'regime' means that several welfare states are characterised by common structural features. This perspective allows us to make statements on a very general basis that apply to specific 'types of societies'. Esping-Andersen distinguishes between the 'liberal regime type', which is found predominantly in the Anglophone groups of countries; the 'conservative regime type', which is found mainly in continental Europe; and the 'social-democratic regime type', which mainly characterises the Scandinavian countries. More advanced studies also distinguish between a Southern European-Mediterranean type, an Eastern European type and a Far Eastern type, but for various reasons we will not discuss these in detail here.[1] The

---

[1] On the one hand, the so-called 'rudimentary welfare regimes' of Southern and Eastern Europe already scored worst in all the indices presented above in the OECD comparison – highest inequality, unemployment, per capita income and value added and correspondingly lowest values in interpersonal trust and well-being. On the other hand, the index scores for the very highly developed regimes of Far Asia, e.g. Japan and South Korea, produced very contradictory results that cannot be attributed to welfare policies in the context of our observations. As mentioned earlier, Japan and Korea, for example, have the highest suicide rates despite high levels of socioeconomic development, and in surveys of subjective well-being, populations often report being unhappy. At this point, cross-national comparative happiness research proves to be limited. Obviously, these countries have culturally different definitions of happiness and well-being, which are not adequately covered by the surveys developed for the societies of the West. In the following, we are therefore only interested in the regime types of Western societies.

8.2 Principle 6: Securing (Basic) Needs and Preventing Potential Social Conflict 235

following is a rough sketch of the three types of welfare regimes found in Western societies (in the following Esping-Andersen, 1990; Fathi, 2013):

The *liberal type* comprises primarily the Anglophone societies such as the USA, Australia, Ireland, England and New Zealand.[2] Typically, all these welfare states have very low welfare state benefits that are means tested. A wage floor is secured by minimum wages; however, market logic and private insurance dominate. The cultural context implies the assumption that all people have the same starting conditions in the Calvinist-Protestant sense and can only achieve personal prosperity on the market through sufficient diligence and work. The best-known myth in this context is that of the 'American dream', in which the dishwasher works his way up to become a millionaire. Accordingly, the provision of state-regulated benefits is low, with a high emphasis on individual liberties and also opportunities to earn income in a very flexible labour market. At the same time, due to the very high dominance of market forces, there is the commodification of work – that is, the 'commercialisation', or in other words, the 'commodification' of work. The increasing pressure of the world market since the 1990s has had an impact on the liberal welfare state, especially on wages and benefits, resulting in the emergence of a class of 'working poor'. As a result, extensive domestic services (e.g. childcare) from this same stratum are recruited and consumed by upper- and middle-class families (Esping-Andersen, 1990; Fathi, 2013).

The *conservative type* is found in continental Europe and includes Germany, Austria, France, Belgium, among others.[3] In this welfare regime, market forces are more strongly regulated for reasons of state policy and paternalism. In addition to basic welfare, benefits such as pensions or unemployment benefits are added – commodification is therefore lower than in the liberal regime. The cultural context is shaped by the influence of the (Catholic) Church and the paternalistic-conservative state. Against this background, welfare policy primarily has a 'status-preserving' character. Accordingly, social security benefits are paid according to the equivalence principle, i.e. depending on the amount and duration of previously paid contributions on wage labour (Esping-Andersen, 1990). Since the 1990s, the increasing pressure of globalisation has led conservative welfare states to stabilise the core workforce of their industrial firms by reducing the labour supply – usually by promoting early retirement, disability and a massive exclusion of women from

---

[2] Canada varies greatly from province to province and therefore cannot be clearly assigned as a whole to the liberal type. For similar reasons, Switzerland cannot be clearly assigned to a regime type.

[3] The Netherlands is considered to be a hybrid type, which has basic features of all three types and therefore cannot be clearly categorised according to the present model.

the labour market. In the 2000s, the response to comparatively high but stable unemployment was to expand the low-wage sector (Fathi, 2013). In the conservative welfare state, the family usually plays a special role. Single-earner or main-earner households occur relatively more often than in the other types, and social services (such as childcare) are mainly provided by private households themselves – usually by women (Esping-Andersen, 1990; Fathi, 2013).

The *social democratic welfare regime* is found in the Nordic societies, i.e. Sweden, Norway, Denmark, Finland and Iceland. Access to social entitlements is universal and tax-funded social benefits are at a comparatively high level. In addition to financial protection, there is a dense network of social services, from active labour market policies to very good childcare, and there is a high degree of protection from market forces, i.e. high 'decommodification'. The cultural background results from a mixture of Protestant individualism and a democratic principle of comprehensive social civil rights. Thus, welfare policy is characterised above all by a very high degree of equality. However, tax revenue and bureaucratic expenditure are relatively high (Esping-Andersen, 1990). Social democratic welfare states have countered the increasing pressure of globalisation since the 1990s with a massive expansion of public services and an active labour market policy. In addition, increasing deregulation can be observed in some countries; in Sweden, for instance, the share of private insurance is being expanded (Fathi, 2013).

In summary, the three welfare regimes of Western societies comprise the following key distinguishing features (Table 8.4):

8.2.2 Conflict Potentials of Welfare Regimes

The three welfare regimes are characterised by potential conflicts that are described in the 'trilemma of the service economy' (Pierson, 2001), with the assumption that every welfare state is faced with three central challenges that it cannot meet in equal measure. These are ensuring the highest possible employment rate, income equality and cost efficiency. Any attempt to fully satisfy one of the three aspects neglects another. Transferred to the welfare state typology presented above, this means that different problems arise and thus lead to conflicts in the distribution of wealth (in the following Esping-Andersen, 1990; Fathi, 2013).

The *liberal welfare regime* does not have high public sector spending. The low-wage sector is strongly promoted, which creates work, but at the price of strong wage inequality. The liberal states are therefore very conflicted about the very unequal distribution of income, which takes on a particular dimension in the USA

8.2 Principle 6: Securing (Basic) Needs and Preventing Potential Social Conflict

Table 8.4 Essential criteria of welfare regimes in the West according to Esping-Andersen (1990)

Welfare regime Distinguishing features	Liberal (Anglophone societies)	Conservative (Continental Europe)	Social Democratic (Nordic companies)
Access to social benefits	Means tested	Means tested	Universal
Quality of social services	Low	Medium	High
Decommodification	Low	Medium	High
Focus	Greatest possible freedom (principle of personal responsibility)	Preservation of status (equivalence principle)	Greatest possible equality (universal principle)
Type of financing	Taxes	Payroll taxes	Taxes

in the so-called gap between the rich 'one-percent' and the rest of the '99 percent' of the population (Esping-Andersen, 1990). The Occupy Wall Street movement is a consequence of this. The potential conflict of the liberal welfare regime is thus that it excludes a large part of the population from potential social benefits and creates a large gap between the rich and the poor. Moreover, this regime stands out as relatively resilient to demographic or labour market changes, as it can be maintained with comparatively low expenditures (Fathi, 2013).

The *conservative welfare regime* has higher expenditures than the liberal regime. However, due to high taxes and regulations on the factor labour, the wage gap is lower than that of the liberal regime and the low-wage sector is not strongly expanded. By contrast, the conservative welfare regime usually has a higher unemployment rate. Since social security is directly related to (permanent) employment relationships, there is also a vulnerability to conflict when unemployment increases (Esping-Andersen, 1990). The conservative welfare state classically leads to a sharp divide between insiders and outsiders, i.e. people in dependent wage employment and people without work or in precarious employment – and this proportion is rising rapidly. Another potential conflict is the generational conflict: Due to the changing age structure of the population, fewer and fewer employees subject to compulsory insurance are paying contributions, while the number of pension recipients, who used to be contributors themselves in the past, is steadily increasing. In sum, the conservative regime is characterised by social conflict prevention potential with regard to inequality. However, it is highly vulnerable in the face of changes in population trends and on the labour market as well as being relatively costly and bureaucratic (Fathi, 2013).

The *social democratic regime* has the comparatively highest income equity and relatively low unemployment. In this respect this type of regime seems at first glance to have a relatively high preventive character towards social conflicts. Its weakness lies in its high level of regulation and bureaucracy. Moreover, there is a risk of budget overload, as jobs are created in the public sector, not in the low-wage sector (Esping-Andersen, 1990). There is a risk of generational conflict here, as the social democratic regime cannot rely on high government spending for long. If too much money is spent in the present generation, no provision is made for future generations. Thus, in systemic terms, the high bureaucratic and financial cost of the social democratic regime makes it very vulnerable to demographic fluctuations. However, it is less vulnerable to fluctuations in the labour market than the conservative regime, which is financed by payroll taxes (Fathi, 2013).

In summary, the welfare regimes in the OECD area respectively exhibit different conflict potentials. There are also correlations with many of the country-comparative indices presented above (Chap. 5, especially Sect. 5.2.4). The following indicators appear to be among the most meaningful:

- Psychological well-being and similar indicators such as life satisfaction and happiness indicate the extent to which the population perceives its needs met in society. Great dissatisfaction is reasonably likely to fuel social protests. Interestingly, as indicated above, in several studies of the last two decades, the global top ten is led by the cluster of Scandinavian societies, mostly Norway. Also high in the top ten are the Anglophone societies, led by New Zealand, Australia and the USA. There is a clear difference to the cluster of conservative welfare states, such as Germany, France and Austria (see also Fathi, 2013).
- Income equality: This indicator reflects social equality in society and, as Wilkinson and Pickett's (2010) study among others suggests, harbours enormous potential for social conflict. As expected, the liberal welfare regimes of Anglophone societies show the most critical values here.
- Employment: According to a variety of studies mentioned in Chap. 5, high unemployment is one of the most serious potential sources of conflict. Here, the societies in the cluster of conservative welfare regimes have the relatively most critical values, while the Anglophone societies, which belong to the relatively less regulated liberal welfare regime, have the best values. The Scandinavian societies are located in between – they are similarly regulated as the societies of the conservative welfare regime, but have a more effective job placement.

As shown above, these factors have cross-connections to other meaningful indices that also correlate strongly with the respective clusters. These include trust, social

8.2 Principle 6: Securing (Basic) Needs and Preventing Potential Social Conflict

behaviour and bad experiences. A summary presentation can be found here. It has not significantly changed in the 2010s (Table 8.5).

Taking all the indicators from comparative welfare, happiness and development research together, one could come to the conclusion that the social democratic welfare regime that prevails in the Nordic societies seems to have the least potential for social conflict in non-crisis times. It provides high general well-being through comparatively good socio-economic integration. However, it is very costly and bureaucratic to maintain and therefore vulnerable in systemic terms. This was demonstrated during the refugee crisis of 2015/2016: The relatively bureaucratic and highly regulated social democratic regimes proved to be overburdened relatively quickly and thus poorly able to cope with sudden crises, instead relying on stability and unchanging conditions (Benedikter & Fathi, 2021). The liberal welfare regime has massive potential for conflict, especially with regard to social inequality. However, because of its low costs, it is relatively easy to maintain, even in the face of change. In this regard, it proves to have the relatively highest resilience. Surprisingly, the liberal regime, despite relatively low socio-economic integration and high inequality, achieves high levels of general well-being and even the highest levels of pro-social behaviour. However, as shown in 2020 during the Coronavirus crisis, liberal welfare regimes, notably in the USA, characterised by low social support and high inequality, appeared to be particularly challenged when it comes to the so-called 'Pandemic-Inequality-Feedback-Loop', which means: 'Declining economic status leads to rising rates of chronic illness. That, in turn, further depresses productivity and raises healthcare costs, leading to more poverty, which leads to more disease' (Fisher and Bubola 2020). In contrast to the liberal regime, the conservative type has a much more elaborate social security system, which is lower in quality and almost comparable in cost to that of the social democratic regime. Strikingly, it even comes to relatively worse results in the average subjective well-being of the population than the other two regime types (see also Fathi, 2013).

All these very general conclusions do not claim to replace more specific comparative country analyses. Notwithstanding this simplification, the following general requirements for a sustainable welfare concept can be derived from this overview:

The results suggest,

1. That no regime is free of crisis potential,
2. That the social democratic regime has comparatively the least potential for conflict,

Table 8.5 Different crisis potentials of all welfare regimes at a glance. (cf. Fathi, 2013, p. 64)

Welfare regime / Selected indicators	Liberal (Anglophone societies)	Conservative (Continental Europe)	Social democratic (Nordic Companies)
Main conflict line	• Poor versus rich	• Insider versus outsider	• Bureaucracy versus citizens
Wellbeing			
• Wellbeing/happiness	• High	• Medium	• High
• Positive experiences	• High	• Medium	• High
• Negative experiences	• High	• Medium	• Low
Family stability			
• Absence of child poverty	• Low	• Medium	• High
Income			
• Absolute income	• Very high	• High	• High
• Relative income (equality)	• Low	• Medium	• High
Employment			
• Employment rate	• High	• Low	• High
Social environment			
• Interpersonal trust	• High	• Medium	• Very high
• Tolerance	• Very high	• Medium	• High
• Anti-social behaviour	• Medium	• Medium	• Medium
• Pro-social behaviour	• Very high	• Medium	• High
Freedom			
• Economic freedom	• Very high	• Very low	• Low
Structural resilience			
• Resilience to fluctuations (demographics/labour market)	• Very high	• Very low	• Very low

3. That the liberal regime, with its comparatively highest focus on personal freedom values (emphasising the motto: 'making its own luck'), and despite the lowest quality of social benefits, seems gradually likely to have less crisis potential than the conservative regime. The conservative type is characterised by higher-quality social benefits but relatively high bureaucratic and financial costs and discrimination.

This third thesis is all the more astonishing as one could have gained the impression from a comparison of the distinguishing criteria of the welfare regimes that the liberal type has by far the greatest potential for conflict. However, if we include other factors, such as general subjective well-being, the conclusion suggests that an elaborately designed welfare state does not per se promise less conflict potential,

8.2 Principle 6: Securing (Basic) Needs and Preventing Potential Social Conflict

and that the citizen's personal freedom to develop and shape his or her own life is of essential importance – maybe more important than social security and equality.

8.2.3 Universal Basic Income: A Suitable Approach for a Sustainable Welfare Concept?

Given the different advantages and disadvantages of all three types of welfare regimes in preventing social conflicts, the question arises as to whether and to what extent the respective advantages can best be integrated into a more sustainable welfare concept. Ideally, it would have to be characterised by a high level of social security while preventing high income gaps and high unemployment at the same time. The concept would also have to be relatively simple and unbureaucratic to finance and should offer citizens the greatest possible freedom in shaping and developing their own lives. Against the background of these challenges, the debate on a universal basic income is gaining new momentum as it offers possible answers to several requirements of social crisis prevention:

- How can a society ensure the greatest possible coverage of the welfare state principles of 'freedom' and 'equality' without at the same time creating an overly bureaucratic and costly welfare system?
- How can it alleviate the economic pressure to perform and thus contribute to greater health, especially mental health?
- How can the greatest possible freedom to exploit innovation potential and creativity be guaranteed?

In post-industrial societies, universal basic income concepts are also seen as a solution to compensate for the loss of millions of jobs expected in the near future. This is because as automation advances, 'simple tasks' such as factory production and delivery of goods will increasingly be performed by robots. At this point, proponents of the universal basic income demand that the money generated by machines be distributed fairly, allowing for a corresponding basic income for all. This in turn would lead to people voluntarily doing more socially useful work to escape 'welfare unemployment'.

What does the universal basic income concretely imply? The universal basic income concept is based on the basic idea of giving every member of society a share in the total income of this society without means testing. In this context, every citizen, regardless of his or her economic situation, would receive a financial allowance from the state that is fixed by law and equal for all – without any consideration in the form of a means test or a requirement of willingness to work. However, all general tax- and contribution-financed social benefits would be abolished, such as social assistance, unemployment benefit or child benefit. The extent to which special social need is separately taken into account is considered differently in the various models. Most models provide for a basic income of between €500 and €1500 per month (Blaschke, 2007). Furthermore, there is still the possibility of earning more income through gainful employment in addition to the universal basic income. Usually, earned income and basic income should not be in competition with each other. In order to achieve greater social justice, some models provide for a reduction in the basic income as earned income increases. The most common approach in this context is the negative income tax. Alongside the consumption tax and the taxation of monetary transactions (Tobin tax), it is currently the most widely discussed approach to financing basic income.

The negative income tax was developed in the 1940s by the British politician Juliet Rhys-Williams and popularised 20 years later by the US economist Milton Friedman (Friedman, 2002). Most negative income tax models assume a constant tax rate, a so-called flat tax. If the citizen's household income is above the basic income, the tax rate is positive, which means that an amount has to be paid. If the income is below, the result is a negative tax rate, i.e. an amount paid to the household by the welfare state. The tariff curve is thus completely defined by the two parameters basic income and tax rate (van Almsick, 1981). The following calculation example illustrates the tariffs for households with different market incomes. The universal basic income and thus the transfer limit (i.e. the income level at which the tax liability turns from positive to negative) is set here at €1000 and the tax rate at 50% (Table 8.6):

The concept of a universal basic income, which has been debated for decades, is gaining more supporters after the Corona crisis. According to an opinion poll in 2020 by Oxford University's professor of European Studies Timothy Garton Ash, after the Corona crisis and its economic fallout on small and medium enterprises,

Table 8.6 Calculation example: Universal basic income and negative income tax (Apolte, 2004, p. 8)

Basic income	Acquisition income	Income tax	Tax liability	Available income	Effective tax rate
1000	0	0	−1000	1000	−∞
1000	1000	500	−500	1500	−50%
1000	2000	1000	0	2000	0%
1000	3000	1500	500	2500	17%
1000	25,000	12,500	11,500	13,500	46%

8.2 Principle 6: Securing (Basic) Needs and Preventing Potential Social Conflict

almost 71% of Europeans were in favour of introducing a universal basic income (Benedikter & Fathi, 2021). Even before the Corona crisis, the universal basic income concept has been discussed worldwide and across party lines. Even in the liberal Anglophone world there has been a lively debate on the concept in the past. In the USA, the concept has been discussed under the name Basic Income Guarantee (BIG) and in Australia, Great Britain, Canada and New Zealand as 'Social Credit' since the 1920s. Initial pilot projects have been and are being implemented in several countries worldwide, such as for example Namibia (Basic Income Grant Coalition, 2009), Macau (Lau, 2013), or Kenia (Bakir, 2017). Regionally, the universal basic income debate is primarily focused on Europe. One of the most important associations is the 'Basic Income European Network' (BIEN), which has existed since 1986 and has expanded its activities worldwide since 2004. In Europe, the universal basic income is being discussed more intensively in Germany, Switzerland, Austria, Luxembourg, France and Spain, among other countries. In Germany alone, there are over a dozen models. Among the best known are the Ulm Model, the Solidary Citizen's Income of the CDU politician Dieter Althaus, which is largely based on it, and the model of the initiative 'Unternimm die Zukunft' ('Undertake the Future'), founded by the entrepreneur Götz Werner. Most recently, in Finland, a basic income of EUR 560 per month was paid to 2000 randomly selected unemployed people for the years 2017 and 2018 in lieu of unemployment benefits. The evaluation of the experiment showed a positive development in all test persons with regard to their health values, but no significantly greater success with regard to successful job placement (kela, 2018).

This is also the main criticism of the universal basic income in the discussion to date. According to the critics the introduction of the universal basic income would result in the loss of any incentive to work, which would lead to a dramatic reduction in the value added and competitiveness of society. Other criticisms focus on some uncertainties that would accompany the introduction of the universal basic income. These include, first, the question of price developments and the possible risk of inflation; second, the question of the risk of social immigration from other countries that have not introduced a universal basic income; third, the question of the undesirable side-effects of financing the universal basic income (e.g. a financial transaction tax or taxation of companies could lead to a corresponding emigration); fourth, the question of how the important but unloved professions can be filled if they cannot be reliably replaced by robots (e.g. nursing professions, refuse collection etc.) (see e.g. Flassbeck et al., 2012).

Proponents of the universal basic income point out that people would never give up work solely for reasons of personal meaning, but that universal basic income would enable a new understanding of work. The universal basic income creates the prerequisite for individual freedom and self-realisation even in activities that are not remunerated as gainful employment – it would thus have a strong decommodifying effect. In general, it would also increase individual risk-taking, self-employment and entrepreneurship – thereby promoting innovation and flexibility. In addition, it would also make workers more independent, no longer 'sticking to their jobs' for reasons of existential fear – thus, it could contribute to a reduction of intra-company competition (including bullying) and a decrease in negative stress and possibly other work-related stress-diseases. Moreover, the improvement in social security would also encourage the possibility of alternative life plans, such as educational phases, which would positively contribute to life-long learning. Another argument is that previously poorly paid but necessary work would be better paid or made more attractive. In addition, the stigmatisation of the unemployed, which is unavoidable for a large number of people when unemployment is inherent in the system (as in conservative regimes), would be eliminated. Finally, it is pointed out that the introduction of the universal basic income and the accompanying reduction in administrative bureaucracy would result in enormous cost savings and that the universal basic income would therefore be not only financeable, but cost-efficient (see e.g. BIEN).

Herbert Haberl, co-founder of the Berlin Empathy Academy (today: 'Fachgemeinschaft für Lebensdienlichkeit und Wohlergehen' [in English: Professional Community for Life Service and Wellbeing], locates the deeper contradictions in this debate in different logics of justice. According to this, the universal basic income critics start from the reciprocity principle, while the ethical principle advocated by the universal basic income proponents corresponds to the 'civil rights principle'. The reciprocity principle is characterised by a logic of justice that seeks an immediate balance between taking and giving. Accordingly, it would be unjust to receive without having given anything in return. Haberl states that this logic can also be found in many solidarity contexts, such as the logic of the intergenerational contract. According to this, one generation provides for the next generation and can expect to be provided for by it in old age. This corresponds to a temporally staggered principle of 'giving and taking', because here, too, taking and giving are placed in a direct relationship. By contrast, the logic of justice advocated by universal basic income proponents is based on a principle in which taking and giving are decoupled from each other in time and space. A reference to this approach can be found, among other things, in the principle of civil rights. Accordingly, by the very fact of being a citizen, a person possesses rights associated with his/her dignity, which must be protected. This is precisely the revolutionary approach of the civil or human rights principle, according to which the protection of civil rights is not linked to any conditions.[4]

Another deep contradiction inherent in the debate refers to different assumptions about basic human needs in the context of the principle of 'taking and giving'. On the one hand, it is assumed that humans prefer to take than to give. According to this, humans are lazy by nature and would rather not work voluntarily if their financial existence did not depend on it – a rather pessimistic assumption that the critics of the universal basic income see confirmed in reality. In contrast to this is the assumption shared by the proponents of a universal basic income that people have a strong need to give, to work, to develop in their lives and to create something – an 'idealistic' assumption at first glance that cannot be dismissed in view of the empirically proven connection between social conflicts and unemployment. According to this assumption, people do not have to be forced to work because their willingness to work arises from a need to find meaning and thus from an incentive that goes far beyond the mere need for material subsistence.

Besides these contradictions, which reach into deeper philosophical-ethical assumptions, such as the understanding of work, meaning of life, needs, the nature of man and civil rights, several points should be clear: First, it is noticeable that the pro-contra debate is highly speculative – further social experiments, e.g. via lighthouse projects, are needed to deduce more far-reaching conclusions and, where appropriate, to create opportunities to incrementally improve existing welfare systems. Incremental innovation steps are also recommended because, second, the existing welfare systems of all societies have grown historically and, if the introduction of the universal basic income proves useful at all, are likely to lead to a profound transformation. Third, the results suggest that the universal basic income is unlikely to replace activating labour market policies. Thus, job placements and, above all, further training opportunities that contribute to lifelong learning and ongoing competency development would remain an integral part of welfare policy and would need to complement the universal basic income system accordingly. Competency development and an activating labour market policy would also be crucial components of a sustainable integration policy. In order to prevent the development of parallel societies, an orientation of integration policy according to the 'promote and demand' principle would be conceivable. However, it could be criticised that this principle runs counter to the underlying core idea of an unconditional basic income guarantee, and would encourage a two-class welfare system that differentiates between nationals and foreigners.

---

[4] Interview conducted in August, 2018.

Regardless of the open questions, it can be concluded that the universal basic income deserves a constructive debate and further incremental testing in social experiments against the background of the fact that all existing welfare types show serious conflict potentials. Moreover, it can be argued that universal basic income combined with a Negative Income Tax may also be the answer to massive losses of jobs that are expected in the course of the automation trend (Brynjolfsson & McAfee, 2014), which has even accelerated during the Corona crisis. The post-industrial post-Corona economies are expected to be 'heavy on technology, but light on labour' (Benedikter & Fathi, 2021).

8.3 Principle 7: Development and Preservation

A further principle of building and safeguarding societal future preparedness is expressed in a balance between the demand for technological and economic development on the one hand and an environmentally friendly orientation on the other. It thus involves nothing less than the integration of the contradiction outlined above between the core claims of the concepts of sustainable and developed society. Related to that are three above-mentioned discourses – green growth, the post-social market economy and the post-growth economy – that attempt to provide an answer to this. There are high hopes for various technological applications, ranging low-tech to high-tech. What future development trends and mutual influences are conceivable?

8.3.1 High Hopes for High-Tech and Low-Tech Applications

Optimistic representatives from the development and sustainability discourse argue that a tendency towards higher resource efficiency is already inherent in the development concept. Development and sustainability would therefore not be contradictory, but be per se compatible. This can be seen clearly in the example of urban development. According to this, cities are not only hubs of development, they are also more efficient than other forms of community, because the dense living space is more space-saving. Although cities produce the most waste in relative terms, it should also be considered that more than half of the world's population live in cities, which in turn account for only about 4% of the world's total area (Horx, 2014). According to this line of argument, cities are not only more efficient and effective forms of community – they are also seamlessly connectable to the concept of sustainability. Sustainable urban development increases the

8.3 Principle 7: Development and Preservation

competitiveness of cities because it delivers a higher quality of life and thus contributes to the further development of cities. The highest form of development of the sustainably developed city is the smart city, which provides for improved satisfaction of the population's needs through real-time communication. Moreover, a smart city correspondingly adapts its framework conditions to inner and (possibly also external) changes. An increased use of green technologies, including renewable energies (green technologies) and circular economy applications (blue technologies) would further reduce the burden on the environment.

In addition, 'smart rural areas' are increasingly becoming the focus of discussion outside the cities. The argument behind this is that many small and medium-sized enterprises are at home in rural areas and that digital technologies and networking could also contribute to a stimulation of rural areas. A research initiative of the Fraunhofer Institute IESE is testing and researching the networking of rural areas in the model region of Western Palatinate within the framework of the pilot project 'Digital Villages'. This includes, for instance, the expansion of the digital infrastructure, which, among other things, enables better medical monitoring of patients and thus faster response in emergencies. Another example is the area of autonomous driving – here, IESE is currently researching applications with partners from the automotive industry, for example, to transform commuting by car from a necessary evil into a meaningfully used transfer time (IESE).

Beyond the discourse on smart technologies and sustainable urban development, it is currently mainly applications from the low-tech sector that many observers hope will integrate development and sustainability. This viewpoint is shared not only by representatives of the green growth discourse, but also by those of the post-growth economy. This is because low-tech products are typically characterised by the following criteria:

- Simple function;
- Simple production;
- Simple operation;
- Robustness;
- Easy maintenance and
- Easy reparability.

The field of so-called 'frugal innovation' (meaning 'simple' or 'modest' innovation), which focuses on low-tech applications, breaks with the typical growth paradigm of 'more and more, better and better'. In this context, the products do not only represent a resource-saving, simplified solution. They are also seen as having great potential for opening up lucrative growth markets and for the simultaneous development of societies. Typical characteristics of frugal products are:

- More for less: The product should have a high benefit for the end customer relative to the price (Radjou & Prabhu, 2014).
- Asset-light: This includes, for example, satisfying needs that were traditionally satisfied with capital-intensive products with the help of capital-saving services, such as Airbnb, lending circles, etc. (Bhatti, 2012).
- Good enough: the product should only cover application-oriented and target group-specific functions. However, it should fulfil these at a medium to high level.

Frugal products are not limited to durable goods, such as the US$800 electrocardiogram, the affordable JioPhone, or the One Laptop Per Child initiative, which involves large-scale distribution of laptops worth as little as US$100 each. Frugal products also include services, such as the 1-cent-per-minute calls, mobile banking, off-grid electricity, and microfinance (Bhatti et al., 2013).

Frugal innovations have been receiving increased attention from research institutions, companies and consultancies since the 2010s. Greater sales potential is expected above all in the emerging markets. According to a study by the consultancy Roland Berger, the global middle class is likely to triple to 4.8 billion potential buyers by 2030 (Tshidimba et al., 2015; Winterhoff et al., 2014). Meanwhile, emerging market industries have been independently bringing frugal products to markets. The best-known examples are all located in India and include:

- The carmaker Tata developed NANO, the 1600-euro car, with the potential as a prototype for a real 'world car'. Currently, it still has production and market difficulties (Gaur & Sahdev, 2015).
- The industrial company Godrej & Boyce Manufacturing developed the 'Little Cool' – a US$70 refrigerator that lasts a long time even during power outages (Winterhoff et al., 2014).
- Anurag Gupta, a telecom entrepreneur, simplified a smartphone and fingerprint scanner for ATMs in poor regions (Malhotra, 2014).

It is likely to be a matter of time before frugal products from India and other emerging markets, such as China or Brazil, become so well-designed that they also become attractive for Western markets. Representatives of the post-growth economy could criticise that frugal innovations per se do not yet bring about the change in consciousness necessary to preserve ecology. Ultimately, the aim of manufacturing companies is not only to provide consumers with inexpensive, simple goods, but also to introduce customers to higher-quality products in the long term.

8.3 Principle 7: Development and Preservation

It is not foreseeable that the contradiction between the sustainability paradigm, which aims at preservation of planetary resources, and the development paradigm, which focuses on resource efficient growth, will be resolved. At least in the short term, societal future preparedness will not result in the formation of an overall social culture in which conservation and development are integrated without ambivalence. Rather, in this context, it is likely that several cultures will coexist, each with a different emphasis on one of the two key values. These cultures are decisively shaped by the visions of the three discourses: green growth, post-social market economy, post-growth economy.

8.3.2 Possible Development Trends

The three discourses – green growth, post social market economy, post-growth economy – cover the essential aspects in the debate on an integration of sustainability and development. Currently, it can be observed that the discourses and corresponding initiatives are developing in parallel. A central reason for this is that they are driven by different agents and society sectors.

The vision of *green growth* or 'sustainable growth' is currently the most widespread and has received political impetus from the advancing energy transition in the countries of Europe. The Federal Republic of Germany, Europe's strongest economy and, since the nuclear disaster in Fukushima in March 2011, the driving force and reference market for the energy transition, has played and continues to play an important role as a driver. In addition, the aforementioned large-scale high-tech projects (e.g. smart cities, Desertec) and low-tech solutions (frugal products) promise a multi-billion-dollar market in the emerging states of Asia, the investment-rich Arab oil monarchies and, of course, in the EU-MENA region. The determining players are top national politicians currently in office and the private sector, including large and small corporations, insurance companies, banks, etc. The underlying assumption is that resource-efficient growth is possible in such a way that there are no harmful effects on the environment and that no ecological side-effects would have to result from it. A more far-reaching reform of the existing economic order or even a restriction of growth is not foreseen in this vision.

The vision of the *post-social market economy* is being advanced primarily under the auspices of the eco-social market economy in Europe and on the world stage. The spotlight is mainly on high-profile agents from politics and academia who previously held positions in top national politics or are at the head of far-reaching and rather idealistic, but not particularly effective supranational organisations (e.g. the United Nations, the implementation of the Global Marshall

Plan etc.). They advocate profound political reforms that go beyond the social market economy – and beyond the limited time frames of legislative periods. They criticise the short-term approaches of the national policy-makers in office, who have the real power to shape implementation. In addition, there are also many civil society agents and initiatives that attempt to influence those political decision-makers and other members of national politics. All in all, it can be summarised that there is a tendency for the idea of a post-social, above all eco-social market economy to become increasingly widespread, even if this is mainly limited to the EU area. However, a breakthrough in the implementation of relevant reforms remains elusive, as this is primarily in the hands of national policy, which still tends to be oriented towards more pragmatic, short-term approaches in the green growth tradition.

The discourse and vision of *post-economic growth* are mostly driven by civil society initiatives, independent of political or economic decision-makers. In this context, the aim is bottom-up cultural and substantive social change, not necessarily top-down institutional change.. From a technological perspective, more robust, durable goods are of special interest (since these can be frugal products, there are certainly synergies here). The major focus is on social innovations based on decentralised self-sufficiency, including resourceful cultural processes (e.g. lending circles). Since the guiding principle of post-economic growth goes hand in hand with a profound transformation of the existing economic system, it appears that it cannot be communicated in an attractive way to the majority of politicians and citizens. However, there is a steadily increasing environmental awareness and implementation in more and more initiatives like the Transition Towns at the municipal level.

We can only speculate about the future development and the mutual influence of the three visions. Green growth is in full swing in its many forms. In contrast, despite increasing international attention in politics and civil society, no country can currently be identified that has fully institutionalised a post- or eco-social market economy in society as a whole. This could, however, change in the not-so-distant future. Amsterdam, for example, was the first city in the world to officially declare its intention to adopt and implement the Doughnut Economy, a variant of the post-growth concept, in 2020 (Purdy, 2020). It is also becoming apparent that green growth and the post-social market economy do not stand in the way of existing local initiatives to realise a post-growth economy. A coexistence of initiatives of all three visions can be observed, which may be due to the fact that they currently seem to be developing more or less independently of each other at different agent levels. While the concepts of green growth and post-social market economy are discussed and initiated at the level of politics and economics and

8.3 Principle 7: Development and Preservation

would have to be established top-down, experiments with post-growth economic projects are being conducted independently of these tendencies at the level of civil society, i.e. bottom-up. In theory, each of the three guiding concepts has the potential to gain the upper hand over the others in the long term. However, the probability depends on different framework conditions, the occurrence of which cannot be predicted. Three scenarios are conceivable:

- *Green growth prevails:* The fact that in the long term the guiding concept of 'green growth' will prevail in Europe and possibly worldwide without major innovations in the market economy system is likely to depend above all on future technological developments in the relevant green and blue sectors and the realisation of major projects, e.g. eco-cities, renewable energy parks etc. If, for example, the resource and energy problem were to be solved by a technological revolution, societies with high economic dynamism, inconsiderate prosperity, rising energy consumption and more extreme consumer behaviour could emerge. The key determinant variable for this scenario is thus technological innovation and, where appropriate, the occurrence of a technological singularity.
- *Post-social market economy prevails:* The implementation of the post-social market economy will depend on whether, in of the face of mounting ecological disasters, political decision-makers give in to growing pressure, especially from civil society campaigns, and implement the necessary reforms. In some areas this could possibly result in legal restrictions of the development aspirations of the private sector and of the consumption behaviour of the population. In other areas, corporate value creation activities on green and blue technology innovations are conceivable. A possible scenario could be that of the above-mentioned 'Brazilianisation' or eco-dictatorship. A best-case scenario, as outlined by Radermacher and Beyers (2011), would mean that policymakers worldwide succeed in introducing sensible regulation of market activity, social balance and good governance, and sustainable development in the near future. Whether decisive change occurs will largely depend on whether necessary initiatives such as the Global Marshall Plan are implemented multilaterally. To sum up: The occurrence of this scenario will be largely determined by the determinant of institutional innovation and the increasing pressure on policymakers to act globally due to the deteriorating ecological situation.
- *Post-growth economy prevails:* Establishing a long-term post-growth economy might largely depend on the occurrence of ecological collapse or/and resource scarcity and the resulting need for an economic and social order based on sufficiency and subsistence structures. Individual transport would be restricted or abolished because of the high costs involved, and a change in values and

consciousness would lead to a situation in which one's own value is no longer measured in terms of private possessions, but in terms of community input. Therefore: The likelihood of this development is primarily determined by cultural innovation and the occurrence of a catastrophic scenario.

If one of these scenarios does not prevail, we will have to assume the coexistence of several of these visions, each with a different emphasis on hot or cold cultural aspects. An overall social integration of growth (hot culture) and preservation (cold culture) into a uniform culture is unlikely in the nearer future. If we have to assume that different concepts coexist, can they be institutionalised accordingly?

8.3.3 Political Integration of Multiple Concepts: The Panarchism Discourse

One possible answer to this could be provided by the still little-known concept of panarchism. At least it is receiving increased attention in the sustainability discourse under the term 'governance panarchy' (Loorbach, 2007, 2014). The term 'panarchy' is composed of the Greek stems 'pan'- (all) and – 'archy' (form of government), thus denoting a form of government that includes all forms of government. The term was introduced as early as 1860 by the Belgian botanist and economic theorist Paul Emile de Puydt. He drew inspiration from free economic competition to propose a free competition of the forms of government. His motto 'laissez faire, laissez passer' should also apply to government policy. In practical terms, this means that clients (citizens) should choose their government based on personal worldview and financial criteria, with no electoral losers. The continued existence of any government would depend solely on the number of its supporters and its finances. All individuals would be allowed to break away from the previous government and join a new one without having to leave the country. Thus, several parallel governments would exist in one territory, organised in different ways (e.g. monarchy, republic, etc.). The social contract, however, would exist equally with each government, that is, each government would draw taxes and duties from its followers and provide services to them. The legal basis would be akin to an exterritorial right of secession, i.e. it would be a matter of a personal declaration of independence, not a territory. In the big picture, according to de Puydt, the concept would amount to a peaceful coexistence of different political systems, similar to a peaceful coexistence of different religions on the same territory. In concrete political implementation, an 'office of political affiliation' would be set up in each

8.3 Principle 7: Development and Preservation

community, where each citizen could register and choose his/her form of government (de Puydt, 1860).

Puydt's approach was called for by anarchists, such as Max Nettlau (1909), and expanded into a political philosophy. Similar models are the 'multigovernment' concept of Le Grand E. Day (1969–1977) or the 'Functional Overlapping Competing Jurisdiction' (FOJC) proposed in 1997 by the two Swiss economists Bruno Frey and Reiner Eichenberger. In 1995, James P. Sewell and Mark B. Salter assessed the concept as an 'inclusive' and 'universal' form of governance with the potential of true global governance (Sewell & Salter, 1995). Similarly, Paul B. Hartzog describes panarchy as 'a new way of accomplishing global governance' in times of increasing complexity and interconnectedness (Hartzog, 2007, p. 2).

Critically, one could note that the panarchy concept is just another utopia whose usefulness has not even been empirically proven. It would also raise the question of concrete implementation. From another perspective, panarchy, multigovernance, FOJC, and other innovative global governance concepts, such as Simpol or even Paul Romer's 'Charter Cities' approach described above represent necessary considerations in order to do justice to the interconnected global reality in a more complexity-adequate and integrative political way. Despite their differences, an essential commonality emerges that could allow conclusions to be drawn on the open question of how the diversity of different political-economic cultures can be integrated: They all have in common that they shift the focus away from the state-centric level to the municipal level (this is the case, for example, with Charter Cities) or even to the citizen level, and provide a mechanism to effectively influence the next larger levels from there. Simpol, for example, is an approach supported by digital communication technologies to influence state policy makers directly and simultaneously from the citizen level.[5] Charter Cities do not provide for democratic

---

[5] Simpol is a global citizens' initiative initiated around 2000 by the British activist John Bunzl. Its goal is to resolve the international political paralysis on global problems (e.g. climate change) by democratic means. By means of a digital platform citizens use their political voice in elections to put pressure on all political parties to join the global momentum of simultaneous policies of all nations, ultimately leading to binding regulations on a global level that are beneficial to all nations. The initiative thus aims to build political incentives for global cooperation and overcome the vicious circle of international location competition and the associated 'political prisoner's dilemma' (more on this below). The interim goal is to gain bipartisan support in all democratic parliaments. Simpol is inspired, among other things, by the above-mentioned Spiral Dynamics model. Based on this, it would be a declared goal of Simpol to create incentives to direct political measures from the level of consciousness of national egoisms (blue) towards a genuine world-centric level of consciousness (orange).

elections per se, but citizens can vote with their feet, creating a competitive situation between cities.

To sum up: Building societal future preparedness in the twenty-first century will also have to deal with the challenge of integrating the diversity of choices between different political, cultural and economic concepts and possible contradictions between them. A particular contradiction exists between the demand of hot cultures for growth on the one hand and the demand of cold cultures for the preservation of ecology on the other. In the current discussion, political approaches are emerging as a possible response to organise an appropriate coexistence of different forms of governance, with a mechanism that allows citizens to choose between them. Most of these approaches have not yet been empirically tested in the form of social experiments; this would, however, be useful and conceivable in the near future, similar to the already discussed universal basic income.

8.4   Principle 8: Collective Wisdom

Collective intelligence can perhaps be described as the most fundamental property that a collective system, such as an organisation or society, needs in order to cope with complex problems of all kinds. As described elsewhere, collective intelligence (sometimes also referred to as swarm intelligence) results from the effective interaction of as many different individuals as possible. If the collective manages to allow all members to contribute their different perspectives, a group opinion emerges with a decision-making quality that is 'greater than the sum of its parts'. Diverse achievements of a society are the result of such collective efforts. Collective intelligence does not only bring new achievements (e.g. the digital transformation), but also unintended side-effects (e.g. cyberterrorism). Weighing this appropriately and coming to optimal decisions is the task of (collective) wisdom. It thus represents a currently completely neglected counterweight to (collective) intelligence. What is meant by this and what distinguishes it? How can it be developed and promoted in concrete terms? And why is it particularly important in the twenty-first century?

8.4.1   Wisdom: Characteristics and Approaches

Wisdom refers above all to a profound understanding of the interrelationships in life and the resulting ability to make the most coherent and sensible decisions when faced with problems. Classical approaches to wisdom are found in philosophy (translated from Greek: 'love of wisdom') and in the so-called wisdom traditions of the religions. For instance, Plato's famous allegory of the cave (which has influenced the history of philosophy and religion alike) describes wisdom as a knowledge of the real world and a turning away from the deceptions and errors of everyday knowledge, public opinion, and conventional prejudices. In the religious wisdom traditions, there are many parables like the one mentioned below, in which equanimity and problem-solving skills of wise people are illustrated.

> **The story of the old farmer and his horse**
> Among the best-known parables is the following story, which originates from Daoism but has also been received in Mahayana Buddhism: The story tells of an old farmer in a poor village. One day his horse ran away from him. His neighbours mourned how terrible it was, but the farmer only said, 'Maybe'. A few days later the horse returned, bringing with it two wild horses. The neighbours all rejoiced at his favourable fortune, but the farmer again replied, 'Maybe'. The next day, the farmer's son tried to ride one of the wild horses. The horse threw him and he broke both his legs. The neighbours all expressed their sympathy to him for this mishap, but again all they heard from the farmer was, 'Maybe'. The next week recruiting officers came to the village to take the young men to join the army. A war with the neighbouring kingdom was brewing. They didn't want the farmer's son because his legs were broken. When neighbours told him how lucky he was, the farmer replied, 'Maybe'...

The parable tells how an old farmer, through no fault of his own, gets into various situations with serious consequences. These events are spontaneously judged by his neighbours, but the old farmer does not evaluate these situations and thus always remains equanimous. Wisdom is shown in this parable in the quality of perceiving things as they are – without judging. In this context, the famous quote of the spiritual teacher, Jiddu Krishnamurti, fits: 'Wisdom is not stored memory, but the highest form of openness to the real' (Krishnamurti, 2007, p. 229). This openness to the real arises explicitly from mindfulness and meditation practices.

What is Meditation?

Meditation translated from Latin means 'to contemplate' or 'to take into focus'. A later translation is 'to align oneself with one's own centre'. These translations clarify what meditation is essentially about: gaining a clearer view of oneself, the nature of things and achieving more equanimity and centeredness. From a technical perspective, meditation mostly involves mindfulness and concentration exercises. There are hundreds of different applications, with the best known and most common being done as a sitting meditation (see below for a more detailed description of what such an application might typically look like). Advanced practitioners, on the other hand, see meditation not just as a technique, but as a basic attitude of constant awareness. In the context of meditation, mindfulness involves a non-judgmental form of intentional attention that focuses on the present moment, i.e. the now, rather than on the future or the past (Kabat-Zinn, 1982). In Eastern cultures in particular, meditation is considered a fundamental mind-transforming practice.

Example of a Meditation Application (Inspired by Fathi, 2019b*).*

Get into a comfortable position. A seated position is usually recommended, e.g. sitting on a chair, kneeling on a meditation stool or sitting on the floor with the buttocks on a cushion. In the latter case, a cross-legged position is recommended – if your stretch allows, it may also be a half or full lotus position. It is important that your back remains erect and as straight as possible. There are also forms of meditation in which you stand (e.g. in the starting posture common in Qi Gong or Tai Qi) or lie on your back (e.g. in some yoga traditions). I will not go into this in detail below, but in principle the following steps could also be carried out in these positions.

Getting started: In the second step, I recommend breathing exercises to quickly centre yourself and interrupt the flow of thoughts triggered by everyday life. At this point, you can breathe in and out in a relaxed manner for a few minutes (about 4 s per breath). To complement this, you can use your breath as an 'object' and just count it. This is a useful method to practice your concentration and presence. Whenever you are getting distracted, you can come back to your breath.

(continued)

8.4 Principle 8: Collective Wisdom

> **continued**
>
> *Main part:* Now you begin with the actual meditation. Out of the dozens of possible variations, I want to introduce you to the so-called Body Scan which is a classic part of the MBSR program and Vipassana meditation (a classical meditation practice from the oldest Buddhist school, Theravada Buddhism). Here you go through all the parts of your body one by one during meditation. My recommendation for the sequence: starting with the feet, through the lower legs, thighs, hips/buttocks/genitals, lower abdomen, lower back, chest, upper back, shoulders, both arms down to the hands and finally the head and face, and then go back through the entire sequence from top to bottom. You can go through these sequences several times or only once in total, but stay longer in each part of the body. At the end of the meditation, try to become aware of the whole body at once. Throughout the Body Scan, it is important to become aware of the fact that everything (thoughts, sensations, feelings, concepts, identity etc.) is ephemeral, that it is not the real you.
>
> *Notes:* During meditation, feelings and thoughts will come up. This is completely normal. Try as best you can not to be distracted by them. And if you do get side-tracked, just come back to the here-and-now by concentrating on your breathing. There are several tricks for dealing with thoughts that arise during meditation: One of my meditation teachers recommended to 'label' arising thoughts – internally, you can also shout to yourself, 'There's one!' Another trick is to briefly ask yourself during a thought what the next thought might be. From experience, thoughts and feelings that arise are likely to dissipate quickly. The rest is a matter of your mindfulness and presence.

Openness to the real expresses itself in something that can be described as 'here-and-now awareness' or 'present awareness'. The change manager and transformation researcher at MIT, Claus Otto Scharmer, also calls this 'presencing' and uses it to describe a field of attention in which one becomes aware of the immediate moment, i.e. what remains when the moment has passed and the new has not yet arrived. A related insight is that every evaluation and every thought process is related to a moment that lies either in the past or in the future. Accordingly, what is really true is only what arises in the immediate here and now, and this is not revealed through thinking, but through absolute presence, as taught, for example, in the practice of mindfulness. The spiritual teacher Eckhart Tolle describes in his bestseller 'NOW' that only through this very presence in the here and now can freedom from negative judgements and thus also negative feelings of stress arise:

> All negativity is caused by a (...) denial of the present. Discomfort, anxiety, tension, stress, worry – all forms of anxiety – arise from too much future and not enough present. Guilt, regret, resentment, grief, bitterness, melancholy, and all forms of non-forgiveness arise from too much past and not enough present. (Tolle, 2012, p. 80)

Present-moment awareness helps to avoid stress and the associated emotional dispositions in decision-making situations.

Another characteristic of wisdom that arises directly from meditative mindfulness is an improved ability of dealing with chaos and information multiplicity. It is in this context that the famous quote by the US psychologist and philosopher William James can be placed: 'The art of being wise is to know what to overlook'. The mindfulness researcher Ellen Langer describes this ability as 'gentle openness':

> What you want is a soft openness to be attentive to the things you're doing but not single-minded, because then you're missing other opportunities. (...) We have new data and analysis coming at us all the time. So, mindfulness becomes more important for navigating the chaos – but the chaos makes it a lot harder to be mindful. I think chaos is a perception. People say that there's too much information, and I would say that there's no more information now than there was before. The difference is that people believe they have to know it – that the more information they have, the better the product is going to be and the more money the company is going to make. I don't think it depends as much on the amount of information someone has as on the way it's taken in. And that needs to be mindfully. (Langer, 2014)

While (collective) intelligence in this context is characterised by the ability to process as much information as possible, also about the past and possible futures, (collective) wisdom is characterised by a fundamentally different perspective. Wisdom takes a step back and asks what actually matters. It focuses less on what is possible and how innovations can be developed, but rather on what is the deeper aspiration of what we do. Moreover, wisdom asks what is worth striving for. In doing so, it questions the framework that the intelligence approach would simply take for granted. What becomes of the human self if we strive for limitless optimisation of the human body, in terms of transhumanism? Does this actually lead to what we are actually striving for, namely greater happiness and well-being and the end of suffering? What is the self anyway, and what are we as a society? These and other questions typically arise from the wisdom approach. This is less about solving problems, which is, after all, the goal of intelligence, but rather about a strategica and philosophical weighing of what actually constitutes the perceived problems, what new problems might arise through the pursuit of their solution and what, on this basis, is the best decision in the long term. This can also mean not

8.4.2 Collective Wisdom to Guide Development Paths

taking certain development paths or not taking them yet. How this can look in concrete terms is explained in more detail below.

Of all three guiding concepts, sustainable society is likely to have the largest share in the principle of collective wisdom (and the developed society definitely the lowest one). However, societal future preparedness goes far beyond the traditional three pillars of sustainability – the social, the economic and the ecological. It at least requires a consideration of the technological dimension, which in some areas is characterised by uncontrolled exponential development and could lead to undesired man-made rebound effects. To prevent these, technological sustainability based on collective wisdom would be needed.

In the private sector, isolated efforts can be observed to boycott dangerous technologies. For example, the founders of Google Deepmind, the company's AI division, and Tesla founder and investor Elon Musk have pledged not to participate in the development of autonomous AI-powered weapons. In 2017, they joined more than a hundred other researchers and companies in an open letter on the Future of Life Institute's website calling for such a ban (Future of Life Institute, 2017).

Similar initiatives can be found in the scientific context. A current example is a call by 18 researchers from seven countries to ban genome editing, i.e. the modification of DNA in sperm, eggs or embryos to produce genetically modified children, worldwide for the time being. All nations should voluntarily commit to this moratorium, and until there is an international set of rules for dealing with such technologies, all clinical trials should be stopped. The appeal was prompted by advances in so-called 'Crispr-Cas9' technology (Crispr for short), which uses the molecule Cas9 to enable the targeted removal and exchange of genetic material. This molecule acts like a pair of scissors and transports the DNA fragments. This should not only make it possible to optimise seeds, but also humans, with far-reaching consequences. Once engineered and born, gene-optimised babies could change humanity forever at the expense of all those who do not have the technology. The scientists argue for measures to create transparency and enable dialogues with the public. They also advocate cross-disciplinary and cross-sectoral cooperation, including ethicists, human geneticists, politicians, religious groups, people with disabilities, etc. (Lander et al., 2019).[6]

---

[6] A novel that deals with this issue in a very insightful way is 'Helix' by Marc Elsberg (2016).

In the political context, control could be achieved not only through bans or (unfortunately not very promising) voluntary commitments, but also preventively by regulating the allocation of funding. This makes sense especially in research areas whose technological possibilities are incalculable and can solve or even promote existential risks. In this context, collective wisdom could aim to steer technological development in certain areas.

However, according to philosopher and risk expert Nick Bostrom, this is not so simple. For example, if a politician proposes to cut funding for a particular field of research out of concern for the potential risks of a technology, he or she must expect outrage and enormous resistance in the research community. Because even if the anticipated risk seems plausible, scientists and other advocates would argue that it would be futile to stop the development of a technology by preventing research on it. After all, if a technology was feasible, it would be developed somewhere by someone in the world. Bostrom points out here that this so-called 'futility objection' is almost never raised when politicians propose to increase research funding in a field, along the lines of 'please don't give us funding, cut it instead. Researchers in other countries will surely step in for us, the work will get done anyway' (Bostrom, 2017, p. 321). According to Bostrom, this emerging double standard is most likely explained by national interests. According to this, a society would push ahead with the development of a potentially dangerous technology because it assumed that it would be developed anyway and that the expected damage to the world would occur in any case. At the same time, it could at least secure the relatively small benefit of being the first to develop the technology. Even though the futility objection is so widespread and popular, it does not – according to Bostrom – make it plausible why it is not reasonable to control technological development. The issue, he argues, is not whether a technology would be developed, but when, by whom, and in what context. And these circumstances could very well be influenced by the allocation of funding (Bostrom, 2017).

Collective wisdom would arise in this context according to the 'principle of unevenly timed technological developments': 'Delay the development of dangerous and harmful technologies, especially those that could pose an existential risk, and accelerate the development of beneficial technologies, especially those that reduce natural or technological existential risks' (Bostrom, 2017, p. 323). A wise strategy could be measured by 'how much of a time lead it gives to desirable over undesirable technologies' (ibid., p. 323).

The question becomes more complex when one considers that some technologies have an ambivalent effect on existential risks, increasing some and reducing others. This would be the case, among others, with superintelligence: a machine

8.4 Principle 8: Collective Wisdom

superintelligence could generate a significant existential risk, such as the 'paperclip scenario' presented above. In turn, being considerably more capable than human intelligence, it could effectively eliminate many other risks. Natural existential risks such as supervolcanoes, pandemics, asteroid impacts, it could transform into non-existential risks, e.g. by colonising space. It would detect man-made risks early, it would make fewer mistakes in developing potentially dangerous technologies, and it would take the right precautions early and implement them properly. Accordingly, it all comes down to the order of technological development. According to Bostrom, there is a case for preferring superintelligence to other potentially dangerous technologies such as advanced nanotechnology, even if superintelligence proves to be the riskiest of all. For in developing a superintelligence, humanity would only have to deal with the risks involved, as preventively as possible – while it is still being built and programmed. With the realisation of another potentially dangerous technology, humanity would have to deal with the resulting dangers and those of superintelligence. Although there is much to be said for developing superintelligence first, Bostrom argues that there are other good reasons for delaying its development. This is because a later intelligence explosion would give humanity more time to become smarter itself and to find solutions to the control problem[7] (Bostrom, 2017).

According to Bostrom, the strategy of planned development delay depends primarily on the distinction between two types of risk: State and stage risks. *State risks* are time-dependent – the longer a system is in a certain state, the higher the overall risk. This is particularly true of natural risks. For instance, as time progresses, the probability increases (rather than decreases) that a superasteroid will strike or a supervolcano will erupt at some point. The risk of nuclear war also tends to increase as long as the international community has not yet developed global governance or something similar and remains in a state of 'international anarchy'. *Stage risks,* on the other hand, arise directly from the transition itself, not from the duration of a state. For instance, the emergence of a superintelligence is per se a stage risk. The magnitude of the risk here would depend more on whether appropriate precautions could be taken and not on whether the emergence of a superintelligence took 20 seconds or 20 hours. Bostrom concludes from this that in the case of expected existential state risks – such as a meteorite impact – it makes more sense to accelerate one's own intelligence and to promote technology in a targeted manner. For expected stage risks – such as a superintelligence – a slowing

---

[7] Bostrom's in-depth analysis concludes that there is currently insufficient knowledge to effectively control a superintelligence and prevent unintended risks already in programming (Bostrom, 2017).

down of development might be advisablein order to give more generations the chance and more time to prepare control measures. Bostrom puts this view into perspective elsewhere by conceding that stage risks, which are characterised precisely as a 'transition into something new', cannot be prevented by more experience. Here, accelerating one's intelligence and sharpening one's foresight could possibly prove to be a more reasonable strategy. Higher intelligence would also lead to a more effective use of preparation time, e.g. for the transition into the era of machine superintelligence (Bostrom, 2017).

Bostrom's considerations that also draw on similar reflections by Eric Drexler, lead to a sample argument in seven steps. They provide an ideal-typical rationale for dealing intensively, and, above all, in international cooperation with future risks, especially existential risks, and for preparing for them accordingly. This requires a corresponding foresight, which typically results from the wisdom approach (hereafter Bostrom, 2017, p. 335):

1. The risks of X are big.
2. Reducing these risks will require a serious preparatory phase.
3. Serious preparations will not be made until the possibility of X is really considered by broad sections of society.
4. Broad segments of society will not really consider the possibility of X until major research efforts are made to develop it.
5. The sooner serious research efforts are initiated, the longer it will take to develop X (since one must assume a more primitive level of technology).
6. Therefore, the sooner serious research efforts are initiated, the longer the period during which serious preparations can be made and the more risks can be reduced.
7. Therefore, serious research efforts towards X should be initiated immediately.

It follows from this logic that research into new technologies should neither be slowed down nor stopped, but rather intensified with a view to the expected risks and the development of sustainable strategies.

A similar approach, inspired more by the resilience concept (appearing relatively callous), would even accept small and medium disasters in order to systematically become aware of one's own vulnerabilities and further develop precautionary measures. The idea of such a 'shock therapy' would be to infect civilisation with minor catastrophes, similar to a 'vaccination', thereby triggering 'immune responses' that prepare humanity for existential variants of infection. Proponents of an extreme version of this concept would also accept terrible consequences in the hope that they would wake up the population (Bostrom, 2017).

Regardless of the preferred strategy for systematically preparing for future risks, perhaps the most important parameter for wise collective management of man-made risks will depend on the degree of global coordination and cooperation.

8.4.3 Wise Decisions in the Global Prisoner's Dilemma

Today, the rather unfavourable scenario of global competition is emerging, resulting in a race to develop potentially dangerous technologies at the expense of security. By contrast, cooperation would have the advantage of reducing precipitous action, allowing greater investment in security, and encouraging the exchange of ideas on how to solve problems (e.g. the problem of control in the development of superintelligence). Moreover, global cooperation and coordination would significantly prevent the escalation of international conflicts. It would not only be of greater overall benefit, but also morally more desirable.

Even though in theory global cooperation appears to be much more advantageous, in practice forms of competition and national egoisms still seem to occupy a considerable part of international events. Current reporting focuses primarily on the 'race for AI' between the USA and China. In addition, there are other competitive dynamics, such as the trade dispute between the US under former president Donald Trump and China and the EU, as well as multiple arms races all over the world, for instance between the US, the EU and Russia, in the Middle East or in the South China Sea, to name but a few. The prisoner's dilemma derived from game theory illustrates how uncooperative decisions of all kinds arise at the expense of the global common good. Can guidelines for wise action be derived from this?

The prisoner's dilemma models the situation of two prisoners accused of a common crime. Each prisoner is interrogated and has the option to remain silent or confess independently of the other. Here, the individual sentence depends not only on how the individual prisoner decides for himself, but how his decision relates to the other: If both remain silent, they receive low sentences; if both confess, they receive high sentences. If, however, only one of the prisoners confesses, that prisoner goes unpunished as a state witness, while the other receives the maximum sentence. The dilemma now is that each prisoner must choose to either confess (betraying the other) or deny (cooperating with the other prisoner) without knowing the other prisoner's decision. Thus, the sentence ultimately imposed depends not only on one's own decision, but also on the decision of the other prisoner. The dominant strategy of both prisoners would be to follow only self-interest and confess. In contrast, prisoner cooperation would result in a lower sentence for both and therefore a lower overall sentence (Axelrod, 1984). The prisoner's dilemma can be simulated with the help of the so-called Red-Blue-Game with groups of different sizes.

In the simulation of the Red-Blue-Game the players divide into two teams that sit in separate rooms. Another person is the moderator. Ten rounds are played – so a so-called 'recurring prisoner's dilemma' is simulated – and in each round the teams decide independently whether to play 'Red' or 'Blue'. 'Red' corresponds to a decision that considers both self-interest and the interest of the other group, and is equivalent to the 'deny' option in the prisoner's dilemma. 'Blue' is equivalent to the 'Betray' option in the prisoner's dilemma. In each round, the facilitator asks the groups for their decision. After both groups have decided, the facilitator records the results for both groups to see. Only then do the groups find out how the other group has decided. The results are scored as follows: If both teams choose red, both teams win +3. If both teams choose blue, both teams lose −3. If one team chooses red and the other team chooses blue, the team playing red gets −6 and the team playing blue gets +6. These points count double in round 9 and in round 10. The points of all 10 rounds are accumulated for each team.

Group A plays	Group B plays	Group A receives	Group B receives
Red	Red	+3	+3
Red	Blue	−6	+6
Blue	Red	+6	−6
Blue	Blue	−3	−3

The goal of both groups is to score as many points per round as possible. If they score 0 points or less at the end of the tenth round, they have lost. After round 4 and after round 7, the teams can each send a representative to a conference to coordinate. It is the only way both groups can communicate with each other.

Personally, I have run this simulation over a hundred times at over a dozen universities with courses in different disciplines and semesters (Bachelor and Master). In about 70% of all cases, both groups fail and do not achieve a total point value above 0. In 10% of all cases, the simulation ended with only one group as the winner with a point value above 0. The remaining 20% of all cases, both groups manage to win by achieving a point value above 0. In the final evaluation, the participants state that both groups can only agree through their representatives when conferences are called, but otherwise have to speculate about the other side's decision. Here it becomes clear that trust can be built up but also quickly destroyed again. The groups can only successfully go through the game, which lasts ten rounds, if they come to an agreement – by whatever means. By aggressively playing

8.4 Principle 8: Collective Wisdom

blue, at least one of the groups can achieve a high score in the short term, but only if the other group plays red and thus accepts a high negative score. If both groups play blue, they both get a low, but still negative score. Applied to the trade dispute between the US and China, for example, US President Donald Trump's decision to unilaterally impose import tariffs on certain goods from the EU and China would correspond to the blue option in the simulation. The option serves short-term self-interest but negatively affects each other's interests. The EU and China then responded with counter-tariffs on US products, which also corresponds to the blue option. The result was a spiral of escalation with a lose-lose-lose outcome.

In the long run, the groups can only win if they take the interests of the other group into account and all play red. This requires, firstly, trusting the other group to a certain extent, secondly, pursuing a decision-making strategy that builds trust with the other group, and thirdly, following a decision-making strategy that can, if necessary and flexibly, also build up threatening potential and resort to uncooperative behaviour without unsettling the other group too much and thus triggering an escalation spiral. A promising communication and decision-making strategy in such situations is 'tit for tat'. A player using the Tit for Tat strategy begins the interaction with a cooperative ('friendly') move. After that, a Tit for Tat player copies the last move of the other player. This strategy was already formalised in the 1960s by Anatol Rapoport (Rapoport & Chammah, 1970), but it was popularised by Robert Axelrod with the book '*The Evolution of Cooperation*' (Axelrod, 1984). Here he described, among other things, a computer tournament in which two-person competitive constellations were simulated and came to the conclusion that the Tit for Tat strategy proved to be the most functional in evolutionarily terms (Fehr, 2004). Today it is known that Tit for Tat proves to be disadvantageous if the reaction of the interaction partner cannot be correctly recognised or interpreted and is misinterpreted as a rejection of cooperation. The danger then consists in rapid provocativeness and corresponding retaliatory automatism (Dixit & Nalebuff, 1993). According to Martin Nowak, the win-stay, lose-shift strategy has proven superior to the Tit for Tat strategy for such misleading situations. It involves keeping the chosen strategy (cheat, cooperate) if it was successful and switching if it was not. The advantage of this strategy is that players can come back to cooperation from a non-cooperation loop (Nowak & Sigmund, 1993).

According to the current state of knowledge, both strategies prove to be the wisest strategies for dealing with the difficult initial situation of a recurring prisoner's dilemma. What makes the initial situation of the prisoner's dilemma so difficult? It is the fact that the parties cannot communicate directly with one another, which leads to the parties not knowing exactly what to expect from each other. This inspires distrust and fear and encourages uncooperative actions. What makes both strategies wise? They provide the other party with the greatest possible certainty of expectation about their own decisions, but also about the negative consequences to be expected should they decide to act uncooperatively. It can thus be deduced from the prisoner's dilemma that the greater the transparency and the more possibilities the parties have to come to an agreement, the more likely a cooperative dynamic is. If both prisoners (or both groups) were in the same room, they would immediately see which decision the other was leaning towards and they could coordinate their actions even before they were implemented. This realisation is not so new: At the height of the Cold War, when the Cuban Missile Crisis almost led to nuclear escalation in October 1962, a 'hot wire' or 'red phone' was set up between the United States and the Soviet Union. This permanent teletype link was intended to provide a means of preventing a threat to peace from errors, misunderstandings, or delays in communications.

In other words, direct communication and transparent structures are conducive to global cooperation and are likely to prove important leverage points for shaping societal future preparedness on a global level.

Outlook: Five Leverage Points to Trigger Social Change 9

From the considerations in the previous chapters, it can be concluded that all three currently dominant societal guiding concepts – the developed, the sustainable and the resilient society – contribute important aspects that need to be considered and integrated in the face of today's and tomorrow's global challenges. The eight principles outlined above provide a possible orientation for this integration. However, they are not sufficient on their own to trigger the necessary change. What are leverage points to induce societal change? In the following subsections, I outline several leverage points that can be derived from the previous considerations and that could contribute to the realisation of the eight above-mentioned principles. It should be noted that the reflections outlined below do not claim to be exhaustive. After all, the question of whether and to what extent social change can be initiated, shaped and possibly controlled could fill an entire library. And even despite this wealth of theoretical and practical knowledge from a wide variety of disciplines, such as systems thinking (including its sub-disciplines), transition research, more recently the transformative sciences, development research, evolutionary research, action research and change management that has emerged from it, the design of successful systemic societal change is (still) in its infancy.

This is true even in the much more tapped context of organisational change management. For instance, practitioners (e.g. Ashkenas, 2016 or Ewenstein et al., 2015) and researchers (e.g. Balogun & Hailey, 2004; Mosadeghrad & Ansarian, 2014) agree that the average success rate of change projects is only about 30–50%. How much more difficult and complex is it to initiate and steer societal change? The following considerations can therefore only represent initial approximations, which will need to be developed in more detail in further publications. In the con-

© The Author(s), under exclusive license to Springer Fachmedien
Wiesbaden GmbH, part of Springer Nature 2022
K. Fathi, *Multi-Resilience - Development - Sustainability*,
https://doi.org/10.1007/978-3-658-37892-9_9

text of this publication, the focus is initially only on 'initiating' social change and less on the question (which is also exciting for further research) of how change can be sustainably established and anchored.

9.1 Five Leverage Points for Social Change

Several leverage points can be derived from the previous considerations. The main findings can be summarised as follows:

- *Communication:* The *simplify* perspective, especially Systems Thinking, clarifies communication as the basic operation that fundamentally constitutes social systems, such as societies. It is communication that constitutes knowledge exchange between members and social cohesion to form a 'whole'. Communication enables the emergence of cultures, (legal) norms, collective identities and visions, and it underpins institutions and infrastructures. In retrospect, successful communication is also an essential prerequisite for the realisation of collective intelligence and collective wisdom and therefore underlies all three guiding principles. For social change processes, the design of successful communication is of fundamental importance and therefore an important leverage point.
- *Multiple dimensions:* From the *complexify* perspective it can be derived that social systems contain multiple dimensions that cannot be reduced to one another. The distinction between an inner or subjective and an outer or objective dimension of reality alone, as well as the consideration of different system levels, i.e. the micro (individuals), meso (organisations) and macro (societies) levels, result in a matrix of at least six dimensions. All dimensions would be potential windows of intervention to trigger change processes with different initiatives. Experience shows that initiatives at the macro level are likely to be relatively more costly than processes initiated on a small scale. Experience from international conflict research and peacebuilding also suggests that the 'culture' dimension (i.e. the inner or subjective dimensions of the meso/macro system level) is located on a relatively deep 'tectonic layer' and that interventions concerning this dimension are correspondingly slower (Galtung, 1998). Pragmatically oriented social interventions to merely trigger first change impulses are likely to neglect the intersubjective dimensions and tend to concentrate on external interventions, especially at the micro and, above all, meso level. In the long term, however, societal change processes should consider all dimensions.
- *Experiments:* The importance of small-scale changes is not only that results are achieved relatively quickly at this system level. They enable relatively manage-

able projects based on experimental trial and error. This is particularly recommended because social systems (such as organisations, communities and societies) are so complex that the success of change processes cannot be predicted with the help of a simple input-output model. Any change in a social system involves a high degree of unknowns. Thus, complexity-adequate interventions anticipate this circumstance and plan for a protected space in which mistakes can be made. Initiation processes in the sense of lighthouse projects should therefore be experimental. However, in the long term, and this cannot be covered within the scope of this publication, social change processes go far beyond real life experiments. The subsequent challenge is to institutionalise the insights gained from the trial-and-error process. Resilience itself, as noted above with reference to the Cynefin model, also goes beyond experimentation: resilience manifests itself in concrete crises as direct action (linked to introspection) – not as a mere test run.

- *Top-down and bottom-up:* The current debates on collective change have also shown that there are already a multitude of initiatives from different society sectors to promote development, but above all resilience and sustainability. In the context of resilience in particular, it is striking that these initiatives exist largely independently of one another, with spill-over effects in two directions. Civil society initiatives, for example, work 'from the bottom up', while initiatives from the political sphere typically work 'from the top down'.

On the basis of these preliminary considerations, at least the following complementary leverage points can be derived that could trigger social change and contribute to the realisation of the eight principles of societal future preparedness outlined above:

1. Work on yourself.
2. Using critical mass to drive change from below.
3. Experimental prototype projects.
4. Communicative networking of the subsystems.
5. Methods of communicative complexity management.

9.1.1 Leverage Point 1: Work on Yourself

In answer to the question how societal future preparedness can be promoted, this leverage point would start with the individual citizen. The underlying motto corresponds to the famous saying: 'If you want to change the world, start with yourself.'

A typical focus is the promotion of individual competences and mindfulness. As shown in the previous chapters, this is an essential building block of all three guiding principles. Currently, it can be observed that it is also at the centre of diverse initiatives of social change, mostly from the civil society. In most cases, training courses are offered (sometimes free of charge), increasingly also in the form of online courses. These include for example, the 'Center for Human Emergence and Spiral Dynamics Integral' or the 'Integral City Initiative'. The latter was founded by activist Marylin Hamilton. In her doctoral thesis of the same name, she describes her vision of an integral city that is geared towards the greatest possible sustainability and development of the potential of its inhabitants (Hamilton, 2008). A similar approach is taken by Nan Ellin in her book '*Integral Urbanism*' (Ellin, 2006), who aptly states that 'Urban design success should be measured by its capacity to support humanity', and 'an Integral Urbanism offers guideposts along that path towards a more sustainable human habitat' (Ellin, 2006, p. vii). Often, the offers aim to train multipliers, i.e. people who in turn are to influence further people and drive social change. One of the best-known and most representative examples may be the MOOC '*U.Lab: Transforming Business, Society and Self*' provided by the Presencing Institute. Since 2015, it has been offered annually and free of charge via the digital learning platform edX (Presencing Institute). Regardless of the diversity of offerings, it appears that all initiatives focus on the promotion of universal competencies. This involves the development of basic skills that can be applied to a wide range of problems and are incorporated into more specific areas of competence. Typical universal competencies include mindfulness (which can be fostered through regular meditation practice), emotional intelligence or Futures Literacy (as outlined elsewhere). Universal competencies run deep and usually go hand in hand with the development of one's own personality and change of consciousness.

The development of competence and awareness described above goes hand in hand with an already observable trend towards greater awareness of sustainability issues. In consumption, organic products, fair trade, the keeping of a CO_2 balance and a vegetarian or even vegan lifestyle are no longer exotic fringe phenomena. Currently, avoiding plastic and packaging waste is receiving increased attention. It seems obvious that increasing access to the internet and digital communication media has contributed to a protest movement and greater assertiveness of civil society towards the private sector. In response, more and more companies are likely to have incorporated what is known as 'Corporate Social Responsibility' (CSR) into their corporate policies. The concept describes the voluntary contribution of business to sustainable development that goes beyond the legal requirements. Critics might concede that these developments are still far from sufficient for sus-

9.1 Five Leverage Points for Social Change

tainable global policy and crisis prevention. This will require, among other things, a profound change in awareness among the populations of the First and Second Worlds about the global consequences of their consumption behaviour. Thus, it appears clear that the decisions of individual consumers have an indirect influence on many areas of a sustainable world.

A complementary approach focuses more on the material, self-sufficient aspect of future preparedness and is above all a central component in promoting community resilience. Here, too, the idea is to make individuals or households 'fit' for a wide range of unpredictable crises – if only because politicians would not be able to guarantee reliable supplies for the population. Currently, since 2016, the German 'Bundesamt für Bevölkerungsschutz und Katastrophenhilfe' (BBK, translated in English: Federal Office of Civil Protection and Disaster Assistance) is again increasingly issuing general recommendations for voluntary private stockpiling, which are intended for protection in a variety of scenarios. They are intended to provide each household with a guideline for the extent of self-protection measures which can be implemented relatively easily. This includes, but is not limited to:

- Food and water: Food and liquid for 2 weeks. They should be kept without refrigeration and be also enjoyed cold. Per week and person about 14 litres of liquid should be expected, preferably water.
- Hygiene: This includes soap, detergent, toothpaste; water for washing, rinsing and flushing toilets.
- First-aid kit: first-aid kit, prescribed permanent medication, charcoal tablets, painkillers, laxatives, fever thermometer, hot water bottle.
- Power failure: camping stove and suitable fuel, warm clothes, flashlights, batteries, candles, matches, cash.
- Document protection: family certificates, pension, retirement and income certificates, savings books, shares, vehicle title, insurance policies, payment receipts for insurance premiums, in particular pension insurance, other certificates, contracts, land register extracts, last will and testament.
- Emergency luggage: First aid material, battery-operated radio and spare batteries, document bag and valuables as well as identity cards, food for 2 days, torch, sleeping bag or blanket, weather protection clothing, other utensils such as eating utensils and camera.
- Radio: preferably battery operated or even better: crank radio (BBK)

These and similar recommendations, which are familiar from the prepper scene that is also slowly spreading in Europe, contain relatively easy-to-implement measures and help to increase the general disaster capacity of households.

In summary, there are numerous top-down and, above all, bottom-up initiatives that aim to foster future preparedness at the individual or household level. This includes, above all, the promotion of universal competencies, which goes hand in hand with the development of one's own personality and mindfulness, educational work for sustainable consumption, and preparedness.

9.1.2 Leverage Point 2: Using Critical Mass to Drive Change Bottom-Up

When the number of people changing a certain behaviour reaches a 'tipping point', a change can take hold in the entire community. This is described by the term 'critical mass'. The term, which originated in epidemic research, means in game theory that it is enough to convince a certain number of participants of a vision etc. to influence the entire group. Once a certain threshold of participants is exceeded or a critical mass is reached, that vision will become self-sustaining (Gladwell, 2001). In the social sciences, the concept was most notably coined by game theorist Thomas Schelling (1978) and sociologist Mark Granovetter (1978). The term became known at the turn of the millennium with Malcolm Gladwell's book *'The tipping point – How Little Things Can Make a Big Difference'* (Gladwell, 2001). In it, he describes three communicative factors that are typical of critical mass in social contexts.

1. *The Law of the Few:* Individual members have more influence to bring about change than others. This is especially true for leaders, people with more decision-making power than others.
2. *Stickiness:* The presentation of the vision is crucial for motivating the addressees to act. Even small changes can have a big impact. Gladwell cites the example of the children's programme 'Sesame Street', which flopped during initial pilot tests in the USA, but eventually became a success thanks to the addition of the character Bibo.
3. *The Power of Context (environmental conditions):* Human actions are strongly influenced by the respective environmental conditions. As an example, Gladwell cites the Broken Windows Theory, which dominates criminology and was successfully implemented in the 1990s by New York's then-mayor Rudolph Giuliani. This involved the New York Police Department concentrating on tackling what appeared to be petty crimes (this includes mainly vandalism, such as smashed windows), but which affected the quality of life of New York residents, in order to send a signal of 'zero tolerance'. Giuliani's policy led to a significant decrease in crime in New York (Gladwell, 2001).

9.1 Five Leverage Points for Social Change

How big does a critical mass have to be to trigger changes of social significance? This is a controversial question. In the current debate, the famous model of Everett M. Rogers is mainly used. According to this model, at least five factors have a decisive influence on whether people are willing to accept a change and whether an innovation will take hold:

1. Subjective benefit of an innovation (e.g. prestige gain, etc.).
2. Compatibility with existing value systems.
3. Complexity – or the simplicity perceived during the initial contact.
4. Trialability (possibility of experimenting with the innovation).
5. Visibility of innovation.

Not all groups of people act in the same way. Rogers distinguished five groups of people that differ in their innovativeness. The process by which a change takes hold and the innovation is adopted by more and more groups of people moves from the curious innovators (5–10%) and early adopters (10–15%) to an early majority (30%) and late majority (30%), and finally to the relatively sceptical laggards (remaining 20%). When the number of people who adopt and implement the innovation reaches a critical mass, the change catches on with the remaining groups of people (Rogers, 2003).

Today, there are a variety of initiatives, especially at the intersection of the academic and civil society subsectors, that offer awareness-raising and competency-developing services on digital platforms, sometimes even free of charge. The aim and purpose is to train multipliers, i.e. people who in turn initiate social change impulses from below with numerous projects. A representative and popular example is the above-mentioned MOOC *'U.Lab: Transforming Business, Society and Self'*, which has been initiated annually by the Presencing Institute since 2015 and is offered free of charge via the digital learning platform edX. This 8-week online training course is moderated personally and live by MIT professor Claus Otto Scharmer, the originator of the change management method 'Theory U'. Here, participants do not only learn how to apply the method, but also independently develop projects in small groups of five in order to initiate social change from below and increase society's capacity for complexity (Scharmer & Käufer, 2013). According to Scharmer/Käufer, a critical mass of five people per group is sufficient to achieve a significant impact on the overall system. In this context, they quote Nick Hanauer, entrepreneur and long-time member of Amazon's supervisory board:

One of my guiding principles comes from Margaret Mead: 'Never doubt that a small group of committed citizens can change the world. On the contrary, that's the only way change has ever happened.' I totally believe in this principle. With just five people, you can do almost anything. With just one person, it's difficult – but when you put that one person together with four or five other people, you have incredible power. Suddenly a momentum is created and almost anything that is intrinsic and possible can be achieved and realized (quoted in Scharmer & Käufer, 2008, p. 9).

The complexity expert Monia Ben Larbi developed this idea further and applied it to the German initiative 'Dörfer im Aufbruch' ('Villages on the Move'), which she co-founded.[1] Following Everett Roger's model, she describes very succinctly how changes in a community can be initiated from below and what is important in the process:

- I only need 5 people, so 4 besides me, who are willing to jump into the unknown – because the first group is made up of people who really enjoy being pioneers;
- We then need to find 27 people who are enthusiastic about the vision, because the second group are visionaries who have the ability to imagine the future;
- By the time we add even one more person, for a total of 33 people, we have already crossed the innovation chasm, which is considered the point at which change is unstoppable. Now we are already ready to work;
- The next group, i.e. the next 68, are pragmatists – they will be infected when there are first experiences, when the visionaries have already worked ahead and there are pragmatic, visible results;
- The next 68 will join later, when it already feels more normal;
- And the last 32 will always think everything is stupid, we don't have to work on them at all.

So, in my village, 33 people can make a lasting difference – a group size that even fits in one room. The power we have as individuals to reach the critical mass for change is something everyone can only envy. What else can we learn from the theory of innovation diffusion?

- Focus on the innovators and the visionaries. The others will come later. Trying to get them excited from the start is doomed to failure, because they can't warm up to ideas but only to the visible. Give them the time to get used to the change.
- Don't be swayed by the 16% who will never like what you are doing, but develop a respectful immune system towards them. They have the right to reject your project and you have the right to implement it anyway. (Ben Larbi, 2018)

---

[1] Similar to Transition Towns, the German government-funded initiative 'Villages on the Move' assumes that communities can take their development into their own hands and effectively use previously unused resources. It assumes that villages are predestined for innovation, as they contain the diversity of society within a manageable framework (Ben Larbi, 2018).

These and other bottom-up initiatives share the assumption that social change does not require reaching the entire population. Changes can be initiated by a relatively small core group and – if a critical mass of the population is reached – can spread and become established in the longer term. This is usually implemented through experimental pilot projects.

9.1.3 Leverage Point 3: Experimental Prototype Projects

Many civil society initiatives, especially those in the resilience and sustainability fields, are implemented as real life experiments or 'real life laboratories' in which new forms of coexistence are further developed through trial and error. Transition Towns initiatives, with more than 450 communities worldwide, are among the best known. However, the question of how to successfully establish and govern socially innovative communities in today's highly individualised, modern and postmodern societies remains open. According to several studies, currently on average about 10% of community projects would survive the first 5 years. The main reason for this, according to expert Iris Kunze, is a lack of the communities' capacities to manage social conflicts (Kunze, 2010, p. 86). Conflict resolution mechanisms, emotional intelligence, communication skills and all other related competencies become key factors for the success of social innovations.

An experimental approach is also recommended for the more top-down policy measures. This concerns, for example, approaches to systematically test the improvement welfare policy (such as the universal basic income) or political governance (e.g. panarchy, e-democracy). The fundamental insight here is that in the face of increasing social complexity, there will be no simple or perfect solutions, but only approximations that need to be adapted in a process. What sets the right incentives (and what not)? What works (and what not)? What can be learned from other experiences in the past and/or other cultures/communities/societies? It is these systematic questions that enable early adjustments and pragmatic institutional learning. Here are two examples:

As illustrated elsewhere, the development of the welfare system is one of the most essential building blocks for societal future preparedness. Since the high socio-economic complexity requires a learning approach, regionally limited pilot projects can already be observed today – e.g. on the advantages and disadvantages and the adjustment potential of the universal basic income. A fundamental controversy is the question of the appropriate relationship between state regulation and the dynamics of a market economy. In his book *'Das Megatrend-Prinzip'* (The Megatrend Principle), futurologist Matthias Horx (2014) cites the example of a successful measure implemented by the British government. This government had set up a lavishly endowed social fund that promotes private initiatives with market economy criteria and at the same time also controls and 'bench markets'. In this case, the aim was to reduce the recidivism rate in a prison for juvenile offenders. Private companies that felt they could do a better job of rehabilitating offenders were contracted to reduce recidivism rates. In return, there is an amount of money that the state could save on its own programs. If the private group succeeded in lowering the recidivism rate below a certain level over time, there was a bonus payment and new contracts at other facilities (Horx, 2014). Such approaches do not only blur the line between state and private sector, they are also tied to concrete quality and success criteria (rather than the cheapest price when bidding for contracts) and allow for incremental improvement.

Another example shows the experimental influence on the decisions of the citizen. The approach is called 'nudging'. The term translates from English as 'nudge' and was coined by economist Richard Thaler and legal scholars Cass Sunstein and their book *Nudge: Improving Decisions About Health, Wealth, and Happiness* (Thaler & Sunstein, 2008). In it, they advocate the idea of 'libertarian paternalism' and discuss 'nudge' as a method of influencing people's behaviour without resorting to prohibitions and commandments or changing economic incentives. Typical examples already implemented today are:

- If an image of a fly is placed in urinals, 80% less urine ends up on the floor because men aim for the fly when they urinate.
- If fruit is presented close to hand at a canteen buffet, while Danish pastries are presented further away, users will reach for fruit more often (Thaler & Sunstein, 2008).

Another example is the 'Speed Camera Lottery', which was tested in Freiburg, Germany, by the initiative 'The Fun Theory' in a real-time experiment and put up for discussion in the community. Here, in a traffic-calmed residential area, all motorists were informed by signs that they were taking part in a lottery that drew the speeding drivers's fine to the correct drivers. As a result, the average speed went down by 22% (Mörchen, 2015). There are over 80 project groups around the world working to find ways to use nudge theory to improve government policy and services. Among others, these so-called 'behavioural insights teams' can be found in England, the United States, Australia, and, since 2014, even in Germany. Among other things, the project groups are investigating ways to increase people's willingness to pay taxes, donate to charitable organisations, avoid prescription errors, and increase voter turnout. In its publication '*Behavioural Insights and Public Policy:*

Lessons from Around the World', the OECD describes over a hundred different applications in politics (OECD, 2017).

Due to their high social complexity, large-scale initiatives with a focus on society as a whole will require an experimental and incremental approach. Flagship projects in the urban sphere seem to be predestined for this, as mentioned above with the examples of smart cities and charter cities. They are relatively costly and require international cooperation with different agents – mostly from politics, the private sector and applied science. The optimistic representatives of these projects also expect a high impact, i.e. a high spillover effect of the realised urban project on society as a whole, if not on all societies involved in the project. Societal future preparedness will certainly also require an in-depth examination of lead projects of this magnitude. This is not only about the pragmatic realisation, but also about weighing the critical implications of these projects. These can include undesired new vulnerabilities through technological dependencies, the high dominance of entrepreneurial interests (both concern particularly smart cities), or neo-colonialist implications of charter cities.

In sum, social change will develop primarily through a gradual, pragmatic and experimental approach. Two things are striking here: Firstly, as we have seen above, these pilot projects and real-time experiments are not only initiated by politicians, but above all by other society sectors, such as the civil society, the private sector and even the science sector. Secondly, this leads to the fact that a mixture of bottom-up and top-down initiatives can already be observed today, from which, according to sustainability researcher Derk Loorbach, a multi-dimensional transitional order of 'governance panarchy' could emerge that drives the social change (Loorbach, 2007, 2014). Third, it is possible to observe not only a juxtaposition of initiatives, but above all collaborations between the society sectors involved – Transition Towns, for example, often involve cooperation between civil society and science, or a learning social policy, as described above, will often involve cooperation between policy-makers and the private sector. This requires communicative 'bridges' between the society sectors.

9.1.4 Leverage Point 4: Building Bridges Between Society Sectors

Societal future preparedness will depend to a large extent on how mankind is able to shape successful communication – not only between societies within the framework of global cooperation, but above all within societies (i.e. between the inherent society sectors). In social terms, this involves nothing less than a 'dialogical refoundation of science, economics and politics' (Scharmer, 1995). Knowledge transfer and networking of functional society sectors are a central criterion (as described in detail in Principle 3) for collectively anticipating unintended side-effects – especially in the development of dangerous technologies. Cross-sectoral collaboration, preferably global in scope, provides the infrastructure for collective intelligence (Principle 4) and collective wisdom (Principle 8). What are the leverage points for promoting such collaborations? The examples presented above show that initiatives from the civil society and political sectors can act as 'bridge builders' to the others. This is also possible from the other sub-sectors, science and the private sector.

In the science sector, it is currently above all the contribution of the director of the Wuppertal Institute, Uwe Schneidewind, who draws attention to the potential of the science sector as an agent of such communicative bridge-building. According to Schneidewind and Singer-Brodowski (2013), however, this presupposes that science not only adopts a transdisciplinary orientation and systematically networks with key persons from the other sectors, e.g. politics, civil society and the private sector, in order to investigate complex societal challenges and develop solutions. The approach also includes a new self-image of science, not only as a 'neutral analytical observer' of society, but also as a normatively oriented active co-designer. The beginning of this debate was already initiated by the concept of a so-called 'mode 2' science coined by Helga Nowotny and Michael Gibbons in the 1990s (Nowotny et al., 1994). Since the second half of the 2000s, Schneidewind and Singer-Brodowski (2013) have even spoken of a 'mode 3' science.

- *Mode 1 science* refers to conventional 'normal science' with a strong reference to systems knowledge, a homogeneous knowledge base (primarily from scientific institutions), little inclusion of societal perspectives, hierarchical organisational structures, and a disciplinary and currently interdisciplinary orientation (Gibbons et al., 1994; Nowotny et al., 2001).
- *Mode 2*, by contrast, stands for a 'context-sensitive science', i.e.: a science that is aware of the close feedback with society and the developments of a reflexive modernity (Nowotny et al., 2001). It therefore sees society as a central component of knowledge production and is based on a correspondingly heterogeneous knowledge base from different institutions and heterarchical organisational structures. The transdisciplinary orientation of knowledge makes it possible to grasp complex problems that go beyond the specialised perceptions of the respective society sectors (Gibbons et al., 1994; Nowotny et al., 2001).
- *Mode 3* underlines the need for a 'third-order change' – this involves not only reorientations in terms of content (mode 2), but also institutional change. In this context, science is aware of its educational and mediating function and actively

develops society in a learning process (Schneidewind & Singer-Brodowski, 2013). The orientation of knowledge is therefore not only transdisciplinary, but also 'transformative'. Science would thus have no less than the role of a 'catalyst for social change processes' (cf. on the definition also WBGU, 2011). The knowledge base would not only be 'heterogeneous' (as in mode 2), but 'heterodox' based on experimental knowledge. In this context, society would not only be a 'component of knowledge production' (as in mode 2), but an 'agent in knowledge production and institutional science organisation' based on cooperative organisational structures (Schneidewind & Singer-Brodowski, 2013).

According to Schneidewind and Singer-Brodowski (2013), there are four agents in particular that are contributing to the development of a so-called 'mode 3' science in Germany (and correspondingly maybe in many other Western modern societies). However, the process of cross-sectoral bridge-building is already in its infancy:

- *Organised civil society* includes, for example, environmental associations, trade unions, churches, which increasingly recognise the importance of science policy as a central policy field to promote their concerns. One example is the initiative 'Civil Society Platform for the Research Transition'.
- *Foundations* and alternative science funders that trigger impulses in the science system with new initiatives and programmes. One example is the Heinrich Böll Foundation's funding of the doctoral focus 'Transformation Research'.
- *Innovative state policies* as pacesetters and drivers for a new science policy. According to Schneidewind and Singer-Brodowski, Baden-Württemberg and North Rhine-Westphalia are currently considered pioneers in this area.
- Innovative strategies on the part of *pilot universities* and research institutes as well as new networks. One example is the Leuphana University of Lüneburg as a pilot university for sustainability (Schneidewind & Singer-Brodowski, 2013).

The potential of the private sector as a bridge builder is significantly underestimated in the current debate. Two heterogeneous approaches and corresponding discussions are to be highlighted as examples. These include, first, transformative entrepreneurship, which in turn includes the much-discussed social entrepreneurship and the relatively little-developed 'village entrepreneurship' and, second, different variations of multi-stakeholder collaboration, which include, for example, multi-stakeholder initiatives (MSI) and collaborative networks (CoIN). The first, *transformative entrepreneurship* generally involves society-changing business innovation. Examples include social entrepreneurship and, more recently, village entrepreneurship.

- *Social entrepreneurship* is an entrepreneurial activity that is innovatively and sustainably committed to solving social problems and, in the broadest sense, strives for positive change in society. Areas in which a social entrepreneur engages include education, environmental protection, job creation for people with disabilities, poverty alleviation or human rights (Scheuerle et al., 2013). In an anthology entitled *'Social Entrepreneurship – Social Business: Unternehmen für die Gesellschaft'* (Social Entrepreneurship – Social Business: Doing Business for Society) experts Helga Hackenberg and Stefan Empter describe it as a conceptually still undefined 'phenomenon in the field of tension between business, the state and civil society'. Social entrepreneurs fill gaps that neither the market nor the less dynamic state activities can cover (Hackenberg & Empter, 2011). Perhaps the most prominent example of social entrepreneurship is Ashoka. Ashoka is an American non-profit organization that promotes social entrepreneurship. In currently around 70 countries, fellows receive financial support, advice and connections to networks in the social sector as well as in business and academia so that they can disseminate their projects (Ashoka).
- The relatively little tapped concept of *'Village Entrepreneurship'* comprises business models based on the principle of systemic sustainability, which lead to a comprehensive development of regions – mostly villages. The above-mentioned initiative 'Dörfer im Aufbruch' (Villages on the move) documents successful cases in German-speaking communities and promotes the exchange of knowledge on successful practices. The most prominent example is the case of the village of Oberndorf in Lower Saxony. Under the guidance of the company BE Solutions & Blue Systems Design GmbH, a biogas plant was built in the 2010s on the basis of 100% liquid manure, the waste heat of which is used for a fish farm. The input material slurry is already separated on the cooperating farms and later processed into fertiliser in a further procedure. The systemic cascading value-added cycle described here results in various products that are marketed. At the same time, 100% of the energy (heat and electricity) is used and no waste is produced. The environment and the groundwater are relieved of nitrate and CO_2 is reduced. The process is shown in detail in the documentary film 'Dreaming of banana trees' (Hubert, 2016).

In addition to transformative entrepreneurship, *multi-stakeholder collaborations* are also important bridge-building private sector initiatives. These are decentralised networks initiated by companies on complex topics. The best-known of these are the multi-stakeholder initiatives (MSI), which are mainly influenced by the sustainability debate, and the collaborative innovation networks (CoIN).

9.1 Five Leverage Points for Social Change

- *Multi-stakeholder initiatives (MSI)* are voluntary associations between civil society, public and private agents with the aim of solving complex societal problems in a cooperative manner. Their focus is usually on promoting sustainable development. Usually, the aim is to better embed corporate social responsibility (Lin-Hi n.d.). The board of an MSI is the highest decision-making body and is usually made up of representatives of the various stakeholders. The work of an MSI is divided into four areas of responsibility: (1) Establishing a common basis for communication and a continuous dialogue between the individual stakeholders. The aim is to identify and jointly solve sustainability problems. (2) Formalising environmental and social sustainability in writing by drawing up production and behavioural standards in codes of conduct. (3) Implementation of agreed standards through targeted training and education measures, certification and accreditation of member companies. (4) Verification that established standards are met through various audits. This is to ensure continuous improvement of MSI as well as social standards. For instance, in the textile industry, there are the Fair Wear Foundation, the Fair Labor Association, the Ethical Trading Initiative, or the Social Accountability International, among others (Both et al., 2012).
- *Collaborative innovation networks (CoIN)* are also characterised by collaborative cooperation between organisations and external stakeholders. The focus here is on the development of innovations, for example in the form of new products, services and business solutions. The term was originally coined by Peter Gloor of MIT Sloan's Center for Collective Intelligence (Gloor, 2006). According to him, CoINs include the following five characteristics:

1. *Dispersed memberships:* The use of digital communication media enables memberships regardless of the physical distance of the stakeholders.
2. *Information transparency:* There is a high level of information transparency, so participation is based on mutual trust.
3. *Common goal:* The participation of all stakeholders is voluntary and intrinsically motivated with a focus on the common goal.
4. *No hierarchies:* CoINs are generally self-organised and decentralised and do not follow any hierarchy.[2]

---

[2] This point must be put into perspective in light of recent findings from research and practice. In their extensive study Heidenreich et al. (2016) show that the hierarchical structure, along with the network, the community and the market, is an established form of governance in CoINs. Their different influences and effects on the quality of knowledge transfer and collective action are discussed in more detail elsewhere (Heidenreich et al., 2016).

5. *Ethical code:* As a rule, they establish common guidelines of conduct to which they adhere in a binding manner (Gloor, 2006).

The industry diversity of CoIN project topics is large, but mostly it is only about technological issues. If a wind energy manufacturer wants to build a turbine in one place, for example, he/she has to coordinate with all stakeholders along the value chain – from various component manufacturers and suppliers to the local agents concerned (Heidenreich et al., 2016).

In summary, every society sector has the potential to promote cross-sector collaboration through appropriate initiatives – some of which are already being implemented today. According to current knowledge, up to four different types of governance and design of collaboration can be distinguished: Hierarchies, markets, networks, communities. They differ in their characteristics, first in the way they organise the exchange of knowledge between agents (on this, Heidenreich et al., 2016), and second, in the mechanisms they provide for regulating social complexity. The following is a brief overview:

Market: In the market, knowledge exchange between different agents is regulated within the framework of a 'buyer-seller' relationship. The buyer (e.g. a company) acquires certain knowledge from a knowledge producer. This is usually contractually agreed in advance. He can freely dispose of the purchased knowledge and also has extensive control over the result. The product can take various forms, for example as a patent, license or as the result of knowledge requirements that are advertised on an open innovation Internet platform. A representative example of such a platform is InnoCentive.com. According to an extensive study by sustainability researchers Heidenreich et al. (2016), the main disadvantage is that in innovation projects the knowledge required can be specified precisely in advance. This leads to contractually agreed regulations having to be renegotiated, concretised and interpreted again and again in the course of the project. The advantage, in turn, is that contractual arrangements between the agents create a certain degree of certainty of expectations (Buss & Ortiz, 2016). Beyond the context of knowledge sharing, the market exhibits an efficient mechanism that has proven effective in regulating social complexity. The mechanism stems from the supply-demand structure, which is self-regulating and allows for a wide range on the one hand, and a high quality of supply on the other, in order to satisfy the needs of the demanders. The example of panarchy shows that the market, at least in theory, allows an almost unlimited variety of offers (here: Forms of rule) to co-exist. In addition, the market has an efficient selection mechanism resulting from competition between suppliers for the favour of the demanders. Thus, it not only creates a diversity of supply, but also includes incentives to create a supply that is attractive to as many market participants as possible. At this point, the example of the learning welfare state shows that, in the long run, it is above all those offers that are most likely to meet the needs of the demand that will prevail.

Hierarchy: In the context of knowledge transfer, this governance type provides for a hierarchical structure in which one agent presides and transfers external knowledge from the other agent(s) into its own structures. As with the market type, agents have extensive control over the bodies of knowledge they wish to integrate. The disadvantage is that the way of learning processes cannot be regulated hierarchically. This can be seen, for example, in the fact that a large proportion of takeovers and mergers between organisations fail because they do not sufficiently take into account and integrate the acquired organisation's knowledge. Since the willingness to collaborate can only be enforced to a limited extent, the mechanisms of hierarchical coordination may prove to be a barrier to learning processes (Cartwright & Schoenberg, 2006). Beyond knowledge transfer, hierarchies may prove to be effective mechanisms for regulating social complexity. This applies in particular to subsidiary structures (see Principle 3), which, on the one hand, enable knowledge exchange and also chains of command or at least directive relationships between the subsystems, and, on the other hand, ensure decentralised decision-making by the agents at the lower levels of the hierarchy. A subsidiary hierarchy – similar to what the EU is already demonstrating at regional level – could be the basis for a complexity-adequate global governance, as will be described in more detail below.

Networks: As an instrument of knowledge exchange, the network functions in such a way that several agents pool their capacities. Hereby, cooperation is usually limited to a specific joint development project. Unlike the previous types of governance, the network aims to generate new knowledge through collaboration (Weyer, 2011). This means that the actors involved can influence the process of knowledge generation, but cannot use it exclusively – as is the case, for example, with a purchaser. The most common forms are development partnerships or strategic alliances (Heidenreich et al., 2012). The network is particularly suitable when the complexity of the project appears to be high, the partners involved have specific competencies, their respective performance outcomes can hardly be determined in advance, and the partners mutually trust each other. As an example, the Blue Brain project mentioned above can be understood as a successful network application. The goal of this international research project was the virtual simulatation of the human brain by building elaborate computer models. Earlier than planned, the first breakthroughs were achieved on the basis of effective international cooperation (Markram, 2006). Ultimately, the central challenge of the network is that the partners involved can be very different in terms of their communication routines and interests, which can make joint learning processes and centralised coordination difficult. As a mechanism for regulating social complexity, the network therefore proves to be a comparatively 'soft' form of coordination. This allows for flexible, process-open cooperation in complex projects, which can hardly be planned in advance anyway. This is especially true for software development, for instance in the cross-national development of ground-breaking, possibly also dangerous technologies. The disadvantage of the network is the great uncertainty about the mutual interests of the agents involved (Heidenreich et al., 2016).

Community: Similar to the network, the community provides for an exchange of knowledge in which the participating agents pool their capacities and their knowledge. This leads to the creation of new knowledge. Here, too, knowledge production is completely open to processes and none of the agents involved can claim this new knowledge exclusively for themselves. The products of the community are therefore to be understood as public goods. Thus, a company that is a member of an innovation community cannot expect to transform the 'free' knowledge assets into a complex product that can then be marketed. Similar to the network type, collaboration with communities is suitable when specific knowledge is required. Usually, this is specific knowledge about a particular industry and in a particular region (Heidenreich et al., 2016; Isaksen and Asheim, 2002). If, for example, a company wants to build an energy generation plant or launch frugal products in a certain region, it will also need to partner with local actors, at least to gain knowledge about local specifics. Similar to the network, the community is a relatively soft form of cooperation, which means that the participants have limited influence on the direction of development processes within the community. Unlike the network, but identical to the market type, the product is relatively clearly defined in advance. Another special feature is that communities are more binding than networks in that members are expected to comply with the rules and manners of the community and must also make their own contribution in order to have permanent access to this knowledge. The best-known form of innovation communities would be open-source communities (Hanekop & Wittke, 2009; O'Mahony, 2006).

All four types of governance described above complement each other and are important to consider for effective collective complexity management. In this context, the term global governance refers to the totality of coordination processes of different agents in order to be able to cope with the current, complex global challenges (Messner, 2000). It can be used largely synonymously with the term 'Weltinnenpolitik' (world domestic policy), which is widely used in German-speaking countries (Bartosch & Gansczyk, 2009), and is understood as a political programme for cooperative, multilateral management of globalisation (Hauff, 1987). In contrast to the concept of 'global government', the 'governance' concept does not presuppose a centralised hierarchy (CGG, 1995). Rather, it emphasises collective regulation of societal activities, which can be organised in a market-like manner (e.g. within the framework of a free trade zone), in a network-like manner (e.g. within the framework of a cooperation to develop a universal technology), or in a community-like manner (e.g. within the framework of a regional cooperation project such as a charter smart city). All this does not exclude subsidiary hierarchy. As mentioned, the EU today represents a typical regional form of governance in which supranational EU law and national law are integrated. This means that the nation states retain the greatest possible sovereignty and are at the same time integrated into a supranational structure with far-reaching rights and obligations. The EU itself can be seen as a process-open experiment in the economic and political integration of states and is the most advanced in the world this respect (Grande, 2000). Despite all the criticism of the EU, the fact that it has been possible to achieve a relatively stable peace order between the member states since World War II speaks in favour of this project. The spectrum of positions in today's global governance debate varies from the voluntary cooperation between states to the position of world federalism (World Federalist Movement) – which envisages that nation states must severely restrict their competences in favour of global rule-of-law institutions – to the position of world government – here the aim is to largely abolish the nation state and regard all citizens as global citizens (World Citizenship Movement).

9.1.5 Leverage Point 5: Methods of Communicative Complexity Management

Collective intelligence and collective wisdom may be among the most fundamental collective properties. This in turn is only made possible by successful communication processes between all members from different disciplines and society sectors. Typical examples of such collective processes are agile development teams or quality circles, policy think tanks, MSIs or CoINs, panels with policy makers or interdisciplinary conferences or research initiatives. The goals of these gatherings may vary in detail; e.g. they may be about creating a complex product, developing new insights or solutions to a complex problem, or complex political decision-making. Regardless of these different goals, the same criteria for success are always at work: they are jointly supported processes in which people with different backgrounds of experience communicate and try to bring together their different perspectives in order to produce results on this basis. This proves to be a domain that is primarily supported by direct personal exchange.

Direct communication, as Heidenreich et al. (2016) confirm in their case studies, can be demonstrated in all governance types (i.e. communities, networks, markets and hierarchical structures). They explain this in many ways: direct communication makes it possible to agree on common procedures and standards in forms of cooperation that are already very difficult to control, such as communities or networks. Moreover, it is the most important means of resolving misunderstandings and solving problems that affect the communication process as such. This proves to be particularly central in collaborations between experts with different professional and academic backgrounds – where misunderstandings are more frequent and this therefore requires increased and improved direct communication. Another advantage of direct communication is that tacit knowledge – i.e. implicit knowledge in the form of personal experience that is not so easily conveyed in writing – can be better passed on (Heidenreich et al., 2016). Designing direct communication in such a way that it enables effective and time-efficient knowledge sharing and, if necessary, leads to joint decisions, is not so easy. The following is an example scenario:

> **Example Case About Typical Problems of Interdisciplinary Conferences**
> At an interdisciplinary conference on the topic of 'climate change', renowned experts from different disciplines and society sectors are invited – sustainability researchers, agricultural scientists, sociologists, engineers, social entrepreneurs, activists, for-profit entrepreneurs, politicians, etc. – to present their different views on the problem. They all present their different views on the problem in talks. At a subsequent panel discussion, they have the opportunity to interact directly with each other. At the end of the event, the moderator concludes that the event was a 'complete success', thanks for the committed contributions and sums up that climate change is one of the most complex global challenges of our time, that there are different perspectives on this topic and that it can only be successfully tackled through interdisciplinary cooperation and global collaboration. An integration of these perspectives or even concrete measures and decisions that could contribute to global cross-sectoral cooperation are completely missing.

This example shows the typical complications that hinder a successful exchange between knowledge and decision makers and lead to suboptimal results. How much better would the outcome of the conference described above have been if the participants had used more effective forms of knowledge exchange and knowledge integration? Since direct communication is inherent to almost all collective decision-making and problem-solving processes, it proves to be a key leverage point for enhancing societal future preparedness.

9.1 Five Leverage Points for Social Change

Based on this basic assumption, I outline several approaches for optimising communication in such processes in another publication (Fathi, 2019a). This is an integrative method set that, on the one hand, allows for situation-adequate interventions and, on the other hand, is generalisable to the extent that it can be applied at all system levels, i.e. organisational and societal. Altogether, the method set comprises six interacting dimensions, which can be summarised as follows (in the following and for more detail, see Fathi, 2019a):

System level: Within the framework of this dimension, the scale and specific features of the respective social systems must be taken into account. The societal level for example includes extra-organisational frameworks in the form of different governance types, legal frameworks, etc. A common mistake, therefore, is to neglect that with each scale of a system there are further level-specific frameworks to consider. It is unlikely to assume that particular workshop or consultation methods, such as Scharmer's above-mentioned Theory U, will be sufficient to bring about social change. At best, they will only be able to make a significant contribution within a more complex context. Again, this contribution should not be underestimated. This is because the system levels are nested, which means that societal problem-solving is always linked to corresponding processes of its inherent units, e.g. organisations and teams. Methods to optimise knowledge sharing at the team level (e.g. in the form of a specific workshop format) are therefore likely to be relevant at all system levels. Thus, a strategy-making workshop with political decision-makers in the societal context, or with organisational decision-makers in the organisational context, can provide important impulses and direction for the collective action of the respective social system.

Intervention type: The intervention type must distinguish between methods, frameworks and guiding concepts. Methods are always implemented at team level. There is a wide variety of intervention approaches, ranging from one-day goal-setting workshops to two-year change interventions. Frameworks are more comprehensive and usually contain a whole set of methods. Design thinking, for example, includes shadowing, stakeholder interviewing, storytelling and prototyping. Scrum includes its own workshop formats, such as Scrum Review or Retrospective. The frameworks differ in their respective problem focus and system level. At the team level, for instance, the realisation of a complex project or product might require a communication design that is oriented towards Design Thinking or Scrum. At the organisational level, preventing unexpected crises might require the implementation of leadership and communication principles that can be realised through the framework of an HRO. Frameworks are typically found at the team and organisational levels. Guiding concepts, such as multi-resilience or sustainability, contain broader worldviews; they may include multiple frameworks and discourses, as detailed above.

Intervention context: The pre-selection of suitable methods, frameworks and guiding concepts is based on the concrete requirements of the respective situation, the available resources and the goals to be achieved. In the area of methods, Nicolai Andler's standard work is a representative example of the classification of over 145 classic methods from the areas of project management, consulting and workshop design. He distinguishes the following categories for classifying different intervention purposes and corresponding methods, such as: definition of the situation, information gathering, information consolidation, creativity, goal formulation, strategy analysis, system analysis, organisational analysis, decision making, and project management (Andler, 2013). In conjunction with the type category of frameworks, a more general and definitely smaller number of classification categories can be formulated. These take into account the fact that different types of complex problems can be distinguished, which in turn require different approaches to communicative complexity management. The agile framework Scrum, for example, aims at the incremental development of products that arise in an unpredictable environment, while the HRO framework, for example, aims at resilient crisis management in chaotic situations.

Framework conditions: Finally, when designing situationally appropriate intervention, the specific framework conditions of the intervention must be taken into account. They affect all system levels differently. The widely used classification scheme PESTLE (Political, Economic, Social, Technological, Legal, Environmental) provides an impression of the variety of framework conditions under which communicative complexity management takes place and which must be taken into account in this very process. In the political context, for example, it is taken into account how directional decision-making is institutionalised in the existing ruling system (e.g. in societies) or management system (e.g. in organisations) at the formal level. This should prove helpful in understanding risk and resilience potentials in organisational and societal systems. Complementarily, the social context, for example, determines the informal relational reality of the members of the social system. This includes, among other things, existing conflicts and further group dynamics. These aspects provide information about the potential for risk and resilience in social systems of all sizes and, moreover, about the extent to which the collective intelligence inherent in the system is used. Furthermore, this dimension contributes to the identification of role and relationship constellations in the social system and corresponding leverage points for influencing them.

Intervention phases: Despite an immeasurable variety of approaches to workshop design and consultation, three typical process phases can essentially be distinguished that underlie any form of communicative complexity management. This distinction allows a rough classification and combination of different methods for an optimal design of communicative processes:

- Orientation: Depending on the concrete situation requirements, this process step typically includes the following communicative activities and inherent purposes: topic collection, definition of the situation, information gathering and information consolidation. In addition, this process phase may also include measures of goal formulation.
- Processing: This process step involves a deeper analysis and processing of the selected topics. Analogous to the U-process, this process step is located on the lower side of the U and can also include measures of letting go and presencing.
- Solution/implementation: This process step is concerned with the development, and if necessary, the selection and decision as well as the testing and implementation of solution options and prototypes.

Even when dealing with chaotic phenomena that require rapid action and leave no room for analysis, these three steps are likely to be found in slightly modified form. Thus, in the practice of military special units, when dealing with immediate crisis situations (e.g. in the form of an immediate response to an armed attack), the following three-step process has proven its worth[3]:

- Orientation: What's going on here?
- Reaction: What do I do?
- Refinement: How do I do it?

Intervention dimensions: In addition, three dimensions can be distinguished, similar to the quadrant model of the *complexify* tradition, which underlie the vast majority of different forms of communicative complexity management: decision-making, understanding, generative communication.

- *Decision-making:* Decision-making is one of the most central operations in any collective problem-solving. It is thus inherent to every collective action of groups, organisations, and societies – whether, for example, politicians and their advisory councils deliberate on draft legislation or specific policies, or think tanks develop policy recommendations, executives weigh their corporate strategy, or performance teams assess project success. In all these and other settings, decisions are made jointly, which often prove inefficient in everyday practice. Meetings often result in endless discussions or in 'lonely' decisions that

---

[3] Personal exchange with a former member of tactical military special forces unit of the German 'Kommando Spezialkräfte' (KSK) in September 2018.

are not supported by the community. These decisions more often have an impact on society as a whole, suggesting that improving decision-making processes is likely to be important for a society's ability to respond and innovate. From this perspective, the systematic use of collaborative decision-making methods could make an important contribution. Such methods are, for example, systemic consensing or the K-i-E concept already described above in the context of (multi-) resilience. The systematic application of such methods contributes to more time-efficient decision-making processes, which are also of high quality because they are able to include the perspectives of all stakeholders in the outcome in a solution-oriented manner.

- *Understanding:* Whenever people with different academic and practical backgrounds communicate with each other, the fundamental challenge of understanding arises. For often these different backgrounds involve different references to reality, conceptualisations and definitions of terms. This becomes clear in the example of a meeting between sociologists and engineers discussing the topic of 'cities and mobility'. Engineers will mainly focus on the material, infrastructural aspects, whereas sociologists have a specialised view on norms, values and socio-techniques. Another challenge is the fact that language is often ambiguous. To date, these challenges have proven so serious that participants of multi-day interdisciplinary conferences on a complex topic do not even manage to agree on common definitions. How can communication between experts from different backgrounds carriers be optimised? How can communicative competences between disciplines and society sectors be developed? Here, too, the targeted application of appropriate methods that can also be combined with each other could contribute to an improvement. This could include, for example, the use of visualisation techniques, symbols/metaphors or/and storytelling. These methods have proven and established themselves especially in innovation promotion, marketing and increasingly also in corporate communication to communicate complex issues in an understandable way (Fathi, 2019a). They could and should also be applied on a broader scale, for instance in political committees, think tanks, at interdisciplinary conferences or in multi-stakeholder dialogues. A complementary, relatively unknown approach would be to use a structural language, such as FORMWELT. FORMWELT is, according to its two developers, Gitta and Ralf Peyn, a

  semantically and formally self-sufficient and Turing-complete linguistic system, a programming language for language(s) and meaning that can be spoken by humans and machines alike (Peyn, 2018, p. 1).

The application centre of FORMWELT is the so-called kernel, which consists of 320 references that can be used to describe and construct any thinkable and perceivable phenomenon. The semantic self-sufficiency should enable the user to acquire basic knowledge without additional material and to formulate her/his own knowledge. The references here correspond less to classical definitions than to concepts of meaning that motivate the user to make the experience on his or her own. They are not definitions, but instructions for action, similar to a programming command or scientific injunction: 'Do this and that, and so have the experience implied by the reference.' A reference is structured like an equivalence: on the left-hand side is the label to be referenced, on the right-hand side the labels with which the reference is to be worked out. Each of the labels used for this purpose are itself part of the kernel and referenced in it accordingly. Example[4]: *Freedom* ⇔ *Form the next possibility*. To the right of the equivalence sign '⇔' are the labels 'form', 'the', 'next', 'possibility'. All these labels in turn have their own reference in the FORMWELT kernel. According to Gitta and Ralf Peyn, FORMWELT can be learned in a few days and can significantly improve the mutual exchange of experiences (for more information and critical reflections on FORMWELT, see Fathi, 2019a).

- *Generative communication:* Direct communication also proves to be a source of innovations and of transformation processes necessary for tackling complex problems. Behind this is the assumption that complex problems cannot be dealt with sustainably with solutions of the same level of complexity. They require 'new', i.e. 'innovative' responses. Innovation results from a recombination of existing knowledge (Schumpeter et al., 2006), in dialogue with different knowledge agents. From another perspective, almost any problem-solving method based on dialogue can be seen as a process of changing systems. From a methodological point of view, every change process goes through the steps of a survey of the current state, a departure from previous problem-solving approaches and an achievement of the target state and the development of new, innovative perspectives. Initial experience shows, especially in conjunction with Theory U by C. O. Scharmer, that three factors essentially determine whether the group succeeds in producing, i.e. 'generating', something innovative: The group's own attention field, the depth of the relationships between the knowledge agents, and the quality of the dialogue. At the heart of the 'attention field' factor

---

[4] This kernel was available for me to view during a multi-day interview in August 2018 (Peyn, 2018) and is being added to a nascent online platform at the time of writing this publication. Therefore, it was not yet allowed to be published as part of this work.

is what is known as 'presencing', i.e. a state of consciousness in the here-and-now that enables the user to optimally engage with creative impulses for generating a desirable future. This usually involves the use of methods of introspection or meditation. The factor 'depth of relationships' between the knowledge agents indicates that the integration of perspectives and thus the outcome of the exchange is often better when the knowledge agents have a deeper, certainly also emotional relationship to each other (...). In addition to this, the factor 'quality of the dialogue' distinguishes four quality levels of the exchange. These depend on the extent to which the participants in the discussion can free themselves from their own ideas and prejudices and engage in the presencing and respond to each other – and not only to the content of the viewpoints of the other participants, but also to their underlying needs and feelings. In terms of stages, a distinction can be made between what is known as 'downloading', 'debate', 'empathic dialogue' and 'generative dialogue'. The latter stage has a particularly high potential for innovation and implies at the same time the highest form of connectedness that can be realised in a dialogue. The targeted use of methods, such as his Theory U, can significantly contribute to a generative, i.e. co-creative, orientation of communication processes.

In addition to the previous considerations, here is a general application example:

> **Case Example of a Method-Mix of Communicative Complexity Management**
> A cross-disciplinary research community addresses the question of the future of the city in 30 years. The exchange takes place according to a classical moderation procedure, i.e. via the steps (1) collection of topics, (2) working on topics in small groups, (3) bringing together, (4) decision on implementation. In parallel, the three dimensions are taken into account to optimise knowledge sharing. Within the framework of generative communication, attention is paid to systematically implementing practices of pausing before and during each round of dialogue. The quality and depth of the exchange is evaluated according to the four levels of conversation. In the context of the comprehension dimension, the researchers systematically use methods to increase their own language and comprehension skills. Thus, they use visualisations in combination with storytelling to present their different positions. This proves to be particularly helpful for the processing and understanding of the respective individual scientific findings on the common topic. In the context of decision-making, the K-i-E method is used to save time by making transparent the extent to which the knowledge agents agree on a particular issue (Fig. 9.1).
>
> *(continued)*

9.1 Five Leverage Points for Social Change

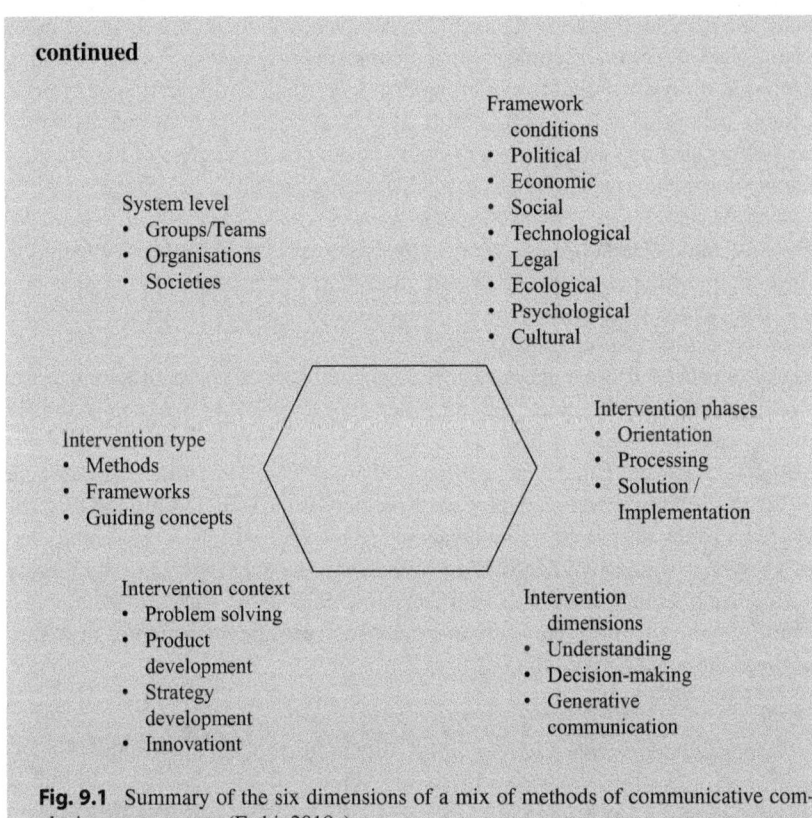

Fig. 9.1 Summary of the six dimensions of a mix of methods of communicative complexity management. (Fathi, 2019a)

It can be argued that the systematic use of methods, frameworks etc. to improve communication processes requires a corresponding mastery, which in turn requires a corresponding training period. At this point it also becomes clear to what extent the leverage points described above are interlinked. Building up the necessary competences would, for example, already be part of Leverage point 1 'Work on oneself'.

However, a high level of communicative competence of all participants is not necessarily a prerequisite. More important is the role of specially trained facilitators who could act as 'complexity workers'. The complexity worker could perform three roles: the role of a knowledge broker and a facilitator.

In the role of *facilitator*, the complexity worker could moderate any dialogue process between the knowledge agents. In this context, even an application of complicated methods, such as K-i-E and Theory U, would be conceivable with participants who are not trained in them, and in completely different settings. Examples would include cross-disciplinary research teams, cross-departmental working groups in the form of quality circles, agile project teams or design thinking teams, change management consulting teams, cross-society sector conferences, workshops or think tanks, and multi-stakeholder dialogues. Regardless of the different application contexts and methods (e.g. Open Space, World Café, Team Syntegrity, scenario planning) or frameworks (e.g. Design Thinking, Scrum, Effectuation, High Reliability Organization), every knowledge exchange potentially contains the three aforementioned starting points of understanding, generative communication and decision-making, which can be influenced by a specially trained facilitator with appropriate knowledge of the methods.

As *knowledge brokers,* the complexity workers 'from below' could support networking with suitable actors from other society sectors. This approach is already widespread in lobbying or political engineering – it could be systematically harnessed for the overarching context of societal problem-solving. Often, this can take on the character of a mediation process for society as a whole. The refugee crisis of 2015, for example, and the social divide which is inherent to the Corona crisis and to a certain extent also to the Russia-Ukraine war, show that agents from different society sectors can be in conflict with each other. Similar to a shuttle mediator, they would support the 'translation' of knowledge content, the mediation of resources between the sometimes conflicting agents.

> **Complexity Work Using the Example of the Refugee Crisis (An Example from Personal Practice [Fathi, 2016a, b, p. 4 f.])**
> At the end of 2015, my co-trainer and I were commissioned by a refugee home operator in a district in eastern Germany to conduct training for the functional and security staff in the refugee facilities as part of our work for the PROTECTIVES[5] competence centre that we founded. The main focus here was on intercultural skills as well as skills in de-escalation and self-protection. However, in view of structural deficiencies, such as lack of space, lack of activities for the refugees, traumatisation, etc., it was foreseeable that this violence prevention measure would only be a drop in the ocean. Nonetheless, my colleague and I used the workshops as an opportunity to think intensively with the staff about the structural challenges and possible resources and potential solutions. Indeed, useful information could be uncovered and recorded.

(continued)

---

[5] PROTECTIVES is a centre of excellence for training and counselling professionals affected by violence in the workplace.

> **continued**
> Obviously, there were untapped resources around the businesses and initiatives concerned with the refugee crisis, especially in civil society. They just needed to be brought together. These included, for example, unused accommodation facilities in the community, activity offers from associations, personnel from refugee initiatives and interest groups (especially in relation to the most often discriminated groups of Sinti and Roma) for translation services, intercultural sensitisation and dispute resolution.
> Unfortunately, the use and pooling of these resources was made difficult by one circumstance. All the actors involved who could have provided the resources were in bitter conflict with the staff in the refugee facilities. The harshest conflict was between the refugee initiatives and the security staff. The former considered the security staff to be racist and incompetent thugs. The latter saw themselves as a buffer for the media in what was in itself an extreme job and perceived the refugee initiatives as 'media-hungry dudes' who were not really interested in the welfare of the refugees but in their own self-dramatisation.
> Our proposal was solution-oriented mediation to optimise communication and knowledge pooling between these disputants. Because from the individual conversations I had already had with all parties, it was obvious that they were all pulling in the same direction and had the refugees' well-being at heart. In the end, all parties involved agreed to this measure, but unfortunately this advance was not approved by the public administration of the district.

9.2 Conclusion and Summarising Theses

9.2.1 Summary

The global situation is characterised by ambivalent developments. Contrary to many pessimistic reports, prosperity and awareness of sustainability issues and man-made ecological and technological risks seem to be increasing more and more. Advancing global development is accompanied by ever greater interdependence and complexity of the overall constellation. This harbours unprecedented risk potentials. Because with ever greater development and complexity also come 'more highly developed' and more complex dangers. Moreover, the structure of needs in today's societies and in the world is more complex than ever before, which does not reduce the general risk of social and psychological crises. With an ever-increasing complexity of the world situation comes an increasing variety of crises – be it climate change, cyberterrorism, energy crises, political conflicts – which are also interconnected as so-called crisis bundles and multidimensional as so-called bundle crises. Especially in the course of rapid and uncontrolled technological development, the danger of man-made existential risks is increasing. In the twenty-first century, societies are thus exposed to increasing complexity, unpredictability and diversity of interconnected risk potentials. Despite, or precisely because of the advancing global development, building up and safeguarding societal future preparedness is more relevant than ever in this century.

The concept of a multi-resilient society has become established as a universal response to this. According to the present study, this is to be regarded as ambivalent. The advantage of the resilience concept is that it can be contextualised in many ways, i.e. it can be applied to different types of crises. Today, it would be hard to find a concept that is similarly flexible and at the same time remains relevant to practice. However, this also carries the risk of diluting this concept, as is already the case in a similar form for the sustainability concept. Resilience research, which is still relatively new and continuing, needs a theoretical foundation, especially with a transdisciplinary orientation. What characterises a multi-resilient society, i.e. one that is equipped to deal with different types of interrelated global problems? At least five systemic principles can be summarised in the context of this book:

- Principle 1: The multi-resilient society is fostered by resilient citizens.
- Principle 2: The multi-resilient society is comfortable with unknowns.
- Principle 3: The multi-resilient society is based on decoupling and knowledge integration of its subsystems.
- Principle 4: The multi-resilient society makes collectively intelligent decisions.
- Principle 5: The multi-resilient society is based on a strong learning culture.

Critically, it should be noted that even the model of the multi-resilient society will not be sufficient to meet the diverse social demands of the twenty-first century. This is especially true with regard to man-made existential crises, such as a superintelligence out of control or a dangerous nanotechnology. These and other scenarios illustrate the great importance of collective wisdom – a collective competence that derives above all from an expanded concept of sustainability. Societal future preparedness therefore requires consistent consideration not only of the concept of multi-resilience, but also of societal development and sustainability. Societal future preparedness should thus include at least the following generalisable orientation principles:

9.2 Conclusion and Summarising Theses

- Principle 1: Development of universal personal key competences.
- Principle 2: Complexity-adequate problem-solving that considers knowns and unknowns.
- Principle 3: Decoupling and knowledge integration of the subsystems.
- Principle 4: Collective intelligence.
- Principle 5: Learning culture.
- Principle 6: Securing (basic) needs and preventing social conflicts.
- Principle 7: Development and preservation.
- Principle 8: Collective wisdom.

In juxtaposition, all three guiding concepts contain several points of intersection (e.g. they presuppose individual competence development and collective intelligence development) and mutual points of complementarity. However, integration also requires that mutual contradictions be overcome. The most serious contradiction is the one between the demand for economic and technological development on the one hand and the demand for control of development and the (economically inefficient) build-up of redundancies on the other.

Moreover, societal future preparedness will ultimately depend above all on the extent to which it succeeds in promoting effective cross-sectoral and global cooperation. This is also one of several possible leverage points to initiate necessary social changes. In essence, the aim will be to promote collaboration between social subsystems, including civil society, the private sector, politics and science, both within and beyond society, and to incrementally test and modify transformation processes on a small scale, in a protected environment. To this end, there are already numerous initiatives and a growing methodological know-how for designing successful communication. In summary, at least five interacting leverage points can be listed:

1. Work on yourself.
2. Using critical mass to drive change from below.
3. Experimental prototype projects.
4. Communicative networking of the subsystems.
5. Methods of communicative complexity management.

9.2.2 Five Milestones for the Further Political Practice

The continuing debate on building societal future preparedness will have to address a multitude of as yet unresolved questions: To what extent can social change be controlled and shaped? What is the desired direction of development and – bearing in mind unforeseeable risks – the desired speed of development? What does multi-resilience, especially evolutionary multi-resilience, mean in a broader framework of space and time, if we assume, for example, in the spatial context, the multi-resilience of the 'global society' (as distinct from the multi-resilience of national societies) and in the temporal context, the multi-resilience of the human species (given the ongoing technological developments that foster the convergence of humans andmachines)?

These are just a few questions that remain open for further research. For political practice, at least five suggestions for promoting national and global future preparedness can be derived from the reflections in this book, which could provide a rough orientation for political practice. These suggestions, which can be seen as important milestones, are not necessarily only addressed to political decision-makers, but also to agents from the other society sectors, especially the science and the civil-society sector. The first three suggestions refer mostly to the national, the last two to the global level:

1. Fostering integrative political practice by adequately considering multi-resilience, sustainability and development as orienting dimensions for the political agenda.
2. Fostering collective intelligence and collective wisdom through cross-sectoral and cross-disciplinary networking and knowledge distribution.
3. Strengthening problem-anticipation and decision-making by improving the quality of communication.
4. Fostering international and cross-sectoral cooperation.
5. Fostering real multilateralism and global governance.

Achieving all milestones requires a lot of time and is likely to go far beyond the framework of a legislative period set in democratic open societies. Political decision-makers are particularly challenged here not only to follow short-term party-strategic calculations that focus primarily on maintaining their own power. There is much more at stake, namely ensuring the survival of future generations. They are therefore called upon to make eventually unpopular decisions that involve investing a lot of time and financial resources, especially in preventive (multi-)re-

9.2 Conclusion and Summarising Theses

silience and sustainability. A general recommendation for addressing the milestones drafted below would be a two-track approach. On track 1, this means setting up competent working groups and advisory staffs to prepare political decisions to deal with the day-to-day political business and all tasks where standardised and best practices need to be applied. On track 2 the aim is to prepare and implement political decisions to realise long-term goals and to develop innovations and new solutions. Highly competent advisory staff, working groups and brain trusts that are interdisciplinary composed, skilled in systemic thinking and equipped with sufficient resources should be assigned just for this purpose. As such, the aim is to induce change via the above-mentioned leverage points (Sect. 9.1). This can mean, for example, testing in lighthouse projects (see Sect. 9.1.3.) little-tapped but promising concepts, such as governance panarchy, the Doughnut Economy or universal basic income, concrete measures to resolve the global prisoner's dilemma, international peace and a multilateral world domestic policy. What do the milestones specifically entail?

9.2.2.1 Milestone 1: Fostering Integrative Political Practice by Adequately Considering Multi-Resilience, Sustainability and Development as Orienting Dimensions for the Political Agenda

It has been outlined that the three societal concepts – development, sustainability, (multi-)resilience – are crucial for future preparedness in the twenty-first century. Current political agendas of many countries appear to focus mostly on economic development and – in the wake of the Corona crisis and the Russia-Ukraine war – increasingly on some narrow aspects of resilience, particularly in terms of robust and decentralised supply chains and military defence capacities. This is not enough.

Within the context of societal development, it is important to consider that economic development is not an end in itself, but rather a means to achieve maximum needs satisfaction of the population. The actual goals of societal development are 'beyond GDP' such as the HDI (including health) and, in the broadest sense, social and psychological well-being (including happiness). They should receive more attention in the current political agendas.

The context of preventive societal resilience is not sufficiently addressed in the political agendas of all societies in the world. In the vast majority of cases, resilience only comes into focus once the crisis has already occurred. Future preparedness requires not only that societies prepare for all conceivable crisis developments, but also that they are able to response appropriately to multiple crises that occur simultaneously and in complex bundles.

Even more than resilience, the context of sustainability is completely neglected in current national and international policy. More efforts could be made to avert the impending global overshoot. More could also be invested in creating fair trade relations with the global South to prevent terrorism, wars and refugee flows.[6] The wealthy countries of the First World are not meeting their global obligations. Germany already used up its share of planetary resources for 2022 on 4 May (Earth Overshoot Day, 2022). In 2019, Germany does not yet pay the required 0.7 per cent of its GDP in development aid. To date, no country is achieving its contributions to the Sustainable Development Goals (SDG) (Bertelsmann Stiftung, 2019). Since 2019, country performance has been declining due to the economic impact of the COVID-19 pandemic:

> For the first time since the adoption of the SDGs in 2015, the global average SDG Index score for 2020 has decreased from the previous year: a decline driven to a large extent by increased poverty rates and unemployment following the outbreak of the COVID-19 pandemic (Sachs et al., 2021, p. vii).

The sustainability concept is indispensable to avoid undesirable side-effects of the policies derived from the development concept. This applies not only to man-made ecological risks, but also to technological risks such as, for example, a grey goo scenario or a superintelligent AI out of control.

To conclude, future prepared policy needs to fully consider all three concepts. The dimensions and indicators of the three concepts outlined in this book can provide an orientation for the political agenda of any country to enhance its future preparedness.

9.2.2.2 Milestone 2: Fostering Collective Intelligence and Collective Wisdom through Cross-Sectoral and Cross-Disciplinary Networking and Knowledge Distribution

Since the COVID-19 pandemic, it can be positively noted that current crisis policies of many countries seem to be better informed in recent decades, for two reasons. First, the highly inter-connected digital infrastructure enabled effective and fast information transfer in near real-time. Second, a very close cooperation between the political and the science sector could be observed. This enabled governments worldwide to react much more quickly and informedly compared to past pandemics. Similarly, the Russia-Ukraine war can be followed almost in real-time and various data about troop movements and military capabilities and strategic analyses by experts from around the world are freely accessible in the

9.2 Conclusion and Summarising Theses 301

internet. Even more than in the Corona crisis, it has become apparent that so-called Open Source Intelligence (Osint) is of increasing importance. An instructive example: in April 2022, a 22 year-old draftman from Biel, Switzerland, used his laptop to analyse data from Google Maps and OpenStreetMap from his room and was able to track down a Russian military camp hidden about 50 km from the Ukrainian border even before Western intelligence services could do so (Roth & Rüesch, 2022).

Although crisis politics is better informed than ever before (if one disregards the current problem of fake news and deep fake), it is noteworthy that to date mainly experts from very few scientific disciplines are consulted by political decision-makers: During the corona pandemic it was immunologists or virologists, and during the Russia-Ukraine war it is increasingly military and geostrategists. This is not enough in the face of crisis bundles and bundled crises. Experts from a wide range of disciplines – e.g. futurology, sociology, psychology, peace studies, economics, geology, systems thinking – and from all society sectors – e.g. the civil society and the private sector – must be systematically included in the consultation process in order to ensure a complexity-appropriate policy.

This implies cross-disciplinary advisory groups, brain trusts, think tanks, task forces, CoINs, MSIs, but can also include innovative initiatives of digital citizen participation. In the following I will give three examples that could serve as inspiration for the development of good or best practices:

- During the Corona crisis, a large-scale hackathon was used in Germany. A hackathon is a design and programming competition in which participants try to solve tasks within a few days. The #WirvsVirus hackathon was organised from 20–22 March, 2020 by the German government together with a number of digital initiatives and Chancellery Minister Helge Braun as patron. 28,361 participants, as well as the German government itself, worked on over 1500 solutions to problems in the context of the Corona crisis (#WirvsVirus, 2020).
- The German company 'Liquid Democracy' develops channels for dialogue democracy, such as citizen platforms for city councils (Liquid Democracy).
- The introduction of citizens' councils. In this process, randomly selected citizens jointly work out solutions to social challenges within a few days. These are then presented to the public and discussed. In Austria, the citizens' council in Vorarlberg, is considered an inspiration (Land Vorarlberg). In Germany, the first Citizens' Council 2019–2020 was organised by the association 'Mehr Demokratie' and the Schöpflin Foundation and supported by the FMER and the Mercator Foundation. In September 2019, 160 drawn citizens developed 'Recommendations for Strengthening Democracy' (Bürgerrat Demokratie).

To conclude: Complex problem-solving requires the inclusion of as many perspectives as possible. In concrete terms, this means that political decision-makers should consult more scientific disciplines than just those directly related to the problem topic and representatives from all society sectors.

9.2.2.3 Milestone 3: Strengthening Collective Problem-Anticipation, Problem-Solving, and Decision-Making by Improving the Quality of Communication

Collective problem analysis and political decision-making take place at almost all system levels. An important implication is: the better the communication quality, the better the integration of perspectives and the higher the collective problem-solving capacity. In practice, this means that in almost every area where experts and decision-makers come together, the more systematic use of appropriate communication tools is highly suggested. Examples for analogue tools are systemic consensing, team syntegrity and scenario planning. Some of these tools have been described in this book, more detailed information can also be found in another publication (see Fathi, 2019a). Digital communication technologies can contribute to better communication as well. In the following, I will give three examples which could serve as inspiration for developing good or best practices:

- At the MIT Center for Collective Intelligence the 'Deliberatorium' was developed – a platform that attempts to make debates on the internet, for example on the topic of 'proposed solutions for the climate crisis', more structured by dividing the individual postings into categories depending on whether someone is contributing a new idea or commenting on an existing idea or commenting on comments (Klein, 2020).
  - The 'Prototype Fund', a support programme for open software and digital-social innovation funded by the German Federal Ministry of Education and Research (FMER), is developing, among other things, digital tools for citizen participation in order to improve the quality of political exchange and opinion-forming in social media, which is currently often characterised by unproductive 'agitated discourses' (Prototype Fund).

To conclude, it is highly recommended to apply analogue and digital tools to improve the quality of communication in any collective problem-anticipation, problem-solving, and/or decision-making-process. However, it is important to point our that communication tools as such should not be seen as an end in themselves, but that the aim is to improve communication and thus foster collective problem-solving.

9.2.2.4 Milestone 4: Fostering International and Cross-Sectoral Cooperation

Current and future complex global problems cannot be solved by nations alone. International cooperation is highly required. In the wake of the Russia-Ukraine war and the expanding global tensions between states, this is more difficult than ever in the twenty-first century. What is the basis of international cooperation?

At all times in human history, nations have developed more or less permanent forms of bilateral or multilateral cooperation – mostly related to military security and economic prosperity. To date, there are no fully developed supranational structures on the international stage. However, international cooperation is necessary in view of the fact that global crises in the twenty-first century cannot be stopped by national solo efforts. As it stands today, the European Union has the comparably highest degree of political integration of any group of nations in the world. However, even this confederation of states is still a long way from the evolutionary stage of a fully developed federation in the sense of a 'United States of Europe' model. In the long run, a similar governance system at the global level is desirable – a multi-level governance that includes binding supranational rules as well as national and local sovereignty.

To date, it is the emerging superpower China that is shaping the discourse on 'Globalisation 2.0' by fostering approaches towards regional and global connectivity. Most noteworthy are its 'New Silk Road', or, as it is officially called, the 'One Belt One Road' (OBOR) initiative, and its plan to build a power grid for the entire world. The latter project includes investments of $50+ trillion and envisions global power connectivity, as well as global power generation through renewable energies. A hope is that such a grid will curb or even reduce international disputes and foster a larger sense of global unity among nations, since power generation and distribution would become a transnational undertaking (Gallego, 2022). The OBOR includes infrastructure development and investment in 70–150 countries worldwide (Godehardt & Kohlenberg, 2017). The OBOR is intended as a vision for the development of a comprehensive Eurasian network that promotes economic, cultural and political connectivity and cooperation between countries, regions and cities along the Silk Road. Optimists argue that the OBOR, due to its connectivity and inclusiveness on the one hand and its openness and flexibility on the other, would largely lead to peaceful win-win cooperation for most of the countries concerned (The World Bank, 2018; Zhang, 2018). Critics, however, tend to see China's initiatives as a threat and emphasise that while each participating country is free to develop its own relations and influence within this network, China clearly represents the core of both the OBOR and the global grid and increasingly binds participating countries to its own interests, which must always take precedence due to its size (Godehardt, 2016; Godehardt & Kohlenberg, 2017). Overall, then, both initiatives can be seen as China's proactive strategic response to the question of how international politics might be reorganised or even de-organised in an increasingly complex and multipolar future. Critical observers emphasise that the hidden incentive of these initiatives is to place China at the forefront of new global rule- and norm-setting, by parallelising and thus dissolving existing international rules and regulatory structures. It can thus be assumed that these initiatives are ambivalent in their impact and may contain both the potential for multilateral win-win cooperation and a national strategy gaining the upper hand in the Great Game.[7]

While the emerging superpower China offers a vision related to development (with emphasis on economic growth) and sustainability (with emphasis on renewable energy supply) that is quite appealing to many societies, the Western world has largely failed to develop appealing alternatives or complementary approaches to innovative multilateral cooperation.

Cooperation-based multi-resilience and sustainability, which would be urgently needed in anticipation of further crisis bundles and bundled crises for the global community, could offer a unique opportunity to create a unified and timely vision for the global future and foster international cooperation. In this context, in addition to existing economic and political institutions, agreements and forms of bilateral and multilateral cooperation, the guiding principle of 'global future preparedness' or 'global multi-resilience' could be an additional motto and incentive to further promote the necessary global political integration and local self-reliance. As such, the EU itself would be predestined to take a leading role in promoting the vision of a multi-resilient world community due to its good global relations, its soft power and its globally unique degree of complexity of political integration.

A possible starting point would be to reflect, design and communicate a vision of global multi-resilience, and, second, to actively initiate transnational mechanisms of global crisis coordination at the formal level of existing supranational institutions (i.e. at EU and UN levels). Inspired by the OBOR, this could also be

---

[6] As a widely discussed example, the European market is effectively closed to Africa at present. According to Roland Süss of the anti-globalisation group Attac, 'subsidized agricultural products from Europe are flooding African markets and destroying local smallholder structures', […] The EU forecasts that its agricultural subsidies will total €365 billion ($423 billion) between 2021 and 2027, while Brussels' entire European-Africa policy is worth just €39 billion a year.' (DW, 2018; see also Schmidt, 2019)

supported by informal bilateral level discussions and cooperation with states at the informal level. Initiatives could include active bilateral offers of support and security partnerships including joint scenario-planning in the face of complex crises. This should not only involve politics, but also initiatives from other sectors of society, e.g. civil society (e.g. the above-mentioned Simpol initiative) or science (e.g. joint transdisciplinary research, anticipation and modelling of man-made side-effects and other complex problems). Thus, cooperation should not only be international, but also cross-sectoral. It should be central to any foreign political agenda.

However, here too, cooperation is not an end in itself. It should clearly aim for a win-win outcome, especially in the long term. In this regard, it is important to consider the reality of national rivalries and egoisms. This ambiguity requires a complexity-adequate foreign policy – in a sense, China's above-mentioned global grid project and the OBOR initiative are representative examples of such a foreign political strategy based on 3000 year-old martial wisdom traditions and stratagems[8] (Fathi, 2021a). As outlined in Sect. 8.4.3., the foreign political approach of any country to manage the global prisoner's dilemma effectively should be a foreign political strategy which enables cooperation, but makes the own political reactions, even of non-cooperation, transparent and predictive to the other nations, thus enabling trust. Scientifically proven strategies of 'tit for tat' 'win-stay, loose-shift' could provide a political orientation.

To conclude, cross-national and cross-sectoral cooperation with the goal of long-term win-win should be central to the foreign political agenda, and the West, particularly the EU could foster this on the basis of global future preparedness or multi-resilience. At the same time, foreign policy should consider the reality of the global prisoner's dilemma.

9.2.2.5 Milestone 5: Fostering Real Multi-Lateralism and Global Governance

The national and international responses to the three most prevalent crises in the last 3 years – the Corona crisis, climate change and the Russia-Ukraine war – revealed several resources and political failures that should be considered in future foreign politics. In terms of resources, the political responses to the Corona crisis and the Russia-Ukraine war have shown in a positive way how quickly collective forces can be mobilised; measures in the form of aid packages, legislative resolu-

---

[7] For more detailed information about this discourse and the ambivalence of Chinese foreign politics, see Fathi (2021a).

[8] Sun Tzu's 'Art of War' might be the most popular source.

tions, cooperation and sanctions can be decided and implemented in a timely manner. However, it is also evident that this only applies to problems that are perceived as immediately urgent. Climate change, which seems to many political decision-makers comparatively abstract, lacks the political commitment that was mustered in the other two crises. It can be assumed that many political leaders know about the necessity of global governance and in the widest sense world domestic politics in the face of current and future global challenges. Many concrete plans, such as the above-mentioned Global Marshall Plan initiative and reforms to democratise the UN's decision-making structures that still reflect the realities of the post-WWII era, are on the table and should be adapted to the realities of the twenty-first century. The failure to establish global governance or even an adequate climate policy is less a question of possibilities than of a lack of political will.

From a systemic perspective, the structural cause might be the global prisoner's dilemma, which can only be successfully managed in the long term by fostering direct communication and transparency about foreign political actions as mentioned above. A related important aspect is mutual trust-building. What can mutual trust-building look like? In principle, this approach is not new; for example, the nuclear disarmament process between Western states and Russia in the post-Cold War era can be counted among them. In the context of a global governance to be striven for in the future, this approach would have to be pursued further. The exact form this would take would have to be fleshed out in further political practice. As this is typical for any peacebuilding, this might also imply for all countries (yes, also for the West) to dare to engage in a process of self-critical reflection and take responsibility for the impact of past policy decisions, to leave an under-complexed 'me: good – you: evil' perspective, and do better for the future. Questions for the West could be: How does the West face the accusation of 'arrogance' and 'double standards' that is often shared by the rest of the world? To what extent would Western states, but also Russia and China, be willing to give up their privileged position in the UN Security Council in favour of a decision-making constellation in which all states of the world are fairly and democratically represented and in which genuine multilateralism can be realised?

On the way to multilaterally represented global governance, it will not be a matter of shaping the world according to Western standards. It will be about finding binding regulations for globally particularly important issues, such as the preservation of national sovereignty (i.e. a ban on war aggressions, which is already at the centre of current international law), the averting of globally threatening crises that can also be enforced supranationally, if necessary. Similar to the EU model, autoc-

9.2 Conclusion and Summarising Theses

racies and democratically open societies will continue to exist for the time being in such a multi-level governance system. The necessary next evolutionary step towards a future prepared world community will be that multilaterally defined rules apply to all states and can be enforced against their will, if necessary.

To conclude: Regardless of what happens in the coming years and decades. What is certain is that the world community, which is partly characterised by mistrust, cannot avoid some form of global governance in the long run if it wants to survive. Perhaps it is a matter of honestly facing the reality of the situation and authentically representing this motto: 'We pull together, we can only do it together'.

References

Allen, J., Burns, N., Garrett, L., Haass, R. N., Ikenberry, J. O., Mahbubani, K., Menon, S., Niblett, R., Nye, J. R., O'Neil, S. K., Schake, K., & Walt, S. M. (2020, March 20). How the world will look after the Coronavirus Pandemic. *Foreign Policy*. https://foreignpolicy.com/2020/03/20/world-order-after-coroanvirus-pandemic/
Allianz. (2018). *Allianz Risk Barometer 2018*. Munich.
Allianz. (2022). *Allianz Risk Barometer 2022*. Munich.
Allison, G. (2018). *Destined for war: Can America and China escape Thucydides's trap?* Scribe.
al-Nawawi. (2007). *Das Buch der Vierzig Hadithe: Kitab al-Arba'in. Mit dem Kommentar von Ibn Daqiq al-'Id*. Verlag der Weltreligionen im Insel Verlag.
Alter, P. (1985). *Nationalismus*. Suhrkamp.
Althaus, D. *Homepage*. http://www.d-althaus.de/22.0.html
American Preppers Network. *Homepage*. http://americanpreppersnetwork.com/
Anderson, J. R., Bothell, D., Byrne, M. D., Douglass, S., Lebiere, C., & Qin, Y. (2004). An integrated theory of the mind. *Psychological Review, 111*, 1036–1060.
Andler, N. (2013). *Tools für Projektmanagement, Workshops und Consulting – Kompendium der wichtigsten Techniken und Methoden*. Publicis.
Antonovski, A. (1997). *Salutogenese. Zur Entmystifizierung der Gesundheit*. Deutsche Gesellschaft für Verhaltenstherapie.
Apolte, T. (2004). Negativ-Einkommensteuer-Transfersystem. Wirtschaftswissenschaftliche Fakultät des Institutes für Ökonomische Bildung, Westfälische Universität Münster. Gutachten für den Parlamentarischen Beratungs- und Gutachterdienst des Landtags NRW, Information 13/1089 vom 13.08.2004.
Arafeh, O. (2018). China's Quiet bid for hegemony: The belt and road initiative. 9 November 2018. *McGill Journal of Political Studies*. https://mjps.ssmu.ca/2018/11/09/chinas-quiet-bid-for-hegemony-the-belt-and-road-initiative/

Arbeitsgemeinschaft Kriegsursachenforschung (AKUF). (2012). Kriege und bewaffnete Konflikte 2012. *AKUF Analysen*, Nr. 11.
Arcadis. (2019). *Citizen centric cities – 2018 Sustainable cities index*. https://www.arcadis.com/campaigns/citizencentriccities/index.html
Argyris, C., & Schön, D. A. (1978). *Organizational learning: A theory of action perspective*. Addison-Wesley.
Argyris, C., & Schön, D. (2008). *Die lernende Organisation*. Klett-Cotta.
Aristoteles. (2006). *Politik*. Translated and published by Gigon, O.
Asad, M. (2011). *Die Botschaft des Koran – Übersetzung und Kommentar*. Patmos.
Ashby, W. R. (1956). *An introduction to cybernetics*. Chapman & Hall.
Ashkenas, R. (2016). Change management needs to change. *Harvard Business Review*. https://hbr.org/2013/04/change-management-needs-to-cha
Ashoka. *Homepage*. https://www.ashoka.org/
Axelrod, R. (1984). *The evolution of cooperation*. Basic Books.
Bachmann, A. (2013). *Hedonismus und das gute Leben*. Mentis Verlag.
Bakir, D. (2017, February 11). In Kenia startet das größte Geld-Experiment der Welt. *Stern Online*. https://www.stern.de/wirtschaft/news/bedingungsloses-grundeinkommen%2D%2Ddas-grosse-geld-experiment-in-kenia-7323118.html
Ball, R., & Chernova, K. (2008). Absolute income, relative income, and happiness. *Social Indicators Research, 88*, 497–529. https://doi.org/10.1007/s11205-007-9217-0
Balogun, J., & Hailey, V. (2004). *Exploring strategic change*. Prentice Hall.
Balsinger, P. W. (2005). *Transdisziplinarität*. Wilhelm Fink Verlag.
Bankoff, G., & Hilhorst, D. (2009). The politics of risk in the Philippines: Comparing state and NGO perceptions of disaster management. *Disasters, 33*(4), 686–704.
Bardi, U. (2017). *The Seneca effect. Why growth is slow but collapse is rapid. A report to the Club of Rome, Winterthur/Switzerland 2017* (Springer book series: the frontiers collection). Springer.
Barinaga, M. (2003). Studying the well-trained mind. *Science, 302*(5642), 44–46.
Baron, S., & Yin-Baron, G. (2018). *Die Chinesen – Psychogramm einer Weltmacht*. Econ.
Bartosch, U., & Ganscyzk, K. (Hrsg). (2009). *Weltinnenpolitik für das 21. Jahrhundert. Carl-Friedrich von Weizsäcker verpflichtet (Weltinnenpolitische Colloquien)* (Bd. 1). LIT.
Basic Income Grant Coalition. (2009, April). *Making the difference! The BIG in Namibia. Basic income grant pilot project- Assessment report*. http://www.bignam.org/Publications/BIG_Assessment_report_08b.pdf
Bateson, G. (1987). *Geist und Natur. Eine notwendige Einheit*. Suhrkamp.
Bauer, S. (2008). Leitbild der Nachhaltigen Entwicklung. In *Informationen zur politischen Bildung* (p. 287). Bonn.
Bauer, T. (2018). *Die Vereindeutigung der Welt. Über den Verlust an Mehrdeutigkeit und Vielfalt*. Reclam.
Baum, S., & Barrett, A. (2017). Global catastrophes: The most extreme risks. In V. Bier (Hrsg), *Risk in extreme environments: Preparing, avoiding, mitigating, and managing* (S. 174–184). Routledge. https://ssrn.com/abstract=3046668
BBK. *Homepage: Vorsorge für den Katastrophenfall. Bundesamt für Bevölkerungsschutz und Katastrophenhilfe*. https://www.bbk.bund.de/DE/Ratgeber/VorsorgefuerdenKat-fall/VorsorgefuerdenKat-fall.html
Beck, U. (1986). *Risikogesellschaft. Auf dem Weg in eine andere Moderne*. Suhrkamp.

References

Beck, U. (1996). Das Zeitalter der Nebenfolgen und die Politisierung der Moderne. In U. Beck, A. Giddens, & S. Lash (Hsrg), *Reflexive Modernisierung. Eine Kontroverse.* Suhrkamp.
Beck, U. (2007). *Risikogesellschaft. Auf dem Weg in eine andere Moderne.* Suhrkamp.
Beck, D., & Cowan, C. (1996). *Spiral dynamics: Mastering values, leadership, and change.* Blackwell.
Beck, U., Giddens, A., & Lash, S. (1994). *Reflexive modernization. politics, tradition and aesthetics in the modern social order.* Stanford University Press.
Beckett, A. (2003, September 8). Santiago dreaming. *The Guardian.* https://www.theguardian.com/technology/2003/sep/08/sciencenews.chile
Beer, S. (1959). *Cybernetics and management.* English Universities Press.
Beer, S. (1974). *Designing freedom.* CBC Learning Systems.
Beer, S. (1985). *Diagnosing the system for organizations.* Wiley.
BEH. (2017). *World risk report 2017.* https://www.cbm.de/static/medien/weltrisikobericht-2017.pdf
BEH. (2021). *World risk report 2021.* https://weltrisikobericht.de//wp-content/uploads/2021/09/WorldRiskReport_2021_Online.pdf
Bell, D. (1985). *Die nachindustrielle Gesellschaft.* Frankfurt a. M.: Campus.
Ben Larbi, M. (2018). Warum Dörfer für Innovation prädestiniert sind. Dörfer im Aufbruch (S. 1974). *Designing freedom.* Wiley. http://www.doerfer-im-aufbruch.de/warum-doerfer-fuer-innovation-praedestiniert-sind/Beer
Béné, C., Godfrey-Wood, R., Newsham A., & Davies, M. (2012). *Resilience: New utopia or new tyranny? – Refection about the potentials and limits of the concept of resilience in relation to vulnerability reduction programmes* (iDs working paper no. 405). Institute of Development studies. http://www.ids.ac.uk/fles/dmfle/Wp405.pdf
Benedikter, R., & Fathi, K. (2012). Der Breivik-Prozeß: Beispiel für eine zukunftsfähige Gesellschaft? *Tiroler Tageszeitung.* http://www.tt.com/home/5492894-91/der-breivik-prozess-beispiel-für-eine-zukunftsfähige-gesellschaft.csp
Benedikter, R., & Fathi, K. (2013, October 21). Der Kampf um das menschliche Ich. *Das Blättchen, 16*(2). http://das-blaettchen.de/2013/01/der-kampf-um-das-menschliche-ich-20489.html
Benedikter, R., & Fathi, K. (2021). *The Coronavirus crisis and its teachings – Steps towards multi-resilience.* Brill.
Bergmann, M., & Schramm, E. (Eds.). (2008). *Transdisziplinäre Forschung. Integrative Forschungsprozesse verstehen und bewerten.* Campus.
Bertelsmann Stiftung. (2019, June 19). Viele Worte, wenig Taten: UN-Nachhaltigkeitsziele könnten scheitern. https://www.bertelsmann-stiftung.de/de/themen/aktuelle-meldungen/2019/juni/viele-worte-wenig-taten-un-nachhaltigkeitsziele-koennten-scheitern/
Bhagat, R. S., Segovis, J. C., & Nelson, T. A. (2010). *Work stress and coping in the era of globalization.* Psychology Press.
Bhatti, Y. A. (2012). *What is frugal, what is innovation? Towards a theory of frugal innovation.* Social Science Research Network. https://www.papers.ssrn.com/sol3/papers.cfm?abstract_id=2005910
Bhatti, Y. A., Khilji, S. E., & Basu, R. (2013). Frugal innovation. In S. Khilji & C. Rowley (Hrsg), *Globalization, change and learning in South Asia.* Chandos Publishing.

BIEN. *Homepage.* https://basicincome.org/news/2017/01/bien-affiliated-organisations-definitions-basic-income/
Big Mountain Aktionsgruppe e. V. (Hrsg). (1993). *Stimmen der Erde. Munich.* Raben.
Binswanger, M. (2006). *Die Tretmühlen des Glücks: Wir haben immer mehr und werden nicht glücklicher. Was können wir tun?* Verlag Herder.
Birkmann, J. (2006). *Measuring vulnerability to natural hazards: Towards disaster resilient societies.* United Nations University Press.
Birkmann, J. (2008). *Assessing vulnerability before, during and after a natural disaster in fragile regions* (Research paper no. 2008/50 UNU-WIDER). Bonn.
Birkmann, J., et al. (2011). *Glossar – Klimawandel und Raumentwicklung.* Akademie für Raumforschung und Landesplanung (E-Paper der ARL).
Bishops of Germany Austria Switzerland, et al. (2017). *Die Bibel: Einheitsübersetzung der Heiligen Schrift.* Herder.
Blaschke, R. (2007). Grundeinkommen zwischen Mindest- und Lebensstandardsicherung. Eine Orientierungshilfe im Zahlenlabyrinth. In A. Exner, W. Rätz, & B. Zenker (Eds.), *Grundeinkommen. Soziale Sicherheit ohne Arbeit* (pp. 156–164). Deuticke.
Bloomberg. (2021, July 28). *The Covid resilience ranking. The best and worst places to be as reopening, variants collide.* Available at: https://www.bloomberg.com/graphics/covid-resilience-ranking/
Bockelbrink, B., Priest, J., & David. L. (2022). *A Practical Guide to Sociocracy 3.0.* The Principle of Consent. 26 April, 2022. https://patterns.sociocracy30.org/principle-consent.html
Boffey, D. (2018, April 8). Amsterdam to embrace 'doughnut' model to mend post-coronavirus economy. *The Guardian.* https://www.theguardian.com/world/2020/apr/08/amsterdam-doughnut-model-mend-post-coronavirus-economy
Bohle, H.-G. (2007). Geographien von Verwundbarkeit. *Geographische Rundschau, 59*(10), 20–25.
Bohle, H.-G., Downing, T., & Watts, M. (1994). Climate change and social vulnerability. Toward a sociology and geography of food insecurity. *Global Environment Change, 4,* 37–38.
Bohlmeijer, E., Prenger, R., Taal, E., & Cuijpers, P. (2010). The effects of mindfulness-based stress reduction therapy on mental health of adults with a chronic medical disease: A meta-analysis. *Journal of Psychosomatic Research, 68*(6), 539–544. https://doi.org/10.1016/j.jpsychores.2009.10.005
Bollmann, R. (2006). *Lob des Imperiums: Der Untergang Roms und die Zukunft des Westens.* wjs verlag.
Bönsel, R. (2012). *Atempause jetzt! Spirituelles Stressmanagement nach S.H. Sri Sri Ravi Shankar.* Kamphausen.
Bostrom, N. (2002). Existential risks: Analyzing human extinction scenarios and related hazards. *Journal of Evolution and Technology, 9*(3), 477.
Bostrom, N. (2013). Existential risk prevention as global priority. *Global Policy, 4*(1), 15–31.
Bostrom, N. (2017). *Superintelligenz – Szenarien einer kommenden Revolution.* Suhrkamp.
Bostrom, N., & Cirkovic, M. (2008). *Global catastrophic risks.* Oxford University Press.
Both, I., Domic, K., Eimers, J., Flath, V., Keuler, K., Krings, C., Noltemeyer, O., Reinke, Y., Steckelbruck, L., & Weidental, R. (2012). *Multi-Stakeholder-Initiative – Eine methodische Einführung, Global Workers Protection (GWP).* Mönchengladbach. https://d-nb.info/1044211555/34

References

Boyd, R. (2014). *Panarchy: Implications for economic & social policy.* http://www.humanitystest.com/panarchy-implications-for-economic-social-policy/
Bright, C. et al. (2004). *The state of the world 2003.* The World Watch Institute. Download: https://doczz.net/doc/3395652/sow-chap-4%2D%2D-worldwatch-institute
Bruni, L., & Porta, P. (2005). *Economics & happiness. Framing the analysis.* Oxford University Press.
Brynjolfsson, E., & McAfee, A. (2014). *The second machine age.* GGP Media.
Buckley, C. (2015, Oktober 29). *The New York Times.* https://www.nytimes.com/2015/10/30/world/asia/china-end-one-child-policy.html?mcubz=3
Bundesministerium für Bildung und Forschung. (Hrsg). (2002). *Bericht der Bundesregierung zur Bildung für eine nachhaltige Entwicklung.* https://www.umweltbildung.de/uploads/tx_anubfne/bfne_bericht_kabinettfassung.pdf
Bundesregierung. *The new Hi-Tech-strategy.* https://www.hightech-strategie.de/de/Thenew-High-Tech-Strategy-390.php
Bundesstelle für Sektenfragen. (2021). *Das Phänomen Verschwörungstheorien in Zeiten der COVID-19-Pandemie.* Bericht der Bundesstelle für Sektenfragen an die Bundesministerium für Frauen, Familie, Jugend und Integration, Austria. May, 2021. Vienna
Bürgerrat Demokratie. *Homepage.* https://demokratie.buergerrat.de/
Burgoon, B. (2006). On welfare and terror – Social welfare policies and political-economic roots of terrorism. *Journal of Conflict Resolution, 50*(2), 176–203.
Buss, K.-P., & Ortiz, A. (2016). Im Schatten des Marktes Mikrologiken marktlicher Governance in kollaborativen Innovationsprojekten in der Softwareentwicklung und der Entwicklung von Windenergieanlagen. *Oldenburger Studien zur Europäisierung und zur transnationalen Regulierung, 2016*(25), 45–74.
Caldeira, K., & Bala, G. (2016). Earth's future: Reflecting on 50 years of geoengineering research. *AGU.* https://doi.org/10.1002/2016ef000454
Calliess, C. (1999). *Subsidiaritäts- und Solidaritätsprinzip in der Europäischen Union.* Nomos.
Campbell, P.-H. (2015). *Space race.* Allitera.
Caplan, N., Choy, M. H., & Whitmore, J. K. (1989). *The boat people and achievement in America: A study of family life, hard work, and cultural values.* University of Michigan Press.
Caplan, N., Choy, M. H., & Whitmore, J. K. (1992). Indochinese refugee families and academic achievement. *Scientific American, 266*(2), 36–45.
Carlisle, K., & Gruby, R. (2017). Polycentric systems of governance: A theoretical model for the commons. *Policy Studies Journal.* https://doi.org/10.1111/psj.12212. http://onlinelibrary.wiley.com/doi/10.1111/psj.12212/full
Carter, O., et al. (2005). Meditation alters perceptual rivalry in Tibetan Buddhist monks. *Current Biology, 15*(11), 412–413.
Cartwright, S., & Schoenberg, R. (2006). Thirty years of mergers and acquisitions research: Recent advances and future opportunities. *British Journal of Management, 17*, 1–5.
Castells, M. (2003). Four Asian tigers with a dragon head. A comparative analysis of the state, economy and society in the Asian Pacific rim. In R. P. Appelbaum & J. Henderson (Hrsg), *States and development in the Asian Pacific Rim* (S. 176–198). Sage.
CGG – Commission on Global Governance. (1995). *Nachbarn in Einer Welt. Der Bericht der Kommission für Weltordnungspolitik.* Stiftung Entwicklung und Frieden.

Chapman, et al. (2006). *Open government in a theoretical and practical context*. Ashgate.
Checkland, P. B. (1981). *Systems thinking* (Systems practice). Wiley.
China.org.cn. (2015, March 25). *China unveils action plan on belt and road initiative.* http://www.china.org.cn/china/2015-03/28/content_35181779.htm
Chouard, T. (2016). The go files: AI computer clinches victory against Go champion. *Nature*. https://www.nature.com/news/the-go-files-ai-computer-clinches-victory-against-go-champion-1.19553
Christmann, G., Ibert, O., Kilper, H., & Moss, T. (2011). *Vulnerabilität und Resilienz in sozio-räumlicher Perspektive – Begriffliche Klärungen und theoretischer Rahmen*. IRS Leibniz-Institut für Regionalentwicklung und Strukturplanung. http://www.irs-net.de/aktuelles/meldungen-detail.php?id=206
CIA. (2018). *Die Welt im Jahr 2035: Gesehen von der CIA und dem National Intelligence Council*. Beck.
CIA-World Factbook. *Homepage*. https://www.cia.gov/library/publications/the-world-factbook/
CIFAR. (2015). *Successful societies*. https://www.cifar.ca/research/successful-societies/
Cisco. (n.d.). *Städte und Gemeinden*. https://www.cisco.com/c/de_de/solutions/industries/smart-connected-communities.htmlCoaffee
Coaffee, J., & Wood, D. (2006). The "everyday" resilience of the city. In *Human security and resilience* (ISP/NSC briefing paper, no. 6(1))
Cohen, B. (2011). Global ranking of top 10 resilient cities. *Triple Pundit*. http://www.triplepundit.com/2011/06/top-10-globally-resilient-cities/
Coutu, D. (2002). How resilience works. *Harvard Business Review*. https://hbr.org/2002/05/how-resilience-works May 2002 Issue.
CSS. (2009). *Resilienz: Konzept zur Krisen- und Katastrophenbewältigung. CSS Analysen zur Sicherheitspolitik*. ETH Zürich: Zürich.
Cutuli, J. J., Herbers, J. E., Lafavor, T. L., & Masten, A. S. (2008). Promoting competence and resilience in the school context. *Professional School Counseling, 12*(2), 76–84.
Cycorp. *Homepage*. http://www.cyc.com/
Dahrendorf, R. (1961). *Gesellschaft und Freiheit. Zur soziologischen Analyse der Gegenwart*. Piper.
Dalai Lama. (2013). *The path to peace and happiness in a global society September 13, 2013*. https://www.dalailama.com/news/2013/the-path-to-peace-and-happiness-in-a-global-society
Damásio, A. R. (1994). *Descartes' Irrtum – Fühlen, Denken und das menschliche Gehirn*. List.
Dasmann, R. (1988). *Toward a biosphere consciousness, the ends of the earth*. Cambridge University Press.
Davidson, R., et al. (2003). Alterations in brain and immune function produced by mindful meditation. *Psychosomatic Medicine, 65*, 564–570.
Day, L. G. E. (1969–1977). *The theory of multigovernment*. http://www.panarchy.org/day/multigovernment.1977.html
de la Peña, V., & Giné, E. (2009). *Decoupling: From dependence to independence*. Springer.
de Puydt, P. E. (1860). Panarchy. *Revue Trimestrielle, 27*, 6–246.
Department of Homeland Security. *Plan ahead for disasters*. https://www.ready.gov/
Desertec. *Homepage*. https://www.desertec.org/

References

Deutscher Alpenverein (Ed.). (2004). *Risiko – Gefahren oder Chancen? Tagungsbericht der Ev. Akademie Bad Boll.*

Dharma, K. (2020). *Mahabharata: The greatest spiritual epic of all time.* Mandala Publishing.

Dietz, K. (2002). *Vulnerabilität und Anpassung gegenüber Klimawandel aus sozialökologischer Perspektive. Aktuelle Tendenzen und Herausforderungen in der internationalen Klima- und Entwicklungspolitik.* Berlin.

Dissen, S., Quaas, M., & Baumgärtner, S. (2009). *The relationship between resilience and sustainability in ecological-economic systems* (Working paper no. 146). Lüneburg.

Dixit, A. K., & Nalebuff, B. J. (1993). *Thinking strategically: The competitive edge in business, politics, and everyday life.* Norton.

DJSI. *Homepage. Dow Jones Sustainability Index.* https://www.spglobal.com/spdji/en/indices/esg/dow-jones-sustainability-world-index/#overview

Dorling, D. (2020, October 6). 'Coronavirus: Is the cure worse than the disease? The most divisive question of 2020'. *The Conversation.* Available at: https://theconversation.com/coronavirus-is-the-cure-worse-than-the-disease-the-most-divisive-question-of-2020-147343

Drath, K. (2014). *Resilienz in der Unternehmensführung. Was Menschen und Teams stark macht.* Haufe.

Drekonja-Kornat, G. (2001). Manfred A. Max-Neef (*1932). Entwicklung nach menschlichem Maß. In Deutsche Stiftung für internationale Entwicklung (DSE) (Hrsg), *E+Z Entwicklung und Zusammenarbeit 07/08* (S. 233–235).

Drexler, K. E. (1986). *Engines of creation. The coming era of nanotechnology.* http://e-drexler.com/d/06/00/EOC/EOC_Table_of_Contents.html

Duclos, D. (2012). Stubenhocker der Apokalypse. *Le Monde diplomatique, 9851,* 3.

Dueck, G. (2018). *Schwarmdumm: So blöd sind wir nur gemeinsam.* Goldmann.

DW. (2018, August 9). EU-Africa free trade will create more imbalances, say critics. *Deutsche Welle.* https://amp.dw.com/en/eu-africa-free-trade-will-create-more-imbalances-say-critics/a-45018168

Earth Overshoot Day. (2022). *Country overshoot days.* https://www.overshootday.org/newsroom/country-overshoot-days/

Ebbighausen, R. (2015). Kommentar: Der gute Autokrat von Singapur. DW. https://www.dw.com/de/kommentar-der-gute-autokrat-von-singapur/a-18334401

Economic Activity. (2017, July 6). *Quaternary sector: Definition, background, examples.* https://www.economicactivity.org/2017/07/quaternary-sector.html

Economist Intelligence Unit. (2022). *2021 Liveability ranking.* https://store.eiu.com/product/liveability-ranking-and-overview/

Edwards, A. (2010). *Thriving beyond sustainability: Pathways to a resilient society.* New Society Press.

Eldar, A. (2006). The eight-stage spiral to peace in the Mideast. *Haaretz.* http://www.haaretz.com/print-edition/features/the-eight-stage-spiral-to-peace-in-the-mideast-1.179848

Ellin, N. (2006). *Integral urbanism.* Routledge.

Elsberg, M. (2016). *Helix – Sie werden uns ersetzen.* Blanvalet.

Environment & Society Portal. (n.d.). *Hans Carl von Carlowitz and "sustainability".* https://www.environmentandsociety.org/tools/keywords/hans-carl-von-carlowitz-and-sustainability

Erdheim, M. (1988). *Psychoanalyse und Unbewußtheit in der Kultur. Aufsätze 1980–1987.* Suhrkamp.
Esping-Andersen, G. (1990). *Three worlds of welfare capitalism.* Oxford University Press.
Etheredge, L. S. (1981). Government learning: An overview. In S. L. Long (Hrsg), *The handbook of political behavior* (Bd. 2). Pergamon.
Etheridge, E. (2009, January 14). How 'soft power' got 'smart'. *The New York Times.* https://opinionator.blogs.nytimes.com/2009/01/14/how-soft-power-got-smart/?mcubz=3
Etzioni, A. (2009). *Die aktive Gesellschaft. Eine Theorie gesellschaftlicher und politischer Prozesse.* Verlag für Sozialwissenschaft.
European Commission. (2022a). *What is the 'Beyond GDP' initiative.* https://ec.europa.eu/environment/beyond_gdp/index_en.html
European Commission. (2022b). *Innovation union.* http://ec.europa.eu/research/innovation-union/index_en.cfm
European Committee of the Regions, Volpe, M., Friedl, J., & Cavallini, S., et al. (2016). *Using the quadruple helix approach to accelerate the transfer of research and innovation results to regional growth*, Committee of the Regions. https://data.europa.eu/doi/10.2863/408040
European Parliament. (2012, December 12). *Eleven EU countries get Parliament's all clear for a financial transaction tax.* European Parliament. https://www.europarl.europa.eu/news/en/press-room/20121207IPR04408/eleven-eu-countries-get-parliament-s-all-clear-for-a-financial-transaction-tax
Evans, D. (2019). *Emotion: A very short introduction.* Oxford University Press.
Ewenstein, B., Smith, W., & Sologar, A. (2015). *Changing change management.* McKinsey. https://www.mckinsey.com/featured-insights/leadership/changing-change-management
Ezzati, M., Vander Hoorn, S., Lawes, C. M., Leach, R., James, W. P., Lopez, A. D., Rodgers, A., & Murray, C. J. (2005, May 3). Rethinking the "diseases of affluence" paradigm: Global patterns of nutritional risks in relation to economic development. *PLOS Medicine, 2*(5). https://doi.org/10.1371/journal.pmed.0020133
Fajnzylber, P., Lederman, D., & Loayza, N. (2002). Inequality and violent crime. *The Journal of Law and Economics, 45*(1), 1–40.
FAO. (2021, September 14). *UN report calls for repurposing of $470 billion of agricultural support that distorts prices and steer us away from environment and social goals.* https://www.fao.org/news/story/en/item/1438889/icode/
Fathi, K. (2013). Conflict potentials of different welfare regimes – A metatheoretical perspective. In I. P. Karolewski & A. M. Suszyki (Eds.), *Identity, citizenship and welfare: National and international perspectives* (pp. 41–74). Fibre.
Fathi, K. (2014). Resilienz – Taugt dieser Begriff als "Ein-Wort-Antwort" auf die Häufung von Krisen? *Forschungsjournal Soziale Bewegungen – PLUS, 29*(4). Supplement, 11 pages
Fathi, K. (2016a). Die Flüchtlingskrise – 5 Komponenten zu einer resilienten Gesellschaft. *Forschungsjournal Neue Soziale Bewegungen PLUS, 29*(4). Supplement, 13 pages
Fathi, K. (2016b). Was hilft Führungskräften, Resilienz bei sich und im Kollektiv aufzubauen? In H. Roehl & H. Asselmeyer (Eds.), *Organisationen klug gestalten* (pp. 301–310). Schäffer-Poeschel.
Fathi, K. (2019a). *Kommunikative Komplexitätsbewältigung – Grundzüge eines integrierten Methodenpluralismus' zur Optimierung disziplinübergreifender Kommunikation.* Springer.

References

Fathi, K. (2019b). *Das Empathietraining – Konflikte lösen für ein besseres Miteinander*. Junfermann.

Fathi, K. (2021a). Die EU und Chinas "Belt and Road Initiative" – Multi-Resilienz als Leitbild für eine komplexitätsangemessene Weltordnung im aktuellen Wettbewerb um Diskursmacht über "Globalisierung 2.0". *Larnaca Conferences Publications. Carl Auer*. 19 April, 2021. https://www.carl-auer.de/magazin/larnaca-conferences/die-eu-und-chinas-belt-and-road-initiative?secret=vpgwa7YNIo7BCWOqfRwCbG2msmkfHHEB

Fathi, K. (2021b). Arbeiten im Bonuspunktesystem. *Arbeitswelten 2050*. 21 June, 2021. Haufe. https://www.haufe.de/personal/hr-management/arbeitswelten-2050-arbeiten-im-digitalen-bonussystem_80_545142.html

Fathi, K., & Benedikter, R. (2013). Was ist eine resiliente Gesellschaft? Plädoyer für ein neues Konzept sozialer Zukunftssicherung in Krisenzeiten. In *Forschungsjournal Neue Soziale Bewegungen* (p. 2).

Fathi, K., & Karolewski, I. (2015). Civil revolutions in a trans-cultural comparison: Eastern Europe and the MENA-region. In S. Arjomand (Ed.), *The Arab revolution of 2011: A comparative perspective*. SUNY Press.

Fathi, K., & Osswald, A. (2017). Der E-Faktor und die Digitalisierung. *Forschungsjournal Neue Soziale Bewegungen* (2). http://forschungsjournal.de/node/2995

Faure, E., et al. (1973). *Learning to be – The world of education of today and tomorrow*. UNESCO. http://unesdoc.unesco.org/images/0000/000018/001801e.pdf

Fehr, E. (2004, November 25). Human behaviour: Don't lose your reputation. *Nature, 432*, 449–450. https://doi.org/10.1038/432449a

FEMA. *Homepage*. https://www.fema.gov/

Feynman, R. P. (1992). *Surely you're joking Mr Feynman: Adventures of a curious character as told to Ralph Leighton*. Vintage.

Fisher, A. G. B. (1939). Production, primary, secondary and tertiary. *Economic Record, 15*(1), 24–38. https://doi.org/10.1111/j.1475-4932.1939.tb01015.x

Fisher, M., & Bubola, E. (2020). 'As Coronavirus Deepens Inequality, Inequality Worsens Its Spread'. *The New York Times*, 15 March. Available at: https://www.nytimes.com/2020/03/15/world/europe/coronavirus-inequality.html.

Fjorback, L. O., Arendt, M., Ørnbøl, E., Fink, P., & Walach, H. (2011). Mindfulness-based stress reduction and mindfulness-based cognitive therapy – A systematic review of randomized controlled trials. *Acta Psychiatrica Scandinavica, 124*(2), 102–119. https://doi.org/10.1111/j.1600-0447.2011.01704.x

Flassbeck, H., et al. (2012). *Irrweg Grundeinkommen – die große Umverteilung von unten nach oben muss beendet werden*. Westend.

FMER. (2019). *Horizont 2020*. German Federal Ministry of Education and Research (Bundesministerium für Bildung und Forschung [BMBF]. http://www.horizont2020.de/einstieg-kurzueberblick.html

Forschungsgruppe Ethische-Ökologisches Rating: Frankfurt-Hohenheimer Leitfaden. Goethe-University Frankfurt a.M. http://www.ethisch-oekologisches-rating.org/veroeffentlichungen/frankfurt-hohenheimer-leitfaden

Foundation for resilient societies. *Homepage*. http://www.resilientsocieties.org/

Frankl, V. (2004). *Man's search for meaning. An introduction to logotherapy*. Random House/Rider.

Freitas, R. (2000). *Some limits to global ecophagy*. http://www.rfreitas.com/Nano/Ecophagy.html
Friebe, R., Karberg, S., & Rövekamp, M. (2020, April 27). Welche gesundheitlichen Schäden der Lockdown verursacht [What damage the lockdown causes to health]. *Der Tagesspiegel*. Available at: https://www.tagesspiegel.de/wissen/risiken-und-nebenwirkungen-der-coronakrise-welche-gesundheitlichen-schaeden-der-lockdown-verursacht/25778410.html
Friedman, M. (2002). *Capitalism and freedom: Fortieth anniversary edition*. University of Chicago Press.
Frodeman, R., Klein, J. T., & Mitcham, C. (Eds.). (2000). *The Oxford handbook of interdisciplinarity*. Oxford University Press.
Fujusawa Sustainable Smart Town. *Homepage*. https://fujisawasst.com/EN/
Fukuyama, F. (1992). *Das Ende der Geschichte*. Kindler.
Fuller, G. (2003, June 1). The Youth Factor: The New Demographics of the Middle East and the Implications for U.S. Policy. *Brookings*. https://www.brookings.edu/research/the-youth-factor-the-new-demographics-of-the-middle-east-and-the-implications-for-u-s-policy/
Future of life Institute. (2017). *An open letter to the United Nations convention on certain conventional weapons*. https://futureoflife.org/autonomous-weapons-open-letter-2017/
Gabriel, J. (2013). *Der wissenschaftliche Umgang mit Zukunft: Eine Ideologiekritik am Beispiel von Zukunftsstudien über China*. Springer.
Gallego, J. (2022, April 4). China wants to build a $50 trillion global wind & solar power grid by 2050. *Futurism*. https://futurism.com/building-big-forget-great-wall-china-wants-build-50-trillion-global-power-grid-2050
Gallup. (2008). *State of the world – 2008 annual report*. University Press.
Galtung, J. (1998). *Frieden mit friedlichen Mitteln. Friede und Konflikt, Entwicklung und Kultur*. Leske + Budrich.
Galtung, J. (2008). *50 Years: 100 peace and conflict perspectives*. Transcend University Press.
Galuska, J., et al. (2010). Zur psychosozialen Lage Deutschlands. *Umwelt·medizin·gesellschaft, 2011*(1), 72–73. http://www.psychosoziale-lage.de/
Garrett, B. (2020, May 14). Living with bunker builders: doomsday prepping in the age of coronavirus. *The Conversation*. https://theconversation.com/living-with-bunker-builders-doomsday-prepping-in-the-age-of-coronavirus-136635
Garshnek, V., et al. (2000). The mitigation, management, and survivability of asteroid/cometimpact with Earth. *Space Policy, 16*(3), 213–222.
Gätjen, H. (2005). Glück: Sieben Faktoren, die Lebensfreude bestimmen. Hamburger Abendblatt. https://www.abendblatt.de/vermischtes/journal/article108603394/Glueck-Sieben-Faktoren-die-Lebensfreude-bestimmen.html
Gaur, L., & Sahdev, S. L. (2015, January 2015). Frugal innovation in India: The case of Tata Nano. *International Journal of Applied Engineering Research, 10*(7), 17411–17420.
Gebser, J. (2015). *Ursprung und Gegenwart*. Chronos.
General Assembly of the UN. (1986). Resolution A/41/128 – Declaration on the right to development. 97. Plenarversammlung. http://www.un.org/documents/ga/res/41/a41r128.html

References

General Assembly of the UN. (2017, July 10). Resolution adopted by the General Assembly on 6 July 2017. *A/RES/71/313*. https://ggim.un.org/documents/a_res_71_313.pdf

Gibbons, M., et al. (1994). *The new production of knowledge. The dynamics of science and research in contemporary societies*. Sage.

Giddens, A. (1995). *Konsequenzen der Moderne*. Suhrkamp.

Gidley, J. (2017). *The future: A very short introduction*. Oxford University Press. https://global.oup.com/academic/product/the-future-a-very-short-introduction-9780198735281?cc=au&lang=en&amp

Giffinger, R., Fertner, C., Kramar, H., Kalasek, R., Pichler-Milanović, N., & Meijers, E. (2007). Smart cities – Ranking of European medium-sized cities. Gemeinsame Studie des Centre of Regional Science (TU Wien), Department of Geography (University of Ljubljana) und des OTB Research Institute for Housing, Urban and Mobility Studies (Delft University of Technology). http://www.smart-cities.eu/download/smart_cities_final_report.pdf

Gigerenzer, G. (2007). *Bauchentscheidungen. Die Intelligenz des Unbewussten und die Macht der Intuition*. Bertelsmann.

Gilan, D., Helmreich, I., & Hadad, O. (2021) *Resilienz – die Kunst der Widerstandskraft. Was die Wissenschaft dazu sagt*. Herder

Gladwell, M. (2001). *The tipping point – How little things can make a big difference*. Brown.

Global Challenges Foundation. (2017). *Global catastrophic risks 2017*. https://globalchallenges.org/

Global Change Data Lab. (n.d.). *Our world in data*. https://ourworldindata.org/

Global Destination Sustainability Movement. *Homepage*. https://www.gds.earth/

Gloor, P. A. (2006). *Swarm creativity: Competitive advantage through collaborative innovation networks*. Oxford University Press.

Godehardt, N. (2016). *No end of history: A Chinese alternative concept of international order?* (SWP research paper 2016/RP). 02.01.2016, 24 Pages. https://www.swp-berlin.org/en/publication/no-end-of-history/

Godehardt, N. and Kohlenberg, P. J. (2017). Die neue Seidenstrasse: Wie China internationale Diskursmacht erlangt. *SWP. Kurz gesagt*. https://www.swp-berlin.org/kurz-gesagt/die-neue-seidenstrasse-wie-china-internationale-diskursmacht-erlangt/

Godschalk, D. R. (2002). *Urban hazard mitigation: Creating resilient cities*. Plenary paper presented at the Urban Hazards Forum, Columbia City University of New York.

Goldstone, J. (2011). Understanding the revolutions of 2011: Weakness and resilience in Middle Eastern autocracies. In Council on Foreign Relations/Foreign Affairs (Hsrg), *The new Arab revolt* (S. 329–343). Council on Foreign Relations.

Goleman, D. (2001). *Emotionale Intelligenz*. dtv.

Goodwin, M., Raines, T., & Cutts, D. (2017). *What do Europeans think about muslim immigration?* Chatham House. https://www.chathamhouse.org/expert/comment/what-do-europeans-think-about-muslim-immigration

Gotham, K. F., & Campanella, R. (2010). Toward a research agenda on transformative resilience. Challenges and opportunities for post-trauma urban eco-systems. *Critical Planning Summer, 2010*, 9–23.

Gotts, N. M. (2007). Resilience, panarchy, and world-systems analysis. *Ecology and Society, 12*(1), 24. http://www.ecologyandsociety.org/vol12/iss1/art24/

Government of the UK. *Emergency planning.* https://www.gov.uk/government/policies/emergency-planning
Govindarajan, V. *Homepage.* http://www.tuck.dartmouth.edu/people/vg/
Graf, R. (2018). *Die neue Entscheidungskultur – mit gemeinsam getragenen Entscheidungen zum Erfolg.* Hanser.
Grande, E. (2000). Multi-Level Governance: Institutionelle Besonderheiten und Funktionsbedingungen des europäischen Mehrebenensystems. In M. Jachtenfuchs & E. Grande (Eds.), *Wie problemlösungsfähig ist die EU?* Baden-Baden.
Granovetter, M. (1978). Threshold models of collective behavior. *American Journal of Sociology., 83*(6), 1420. https://doi.org/10.1086/226707
Graßl, H. (2007). *Klimawandel. Die wichtigsten Antworten.* Herder.
Gredler, M. (1994). *Designing and evaluating games and simulations. A process approach.* Gulf Publishing.
Greenfield, A. (2013). *Against the smart city (The city is here for you to use).* Kindle.
Greve, G. (2012). *Organizational Burnout: Das versteckte Phänomen ausgebrannter Organisationen.* Gabler.
Griffiths, J. (2020, April 5). Taiwan's coronavirus response is among the best globally. *CNN.* https://edition.cnn.com/2020/04/04/asia/taiwan-coronavirus-response-who-intl-hnk/index.html
Groh, A. (2004). Kulturwandel durch Reisen: Faktoren, Interdependenzen, Dominanzeffekte. In C. Berkemeier (Ed.), *Begegnung und Verhandlung. Möglichkeiten eines Kulturwandels durch Reise.* Lit.
Grosvenor. (2014). *Global cities – A grosvenor research report.* Grosvenor. www.triplepundit.com/…/top-50-cities-world-ranked-resilience/
Gualda, G. A. R., Pamukcu, A. S., Ghiorso, M. S., Anderson, A. T., Sutton, S. R., & Rivers, M. L. (2012). Timescales of quartz crystallization and the longevity of the bishop giant magma body. *PLoS ONE, 7*(5). https://doi.org/10.1371/journal.pone.0037492
Gulley, A. (2020, April 30). New Zealand has 'effectively eliminated' coronavirus. Here's what they did right. *National Geographic.* https://www.nationalgeographic.com/travel/2020/04/what-new-zealand-did-right-in-battling-coronavirus/
Haase, T. (2017). *Versicherung ersetzt 34 Mitarbeiter durch Künstliche Intelligenz.* Deutschlandfunk Nova. https://www.deutschlandfunknova.de/beitrag/versicherung-in-japan-statt-angestellten-rechnen.
Habermas, J. (1981). *Theorie des kommunikativen Handelns. Band 1: Handlungsrationalität und gesellschaftliche Rationalisierung. Band 2: Zur Kritik der funktionalistischen Vernunft.* Suhrkamp
Hackenberg, H., & Empter, S. (2011). *Social Entrepreneurship – Social Business: Für die Gesellschaft unternehmen.* Springer.
Hagerty, M. R., & Veenhoven, R. (2003). Wealth and happiness revisited – Growing wealth of nations does go with greater happiness. *Social Indicators Research, 64,* 1–27.
Hall, P. A., & Soskice, D. (2001). *Varieties of capitalism: The institutional foundations of comparative advantage.* Oxford University Press.
Halvorsen, K. (2001). *"How to combat unemployment?" Vortrag an der Cost A 13 Konferenz zum Thema "Social Policy, Marginalisation and Citizenship".* Aalborg University.
Hamilton, M. (2008). *Integral city: Evolutionary intelligences for the human hive.* New Society Publishers.

Händeler, E. (2005). Das Gesundheitswesen als Wachstumsmotor. *Außenreport Versicherungsdienste, 2005* (4). http://www.kondratieff.biz/download/artikel_a_v.pdf

Hanekop, H., & Wittke, V. (2009). Kollaboration der Prosumenten. Die vernachlässigte Dimension des Prosuming-Konzepts. In B. Blättel-Mink & K-U. Hellmann (Hrsg), *Prosumer Revisited. Zur Aktualität einer Debatte* (S. 96–113). VS Verlag.

Hansjürgens, B., & Heinrichs, D. (2007). *Mega-Urbanisierung: Chancen und Risiken: Nachhaltige Entwicklung in Megastädten*. Bundeszentrale für politische Bildung (BPB). http://www.bpb.de/internationales/weltweit/megastaedte/64706/urbanisierung-chancen-und-risiken?p=all

Hartmann, J. (2012, September 19). Siemens errichtet spektakulären Glaspalast in London. *Die WELT*. https://www.welt.de/wirtschaft/article109339056/Siemens-errichtet-spektakulaeren-Glaspalast-in-London.html

Hartzog, P. B. (2007). *Panarchy: Governance in the network age. Masterarbeit an der University of Utah*. Wayback Machine. https://web.archive.org/web/20070928104543/http:/panarchy.com/Members/PaulBHartzog/Papers/Panarchy%20-%20Governance%20in%20the%20Network%20Age.pdf

Harvey, D. (2013). *Rebellische Städte*. Suhrkamp.

Hasel, V. F. (2019, December 26). Gebt den Kindern einen Grund zum Lernen [Give the children a reason to learn]. *Die Zeit Online*. https://www.zeit.de/gesellschaft/schule/2019-12/bildung-neuseeland-schulen-lehrer-kinder-lernen

Hauff, V. (Ed.). (1987). *Unsere Gemeinsame Zukunft. Der Brundtland-Bericht der Weltkommission für Umwelt und Entwicklung*. Eggenkamp.

Heclo, H. (1974). *Modern social politics in Britain and Sweden: From relief to income maintenance*. Yale University Press.

Heidelberger Institut für Internationale Konfliktforschung (HIIK). http://www.hiik.de/de/konfliktbarometer/index.html

Heidenreich, M., Barmeyer, C., Koschatzky, K., Mattes, J., Baier, E., & Krüth, K. (2012). *Multinational enterprises and innovation: Regional learning in networks*. Routledge.

Heidenreich, M., Kädtler, J., & Mattes, J. (Hrsg). (2016). *Die innerbetriebliche Nutzung externer Wissensbestände in vernetzten Entwicklungsprozessen – Endbericht zum Projekt "Kollaborative Innovationen"*. Oldenburger Studien zur Europäisierung und zur transnationalen Regulierung, Nr. 25.

Heidrun, A. (2004). *DDC-Sachgruppen der Deutschen Nationalbibliografie: Leitfaden zu ihrer Vergabe*. Bibliothek.

Helfrich, S. (2012). Gemeingüter sind nicht, sie werden gemacht. In S. Helfrich & Heinrich-Böll- Stiftung (Hrsg), *Commons. Für eine neue Politik jenseits von Markt und Staat* (S. 66–69). Transcript.

Herrmann, U. (2015). *Der Sieg des Kapitals: Wie der Reichtum in die Welt kam: Die Geschichte von Wachstum, Geld und Krisen*. Piper.

HEUNI/UNODC. (2010). *International statistics on crime and justice* (HEUNI Publication Series No. 64). Helsinki. http://www.heuni.fi/Satellite?blobtable=MungoBlobs&blobcol=urldata&SSURIapptype=BlobServer&SSURIcontainer=Default&SSURIsession=false&blobkey=id&blobheadervalue1=inline;%20filename=Hakapaino_final_07042010.pdf&SSURIsscontext=Satellite%20Server&blobwhere=1266335656647&blobheadername1=Content-Disposition&ssbinary=true&blobheader=application/pdf

Heylighen, F. (2007). The global superorganism: An evolutionary-cybernetic model of the emerging network society. *Social Evolution & History, 6*(1), 58–119. http://pespmc1.vub.ac.be/Papers/Superorganism.pdf

Himmelreich, L. (2010). US-Ökonom empfiehlt Deutschland als Kolonialmacht. *SPIEGEL Online*. http://www.spiegel.de/wirtschaft/soziales/entwicklungshilfe-us-oekonom-empfiehlt-deutschland-als-kolonialmacht-a-668449.html

Hodgson, A. (2009). *Transformative resilience – A response to the adaptive imperative*. Compilation, based on papers written for the Carnegie UK Trust, Investigations of the International Futures Forum and a research seminar on the 9 December 2009, held at The Boathouse, Aberdour and supported by the RSA Scotland.

Hoffmann, M., & Patton, K. M. (1996). *Knowledge management for an adaptive organization. The tenets of knowledge management* (Report no. 839). Business Intelligence Program, Menlo Park.

Hofstede, G. (1994). *Cultures and organizations: Software of the mind: Intercultural*. HarperCollins.

Hölldobler, B., & Wilson, E. O. (2009). *Der Superorganismus. Der Erfolg von Ameisen, Bienen, Wespen und Termiten*. Springer.

Holling, C., & Gunderson, L. (2002). Resilience and adaptive cycles. In C. Holling & L. Gunderson (Hrsg), *Panarchy. Understanding transformations in human and natural systems* (S. 25–62). Island Press.

Holling, C., Gunderson, L., & Peterson, G. (2002). Sustainability and panarchies. In L. Gunderson & C. Holling (Eds.), *Panarchy. Understanding transformations in human and natural systems* (pp. 63–102). Island Press.

Homm, F. (2016). *Endspiel: Wie Sie die Kernschmelze des Finanzsystems sicher überstehen*. FinanzBuch.

Horton, H. (2016). Microsoft deletes 'teen girl' AI after it became a Hitler-loving sex robot within 24 hours. *The Telegraph*. https://www.telegraph.co.uk/technology/2016/03/24/microsofts-teen-girl-ai-turns-into-a-hitler-loving-sex-robot-wit/

Horx, M. (2003). *Future Fitness: Wie Sie Führungskompetenz erhöhen. Ein Handbuch für Entscheider*. Eichborn.

Horx, M. (2014). *Das Megatrend-Prinzip: Wie die Welt von morgen entsteht*. Pantheon.

Hübenthal, U. (1991). *Interdisziplinäres Denken. Versuch einer Bestandsaufnahme und Systematisierung*.

Hubert, A. (2016). *Von Bananenbäumen träumen*. http://www.vbbt-derfilm.de/

Huddleston, T. Jr. (2015, September 15). It's Jeff Bezos vs. Elon Musk and Richard Branson – In space. *Fortune*. https://fortune.com/2015/09/15/jeff-bezos-space/

Huesmann, F. (2021). Der QAnon-Boom. Der Erfolg der Verschwörungsideologie in Deutschland. In H. Kleffner & M. Meisner (Eds.), *Fehlender Mindestabstand. Die Coronakrise und die Netzwerke der Demokratiefeinde* (pp. 109–116). Herder.

Huffington, A. (2013). Davos 2013: Resilience as a 21st century imperative. *Huffington Post*. https://www.huffingtonpost.com/arianna-huffington/davos-2013-resilience-as-_b_2526545.html?guccounter=1

Human Wrongs Watch. (2012). *Humanity could not survive a nuclear war using even a fraction of existing Arsenals*. Pressenza International Press Agency. https://www.pressenza.com/2012/12/humanity-could-not-survive-a-nuclear-war-using-even-a-fraction-of-existing-arsenals/

Huntington, S. P. (1993, Summer). The clash of civilizations? *Foreign Affairs.* https://www.foreignaffairs.com/articles/united-states/1993-06-01/clash-civilizations

Ibert, O., Müller, F. C., & Stein, A. (2014). *Produktive Differenzen. Eine dynamische Netzwerkanalyse von Innovationsprozessen.* Transcript.

IBM. *The DeepQA research team.* https://researcher.watson.ibm.com/researcher/view_group.php?id=2099

ICRC. (1976). *Convention on the prohibition of military or any hostile use of environmental modification techniques.* https://ihl-databases.icrc.org/applic/ihl/ihl.nsf/INTRO/460?OpenDocument

IDGs. (2021). *Inner Development Goals: Background, method and the IDG framework.* https://www.innerdevelopmentgoals.org/framework

IESE. *Smart rural areas.* https://www.iese.fraunhofer.de/de/innovation_trends/sra.html

Ikenberry, G. J. (2008). The rise of china and the future of the west. *Foreign affairs.* https://www.foreignaffairs.com/articles/asia/2008-01-01/rise-china-and-future-west

ILO. (2016). *A collective challenge – World day for health and safety at work.* www.ilo.org/wcmsp5/groups/.../wcms_473267.pdf

Imbrie, J., & Imbrie, K. P. (1979). *Ice ages: Solving the mystery.* Enslow Publishers.

IMD. (2018a). *World competitiveness ranking 2018.* https://www.imd.org/wcc/world-competitiveness-center-rankings/world-competitiveness-ranking-2018/

IMD. (2018b). *World digital competitiveness ranking 2018.* https://www.imd.org/wcc/world-competitiveness-center-rankings/world-digital-competitiveness-rankings-2018/

IMD. (2022a). *World digital competitiveness ranking 2021.* https://www.imd.org/centers/world-competitiveness-center/rankings/world-digital-competitiveness/

IMD. (2022b). *World competitiveness ranking 2021.* https://www.imd.org/centers/world-competitiveness-center/rankings/world-competitiveness/

Initiative Grundeinkommen Ulm. *Homepage.* https://www.grundeinkommen-ulm.de/

Innovative Mitte. *Homepage.* http://www.innovativemitte.de/

INSEAD/WIPO/Cornell SC Johnson College of Business. *Global Innovation Index 2018.* https://www.globalinnovationindex.org/gii-2018-report

Institute of Physics. (2004). Nanotechnology pioneer slays 'Grey Goo' myths. *ScienceDaily.* www.sciencedaily.com/releases/2004/06/040609072100.html

IRS Leibnitz-Institut. (2013). *Der State of the Art der Resilienzforschung.* http://www.irs-net.de/aktuelles/meldungen-detail.php?id=206

Isaksen, A., & Asheim, B. (2002). Regional innovation systems: The integration of local 'sticky' and global 'ubiquitous' knowledge. *The Journal of Technology Transfer, 27*(1), 77–86.

Ishay, M. R. (2008). *The history of human rights – From ancient times to the globalization era.* University of California Press.

ISO. (2017). *ISO 2316:2017 Security and resilience –Organizational resilience – Principles and attributes.* https://www.iso.org/standard/50053.html

Jackson, M. C. (2003). *Systems thinking: Creative holism for managers.* Wiley.

Jaeckel, M., & Bronnert, K. (2013). *Die digitale Evolution moderner Großstädte. Apps-basierte innovative Geschäftsmodelle für neue Urbanität.* Wiesbaden.

Jäger, W. (2008). Einführung "Humane Marktwirtschaft – Wirtschaftsordnung für eine menschliche Zukunft". Colloquium politicum, Albert-Ludwigs-Universität Freiburg, Vortragsserie Humane Marktwirtschaft.

Jaques, M. (2012). *When china rules the world: The end of the western world and the birth of a new global order by Martin Jacques.* Penguin.
Kabat-Zinn, J. (1982). An outpatient program in behavioral medicine for chronic pain patients based on the practice of mindfulness meditation: Theoretical considerations and preliminary results. *General Hospital Psychiatry, 4*(1), 33–47. https://doi.org/10.1016/0163-8343(82)90026-3
Kahneman, D. (2012). *Schnelles Denken, langsames Denken.* Siedler.
Kaiser, R. (2001). *Bürger und Staat im virtuellen Raum. Elektronische Demokratie und virtuelles Regieren.* VS Verlag.
Kaiser, J., Chmura, T., & Pitz, T. (2006, October 12). The Tobin Tax – A game-theoretical and an experimental approach. *SSRN.* https://papers.ssrn.com/sol3/papers.cfm?abstract_id=936924
Kamei, R. (2017, October 12): How Singapore encourages lifelong learning and workforce resilience. *The Diplomat.* https://thediplomat.com/2017/10/how-singapore-encourages-lifelong-learning-and-workforce-resilience/
Kela. (2018). Basic income experiment 2017–2018. *Kela.* https://www.kela.fi/web/en/basic-income-experiment-2017-2018
Kevenhörster, P., et al. (2010). *Japan: Wirtschaft – Gesellschaft – Politik.* VS Verlag.
Khanna, P. (2009). *Der Kampf um die Zweite Welt – Imperien und Einfluss in der neuen Weltordnung.* Berlin Verlag.
Khanna, P. (2019). *The future is Asian.* Weidenfeld & Nicolson.
Killingsworth, M. A. (2021, January 18). Experienced well-being rises with income, even above $75,000 per year. *Psychological and Cognitive Sciences.* https://doi.org/10.1073/pnas.2016976118
Kim, T. H. (2020, April 11). Why is South Korea beating coronavirus? Its citizens hold the state to account. *The Guardian.* https://www.theguardian.com/commentisfree/2020/apr/11/south-korea-beating-coronavirus-citizens-state-testing
Kirchberg, D. (2007). *Kirchberg: Der Aufstieg der Tigerstaaten im 20. Jahrhundert. Eine historische Analyse.* VDM.
Kleibrink, A., & Schmidt, S. (2015). Communities of practice as new actors: Innovation labs inside and outside government. In E. Commission (Hsrg), *Open innovation 2.0 yearbook 2015* (S. 64–73). Luxembourg Publication Office of the European Union.
Klein, J. T. (1990). *Interdisciplinarity: History, theory and practice.* Wayne State University Press.
Klein, L. (2015). What the hell is Systemtheorie? *Zeitschrift OrganisationsEntwicklung (ZOE), 34*(1), 66–68.
Klein, M. (2020). *Research.* Massachusetts Institute of Technology (MIT) Center for Collective Intelligence. https://cci.mit.edu/klein/research/#deliberatorium
Klein, J. T., & Newell, W. H. (1997). Advancing interdisciplinary studies. In J. G. Gaff & J. L. Ratcliff (Hsrg), *Handbook of the undergraduate curriculum: A comprehensive guide to purposes, structures, practices, and change* (S. 393–415). Jossey Bass.
Kneer, G., & Nassehi, A. (2000). *Niklas Luhmanns Theorie sozialer Systeme.* UTB.
Knieps, F., & Pfaff, H. (Eds.). (2016). *Gesundheitsreport 2016.* BKK.
Knoll, N., Rieckmann, N., & Schwarzer, R. (2005). Coping as a mediator between personality and stress outcomes: A longitudinal study with cataract surgery patients. *European Journal of Personality, 19*, 229–247. http://www.psy.miami.edu/faculty/ccarver/sclGermanBriefCOPE.pdf

Kobasa, S. C. (1979). Stressful life events, personality, and health – Inquiry into hardiness. *Journal of Personality and Social Psychology, 37*(1), 1–11.

Kocka, J. (Ed.). (1987). *Interdisziplinarität: Herausforderung – Praxis – Ideologie*. Suhrkamp.

Koelling, M. (2011, October 30). Made in Japan – Die erste Fertigstadt der Welt. *WELT*. https://www.welt.de/wissenschaft/article13685979/Made-in-Japan-Die-erste-Fertigstadt-der-Welt.html

Kohlberg, L. (1995). *Die Psychologie der Moralentwicklung*. Suhrkamp.

Krall, M. (2017). *Der Draghi-Crash: Warum uns die entfesselte Geldpolitik in die finanzielle Katastrophe führt*. FinanzBuch Verlag.

Kraut, R. (1989). *Aristotle on the human good*. Princeton University Press.

Krishnamurti, J. (2007). *Freiheit und wahres Glück*. Heyne.

Kristof, N. (2019). Why 2018 was the best year in human history! *The New York Times*. https://www.nytimes.com/2019/01/05/opinion/sunday/2018-progress-poverty-health.html

Kuhn, T. (1988). *Die Struktur wissenschaftlicher Revolutionen*. Suhrkamp.

Kungfutse. (1980). *Lun Yü. Gespräche*. Translated and published by Wilhelm, R., Eugen-Diederichs-Verlag.

Kunze, I. (2010). Gemeinschaftsprojekte als Experimente nachhaltiger Ökonomie. In *IFIS Wirtschaft in der Zeitenwende – zur Vision einer Maßwirtschaft der Lebensfülle und Schritte zu ihrer Verwirklichung*. Institut für Integrale Studien (IFIS).

Laloux, F. (2014). *Reinventing organizations*. Nelson Parker.

Land Vorarlberg. *Bürgerräte in Vorarlberg*. https://vorarlberg.at/-/buergerraete-in-vorarlberg

Lander, E., Baylis, F., Zhang, F., Charpentier, E., Berg, P., Bourgain, C., Friedrich, B., Joung, J. K., Li, J., Liu, D., Naldine, L., Nie, J.-B., Qiu, R., Schoene-Seifert, B., Shao, F., Terry, S., Wei, W., & Winnacker, E.-L. (2019). Adopt a moratorium on heritable genome editing. *Nature, 567*, 165–168. https://doi.org/10.1038/d41586-019-00726-5

Langer, E. (2014). Mindfulness in the age of complexity. *Harvard Business Review*. https://hbr.org/2014/03/mindfulness-in-the-age-of-complexity

Laotse (2004). *Laotse: Tao Te-King*. Translated and published by Wilhelm, R. Eugen Diederich.

Larsen, N., Mortensen, J. K., & Miller, R. (2020, February 11). *What is 'futures literacy' and why is it important?* https://medium.com/copenhagen-institute-for-futures-studies/what-is-futures-literacy-and-why-is-it-important-a27f24b983d8

Latour, B. (1996). On actor-network theory. A few clarifications. *Soziale Welt, 47*(4), 369–382.

Lau, S. (2013, November 12). Macau's residents each get 9,000 pataca handout, but critics not satisfied. *South China Morning Post*. https://www.scmp.com/news/china/article/1354764/macaus-residents-each-get-9000-pataca-handout-critics-not-satisfied

Layard, R. (2005). *Die glückliche Gesellschaft*. Campus.

Layder, D. (1997). *Modern social theory: Key debates and new directions*. UCL Press.

Lazar, S. W., Kerr, C. E., Wasserman, R. H., Gray, J. R., Greve, D. N., Treadway, M. T., McGarvey, M., Quinn, B. T., Dusek, J. A., Benson, H., Rauch, S. L., Moore, C. I., & Fischl, B. (2005). Meditation experience is associated with increased cortical thickness. *Neuroreport, 16* (17): 1893–1897. PMC US National Library of Medicine. http://www.ncbi.nlm.nih.gov/pmc/articles/PMC1361002/. Accessed 28 Nov 2005.

Lederach, J. P. (2003). *Conflict transformation, extracts from Lederach's book "the little book of conflict transformation"*. zusammengesetzt von Maiese, M. http://www.beyondintractability.org/essay/transformation/?nid=1223

Lederach, J. P. (2005). *The moral imagination – The art and soul of building peace*. Oxford University Press.

Lee, K. Y. (2000). *From third world to first. The Singapore story: 1965–2000. Memoirs*. Times Publishing.

Legatum Institute. (2018). *The Legatum prosperity index 2018*. https://prosperitysite.s3-accelerate.amazonaws.com/2515/4321/8072/2018_Prosperity_Index.pdf

Leggewie, C. (2017). Begegnungen mit dem Unvorhergesehenen. *EgonZehnder*. https://www.egonzehnder.com/de/insight/begegnungen-mit-dem-unvorhergesehenen. Accessed 1 Jan 2017.

Lenz, C., & Rucklak, N. (2016, April 27). Honduras als Experimentierfeld neoliberaler Utopien. *amerika21.de*. https://amerika21.de/analyse/145288/charter-cities-honduras

Leven met water. *Homepage*. http://www.levenmetwater.nl/home/

Lexas. (2020). *Länderdaten – Weltrisikoindex*. https://www.laenderdaten.de/indizes/weltrisikoindex.aspx

Libbe, J. (2014). *Difu-Berichte 2/2014 – Standpunkt: Smart City: Herausforderung für die Stadtentwicklung*. https://difu.de/publikationen/difu-berichte-22014/standpunkt-smart-city-herausforderung-fuer-die.html

Libet, B. (1985). Unconscious cerebral initiative and the role of conscious will in voluntary action. *The Behavioral and Brain Sciences, 8*, 529–566.

Lieven, O., & Maasen, S. (2007). Transdisziplinäre Forschung. Vorbote eines "New Deal" zwischen Wissenschaft und Gesellschaft? *GAIA, 1*, 35–40.

Lin-Hi, N. (n.d.). *Multi-stakeholder-initiative*. Springer. https://wirtschaftslexikon.gabler.de/definition/multi-stakeholder-initiative-53810

Lipset, S. M. (1960). *Political man. The social bases of politics*. Doubleday.

Liquid Democracy. *Homepage*. https://liqd.net/de/about/

Littig, B., & Grießler, E. (2004). *Soziale Nachhaltigkeit. Informationen zur Umweltpolitik: Bd. 160*. Kammer für Arbeiter und Angestellte.

Lobe, A. (2018, January 1). Filterblasen, in denen wir wohnen wollen. *DIE ZEIT Online*. https://www.zeit.de/kultur/2017-12/smart-cities-google-sidewalk-labs-modellstadt/komplettansicht

Loevinger, J. (1976). *Ego development. Conceptions and theories*. Jossey-Bass.

Loh, J. (2020). *Trans- und Posthumanismus zur Einführung*. Junius.

Lohmann-Haislah, A. (2012). *Stressreport Deutschland 2012 – Psychische Anforderungen. Ressourcen und Befinden*. Bundesanstalt für Arbeitsschutz und Arbeitsmedizin.

Loorbach, D. (2007). *Transition management. New mode of governance for sustainable development*. International Books.

Loorbach, D. (2014). To transition! Governance panarchy in the new transformation. *Drift*. https://drift.eur.nl/wp-content/uploads/2016/12/To_Transition-Loorbach-2014.pdf

Loughborough University. *Homepage*. http://www.lboro.ac.uk/service/publicity/news-releases/2012/72_resilience.html

Lovelock, J. (1991). *Das Gaia-Prinzip: Die Biographie unseres Planeten*. Artemis & Winkler.

Lovelock, J. (1992). *Gaia – Die Erde ist ein Lebewesen*. Scherz.

References

Lu, F. (2019, February 18). Die Demütigung eines uralten Reiches. *Die Zeit*. https://www.zeit.de/kultur/2019-02/china-usa-weltmacht-handelskonflikt-welthandel

Luhmann, N. (1988). *Die Wirtschaft der Gesellschaft*. Suhrkamp.

Luhmann, N. (1993). *Soziale Systeme – Grundriß einer allgemeinen Theorie*. Suhrkamp.

Luhmann, N. (2006). *Beobachtungen der Moderne*. VS Verlag.

Lukesch, R., Payer, H., & Winkler-Riederer, W. (2010). *Wie gehen Regionen mit Krisen um? Eine explorative Studie über die Resilienz von Regionen, im Auftrag des Bundeskanzleramtes*. ÖAR Regionalberatung GmbH. http://www.oear.at/downloads/

Luks, F. (2002). *Nachhaltigkeit*. Europäische Verlagsanstalt.

Luthar, S. S. (2003). The culture of affluence: psychological costs of material wealth. *Child development, 74*(6), 1581–1593. https://doi.org/10.1046/j.1467-8624.2003.00625.x

Luthar, S. S., & Cicchetti, D. (2000). The construct of resilience: Implications for interventions and social policies. *Developmental Psychopathology, 12*, 857–885.

Lutz, A., et al. (2004). Long-term meditators selfinduce high-amplitude gamma synchrony during mental practice. *Proceedings of the National Academy of Sciences, 101*(46), 16369–16373.

Lykken, D., & Tellegen, A. (1996). Happiness is a stochastic phenomenon. *Psychological Science, 7*(3), 186–189. http://www.psych.umn.edu/courses/fall06/macdonalda/psy4960/Readings/LykkenTwinHappiness_PS96.pdf

Maharshi, R. (2018). *Nan Yar? Wer bin ich?* AdvaitaMedia.

Mair, J., Robinson, J., & Hockerts, K. (2006). *Social entrepreneurship*. Palgrave Macmillan.

Maiwald, J. (2018). *Smart decision-making: Systemic Consensing for Managers* (SmarterLife). A-bis.

Malhotra, S. (2014, April 10). Indian companies will have to elevate their status from low-cost producers to innovative businesses. *BusinessToday*. https://www.businesstoday.in/opinion/interviews/story/indian-companies-will-have-to-elevate-their-status-from-low-cost-producers-to-innovative-businesses-47423-2014-04-10

Malik, F. (1992). *Strategie des Managements komplexer Systeme. Ein Beitrag zur Management-Kybernetik evolutionärer Systeme*. Haupt.

Markram, H. (2006). The blue brain project. *Nature Reviews Neuroscience, 7*, 153–160. https://doi.org/10.1038/nrn1848. Accessed 1 Feb 2006.

Markram, H. (2015). Reconstruction and simulation of neocortical microcircuitry. *Cell Press, 163*(2):456–492. https://doi.org/10.1016/j.cell.2015.09.029. Accessed 8 Okt 2015.

Marr, M., & Zillien, N. (2010). Digitale Spaltung. In W. Schweiger & K. Beck (Hrsg), *Handbuch Online-Kommunikation* (S. 257–282). VS Verlag.

Masdar City. *Homepage*. https://masdar.ae/en/masdar-city

Maslow, A. (1954). *Motivation and personality*. Harper.

Masten, A. S. (2011). Resilience in children threatened by extreme adversity: Frameworks for research, practice, and translational synergy. *Development and Psychopathology, 23*, 493–506.

Matheny, J. G. (2007). Reducing the risk of human extinction. *Risk Analysis, 27*(5), 1335–1344.

Maturana, H., & Varela, F. (1982). *Autopoietische Systeme: Eine Bestimmung der lebendigen Organisation* (pp. 170–235). Erkennen.

May, P. (1992). Policy learning and failure. *Journal of Public Policy, 12*(4), 331–354.

MBSR Association. *Homepage*. http://www.mbsr-verband.de/mbsr-mbct/forschung.html

Meadows, D.H., Meadows, Randers, J., & Behrens, W.W. (1972). *The limits to growth – A report for the club of Rome's project on the predicament of mankind.* Universe Books. Download at: https://collections.dartmouth.edu/content/deliver/inline/meadows/pdf/meadows_ltg-001.pdf

Measuring-Progress.eu (n.d.). *Technological achievement index.* https://measuring-progress.eu/technology-achievement-index

Medina, E. (2011). *Cybernetic revolutionaries: Technology and politics in Allende's Chile.* MIT Press.

Medrano, J. (2012). *Interpersonal Trust, WVS Archive.* http://www.jdsurvey.net/jds/jdsurveyMaps.jsp?Idioma=I&SeccionTexto=0404&NOID=104

Meier, A. (2009). *eDemocracy & eGovernment – Entwicklungsstufen einer demokratischen Wissensgesellschaft.* Springer.

Messner, D. (2000). Globalisierung, Global governance und Perspektiven der Entwicklungszusammenarbeit. In F. Nuscheler (Hrsg), *Entwicklung und Frieden im 21. Jahrhundert* (S. 267–294). Stiftung Entwicklung und Frieden.

Michael, L. (2019). *Wir sollten uns vertrauen – Der Aufstand in gelben Westen.* Nautilus.

Michaelson, J., Abdallah, S., Steuer, N., Thompson, S., & Marks, N. (2009). *National accounts of well-being: Bringing real wealth onto the balance sheet.* New economics foundation (nef). https://neweconomics.org/uploads/files/2027fb05fed1554aea_uim6vd4c5.pdf

Miller, R. (2018). *Transforming the future: Anticipation in the 21st century.* UNESCO/Routledge.

MIPEX. (2015). *Migrant integration policy index.* http://www.mipex.eu/

Mittelstraß, J. (1998). *Häuser des Wissens. Wissenschaftstheoretische Studien.* Suhrkamp.

Mohideen, R. (2010). Disaster management: New Zealand, Haiti and the 'Cuban way'. *International Journal of Socialist Renewal.* http://links.org.au/node/1890

Monetative. *Homepage.* http://www.monetative.de/

Mörchen, L. (2015). The speed camera lottery – The fun theory. *JPG Gemeinde Freiburg.* https://www.jpg-freiburg.de/node/376. Accessed 28 Jan 2015.

Morin, E. (2010). *La Méthode. 6 Bände: Bd. 1: Die Natur der Natur.* Turia-Kant.

Morris, I. (2011). *Wer regiert die Welt?: Warum Zivilisationen herrschen oder beherrscht werden.* Campus.

Morrison, D. (2006). Asteroid and comet impacts: The ultimate environmental catastrophe. *Philosophical Transactions of the Royal Society, 364*(1845), 2041–2054. https://doi.org/10.1098/rsta.2006.1812

Mosadeghrad, A., & Ansarian, M. (2014). Why do organisational change programmes fail? *International Journal of Strategic Change Management, 5*(3), 189. https://doi.org/10.1504/IJSCM.2014.064460

Mourlane, D. (2017). *Resilienz: Die unentdeckte Fähigkeit der wirklich Erfolgreichen.* BusinessVillage.

Mühlhans, T. (2018). *Open cities.* https://projektzukunft.berlin.de/projekt-zukunft/services/internationale-kooperationen/open-cities/

Müller-Seitz, G. (2015). Von Risiko zu Resilienz – zum Umgang mit Unerwartetem aus Organisationsperspektive. *Zeitschrift für betriebswirtschaftliche Forschung, 68*(14), 102–122.

Münkler, H. (2007). *Imperien – Die Logik der Weltherrschaft.* Rowohlt.

Murawski, H., & Meyer, W. (2004). *Geologisches Wörterbuch*. Spektrum Akademischer Verlag.
Neapolitan, J. L. (1999). A comparative analysis of nations with low and hoch levels of violent crime. *Journal of Criminal Justice, 27*(3), 259–274.
Nettlau, M. (1909). Panarchie, eine vergessene Die von 1860. *Der Sozialist, 15*(3), 1909.
Neumann, K. E. (1922). *Die Reden Gotamo Buddhos. Aus der Mittleren Sammlung Majjhimanikayo des Pali-Kanons zum ersten Mal übersetzt von K. E. Neumann. Hier in 3 Bänden komplett*. Piper.
New Economic Foundation (NEF). (2012). *National accounts of well-being*. http://www.nationalaccountsofwellbeing.org/explore/countries/gb
Nonaka, I., & Takeuchi, H. (1997). *Die Organisation des Wissens. Wie japanische Unternehmen eine brachliegende Ressource nutzbar machen*. Campus.
Nowak, M., & Sigmund, K. (1993, July 1). A strategy of win-stay, lose-shift that outperforms tit-for-tat in the Prisoner's Dilemma game. *Nature, 364*(6432), 56–58. https://doi.org/10.1038/364056a0
Nowotny, H., Limoges, C., Schwartzman, S., Scott, P., Trow, M., & Gibbons, M. (1994). *The new production of knowledge. The dynamics of science and research in contemporary societies*. Sage.
Nowotny, H., Scott, P., & Gibbons, M. (2001). *Re-thinking science. Knowledge and the public in an age of uncertainty*. Polity.
Nuscheler, F. (2006). *Entwicklungspolitik*. Bundeszentrale für Politische Bildung.
Nuscheler, et al (2007). Globale Verwundbarkeiten und die Gefährdung "menschlicher Sicherheit". In: SEF/INEF (Eds.), *Globale trends 2007* (pp. 9–36).
Nye, J. (2004). *Soft power. The means to success in world politics*. PublicAffairs.
Nye, J. (2011). *The future of power*. PublicAffairs.
O'Dougherty, W. M., Masten, A. S., & Narayan, A. J. (2013). Resilience processes in development: Four waves of research on positive adaptation in the context of adversity. In S. Goldstein & R. B. Brooks (Eds.), *Handbook of resilience in children* (pp. 15–37). Springer.
O'Mahony, S. (2006). Developing community software in a commodity world. In M. S. Fischer & G. Downey (Eds.), *Frontiers of capital* (pp. 237–266). Duke University Press.
OECD. (1972). *Interdisciplinarity. Problems of teaching and research in universities*. OECD.
OECD. (2001). *OECD science, technology and industry scoreboard 2001 – Towards a knowledge-based economy*. OECD.
OECD. (2008, November). *Policy brief – Mental health in OECD countries*. www.oecd.org/publications/Policybriefs
OECD. (2011a). *Society at a glance 2011 – OECD social indicators*. OECD.
OECD. (2011b). *OECD family database*. OECD. www.oecd.org/social/family/database
OECD. (2011c). *Better life index*. http://www.oecdbetterlifeindex.org/de/
OECD. (2012). *Sick on the job? Myths and realities about mental health and work* (Mental health and work series). OECD.
OECD. (2017). *Behavioural insights and public policy: Lessons from around the world*. OECD. https://doi.org/10.1787/9789264270480-en
OECD. (n.d.). *PISA – Internationale Schulleistungsstudie der OECD*. http://www.oecd.org/berlin/themen/pisa-studie/
Oekom: oekom Research AG. https://www.oekom.de/person/oekom-research-ag-293

Online Etymology Dictionary. (n.d.). *Sustain*. https://www.etymonline.com/word/sustain
OPHI. (n.d.). *Policy – A multidimensional approach*. Oxford Poverty and Human Development Initiative. http://www.ophi.org.uk/policy/multidimensional-poverty-index/
Orrell, D. (2007). *The future of everything – The science of prediction*. Thounder's Mouth Press.
Ortiz-Ospina, E., & Roser, M. (2016). Trust. *OurWorldInData.org*. https://ourworldindata.org/trust
Otto, G. (2021, January 8). Corona verschärft die soziale Spaltung. *Wirtschaftszeitung*. https://www.die-wirtschaftszeitung.de/aktuelles/__trashed/
Paech, N. (2012). *Befreiung vom Überfluss – auf dem Weg in die Postwachstumsökonomie*. Oekom-Verlag.
Palin, P. (2011). Learning from Japan: Sources of resilience. *Homeland Security Watch*. http://www.hlswatch.com/2011/03/25/learning-from-japan-sources-of-resilience/. Accessed 25 März 2011.
Palley, T. I. (2000, June). Destabilizing speculation and the case for an international currency transactions tax. *Global Policy Forum*. https://archive.globalpolicy.org/social-and-economic-policy/global-taxes-1-79/currency-transaction-taxes/45990-destabilizing-speculation-and-the-case-for-an-international-currency-transactions-tax.html
Panyasara. *Homepage*. http://www.panyasara.de/
Parsons, T. (1971). *The system of modern societies*. Prentice Hall.
Patel, S., Rogers, M., & Amlot, R., et al. (2017, February 1). What do we mean by, community resilience? A systematic literature review of how it is defined in the literature. *PLoS Currents*
Pautz, H. (2010). The internet, political participation and election turnout. *German Politics & Society, 28*(3), 156–175. https://doi.org/10.3167/gps.2010.280259
PECC/JANCPEC. (2010). *Towards a more resilient society: Lessons from economic crises*. Pacific Economic Cooperation Council (PECC)/Japan National Committee for Pacific Economic Cooperation (JANCPEC).
Pendall, R., Foster, K., & Cowell, M. (2010). Resilience and regions: Building understanding of the metaphor. *Cambridge Journal of Regions, Economy and Society, 3*(1), 71–84.
Perlas. N. (2000). *Die Globalisierung gestalten*.
Perron, W. (2011). Resilienz in der offenen Gesellschaft – das Freiburger Center for Security and Society. In H. Just, H. Kind, & H. Koch (Eds.), *Solidarität: Dem Einzelnen oder der Gesellschaft verpflichtet? Kolloquium 19. November 2010, Schriftenreihe der Ethik – Kommission der Albert-Ludwigs-Universität Issue 5*. Ethik-Kommission.
Pestel-Institut. (2010). *Regionale Krisenfestigkeit. Eine indikatorengestützte Bestandsaufnahme auf der Ebene der Kreise und der kreisfreien Städte*. Pestel-Institut.
Petermann, T., Bradke, H., Lüllmann, A., Poetzsch, M., & Riehm, U. (2010). *Gefährdung und Verletzbarkeit moderner Gesellschaften – am Beispiel eines großräumigen und langandauernden Ausfalls der Stromversorgung. TAB-Arbeitsbericht Bd. 141*. Büro für Technikfolgen-Abschätzung beim Deutschen Bundestag.
Peyn, G. (2018). https://gitta-peyn.de/formwelt_nerdread/. Accessed: 27 März 2018.
Peyn, G. (2019, Mai). *Persönlicher Austausch*.
Pierson, P. (2001). Post-industrial pressures on the mature welfare states. In P. Pierson (Hrsg), *The new politics of the welfare state* (S. 80–105). Oxford: Oxford University Press.
Popper, K. R. (2003). *Die offene Gesellschaft und ihre Feinde*. Mohr Siebeck.

Popper, K. R. (2004). *Alles Leben ist Problemlösen – Über Erkenntnis, Geschichte und Politik.* Piper.
PopTech. *Homepage.* http://poptech.org/the_city_resilient
Posener, A. (2007). *Imperium der Zukunft. Warum Europa Weltmacht werden muss.* Pantheon.
Presencing Institute: u.lab: Leading From the Emerging Future – An introduction to leading profound social, environmental and personal transformation. Massachusetts Institute of Technology (MIT). edX. https://www.edx.org/course/u-lab-leading-emerging-future-mitx-15-671-1x-0
Probst, G. J. B., & Büchel, B. S. T. (1994). *Organisationales Lernen – Wettbewerbsvorteil der Zukunft.* Gabler.
Probst, G.J.B. & Raisch, S. (2004). Die Logik des Niedergangs. *Harvard Business Manager.* März, S. 37–45.
Prognos. (2010). *Prognos Zukunftsatlas 2010 – Deutschlands Regionen im Zukunftswettbewerb.* Prognos.
pronova BKK. (2018, April 10). 87 Prozent der Menschen in Deutschland sind gestresst – Jeder Zweite glaubt, von Burn-out bedroht zu sein. *Presseportal.* https://www.presseportal.de/pm/119123/3912240
Prototype Fund. *Homepage.* https://prototypefund.de/about/team/
Prout. *Homepage.* http://www.prout.org/
Purdy, C. (2020, April 9). The Dutch want to make big economic reforms with a doughnut. *Quartz.* https://qz.com/1835237/amsterdam-adopted-a-new-economic-model-for-life-after-covid-19/
Purvis, B., Mao, Y., & Robinson, D. (2019). Three pillars of sustainability: In search of conceptual origins. *Sustainability Science, 14,* 681–695. https://doi.org/10.1007/s11625-018-0627-5
Putzier, K. (2011, July 12). Warum bauen die Schweizer so viele Bunker? *Welt Online.* https://www.welt.de/politik/ausland/article13481396/Warum-bauen-die-Schweizer-so-viele-Bunker.html
Radcliff, B., & Pacek, A. (2008). Assessing the welfare state: The politics of happiness. *Perspectives on Politics, 6,* 267–277.
Radermacher, F. J., & Beyers, B. (2011). *Welt mit Zukunft: Die ökosoziale Perspektive.* Murmann.
Radjou, N. & Prabhu, J. (2014). 4 CEOs Who are making frugal innovation work. *Harvard Business Review.* https://hbr.org/2014/11/4-ceos-who-are-making-frugal-innovation-work. Accessed 28 Nov 2014.
Raford, N. (2010). *Drawing a better Panarchy diagram.* http://noahraford.com/?p=648
Ramge, T. (Ed.). (2010). *Jetzt neu – wie wir eine kreative(re) Gesellschaft werden.* Stiftung neue verantwortung.
Rampino, M. R., & Ambrose, S. H. (2002). Super eruptions as a threat to civilizations on Earth-like planets. *Icarus, 156,* 562–569.
Rapoport, A., & Chammah, A. M. (1970). *Prisoner's dilemma – A study in conflict and cooperation.* University of Michigan Press.
Raworth, K. (2018). *Doughnut economics: Seven ways to think like a 21st-century economist.* Random House Business Books.

Rehäuser, J., & Krcmar, H. (1996). Wissensmanagement im Unternehmen. In G. Schreyögg & P. Conrad (Eds.), *Managementforschung 6: Wissensmanagement* (pp. 1–40).
Reivich, K., & Shatté, A. (2003). *The resilience factor*. Random House.
Resalliance. *Homepage*. http://www.resalliance.org/index.php/research
Resonance Consultancy: 2020 World's greenest cities index. https://www.bestcities.org/news/2020/04/22/the-worlds-greenest-cities/
Rheingold, H. (2003). *Smart mobs. The next social revolution*. Perseus.
Rickards, J. (2012). *Währungskrieg: Der Kampf um die monetäre Weltherrschaft*. Finanz-Buch Verlag.
Rinke, A., & Schwägerl, C. (2015). *11 drohende Kriege: Künftige Konflikte um Technologien, Rohstoffe, Territorien und Nahrung*. Btb Verlag.
Rist, M. (2017). *Singapur – Vorzeigestaat oder Auslaufmodell?* NZZ Global Risk. http://nzz-files-prod.s3-website-eu-west-1.amazonaws.com/2018/2/27/d04ce686-decb-4e0b-9265-931ea3362e5b.pdf
Ritchie, H., & Roser, M. (2019). Urbanization. *OurWorldInData.org*. https://ourworldindata.org/urbanization
Ritter, F. E., Shadbolt, N. R., Elliman, D., Young, R., Gobet, F., & Baxter, G. D. (2001, May 13). *Techniques for modeling human performance in synthetic environments: A supplementary review*. Human Systems Information Analysis Center.
Robock, A. (2008). 20 reasons why geoengineering may be a bad idea. *Bulletin of the Atomic Scientists, 64*(2), 14–18., 59. https://doi.org/10.2968/064002006
Roehl, H. (2010). *Perspektiven für eine differenzierende Interventionspraxis*. Gabler.
Rogers, E. M. (2003). *Diffusion of innovations*. Free Press.
Romer, P. (2009, July). Why the world needs charter cities. *TEDGlobal*. https://www.ted.com/talks/paul_romer_why_the_world_needs_charter_cities
Rosenberg, M. (2009). *Gewaltfreie Kommunikation: Eine Sprache des Lebens*. Junfermann.
Roth, J., & Rüesch, A. (2022, April 25). – wie ein 22-jähriger Schweizer mit ausländischen Geheimdiensten wetteifert. *Neue Züricher Zeitung*. https://www.nzz.ch/schweiz/ukraine-ein-22-jaehriger-schweizer-spuert-putins-truppen-nach-ld.1679787?utm_source=pocket-newtab-global-de DE
Sabatier, P. (1987). Knowledge, policy-oriented learning, and policy change. *Knowledge: Creation, Diffusion, Utilization, 8*, 649–692.
Sachs, J. (2005). *Das Ende der Armut*. Bundeszentrale für Politische Bildung.
Sachs, J. (2010, March 18). Robin Hood tax's time has come. *The Guardian*.. https://www.theguardian.com/commentisfree/2010/mar/18/robin-hood-tax-benefits
Sachs, J., Kroll, C., Lafortune, G., Fuller, G., & Finn Woelm, F. (2021). *Sustainable development report 2021*. Sustainable Development Solutions Network (SDSN). https://s3.amazonaws.com/sustainabledevelopment.report/2021/2021-sustainable-development-report.pdf
Salzborn, S. (Ed.). (2014). *Klassiker der Sozialwissenschaften – 100 Schlüsselwerke im Portrait*. Springer.
Sarvodaya. *Homepage*. http://www.sarvodayausa.org/learn/10-basic-human-needs/
Sautet, F. (2013). Local and systemic entrepreneurship: Solving the puzzle of entrepreneurship and economic development. *Entrepreneurship Theory and Practice, 37*, 387–402. https://doi.org/10.1111/j.1540-6520.2011.00469.x

Savage, M. (2020, July 23). Did Sweden's coronavirus strategy succeed or fail? bbc. https://www.bbc.com/news/world-europe-53498133.
Scharmer, C. O. (1995). *Reflexive Modernisierung des Kapitalismus als Revolution von innen: auf der Suche nach Infrastrukturen für eine lernende Gesellschaft; dialogische Neugründung von Wissenschaft, Wirtschaft und Politik.* M & P Verlag für Wissenschaft und Forschung.
Scharmer, C. O. (2009). *Theory U: Leading from the Future as It Emerges.* Berrett-Koehler Publishers.
Scharmer, C.O. (2017). *Theory U.* https://www.presencing.com/theoryu
Scharmer, C. O., & Käufer, K. (2008). Führung vor der leeren Leinwand. *Presencing als soziale Technik. Zeitschrift OrganisationsEntwicklung (ZOE), 2,* 4–11.
Scharmer, C. O., & Käufer, K. (2013). *Leading from the emerging future; from ego-system to eco-system economies: From ego-system to eco-system economies. Agency/distributed.* Berrett-Koehler Publishers.
Schein, E. H. (1993). On dialogue, culture and organizational learning. *Organizational Dynamics, 22*(2), 40–51.
Schelling, T. (1978). *Micromotives and macrobehavior.* Norton.
Scheuerle, T., Gunnar G., Rüdiger K., & Then, V. (2013). *Social Entrepreneurship in Deutschland: Potentiale und Wachstumsproblematiken.* CSI der Universität Heidelberg, on behalf of KfW Bankengruppe Research.
Schmelzer, A. (1991). *Die Dreigliederungsbewegung 1919. Rudolf Steiners Einsatz für den Selbstverwaltungsimpuls.* Freies Geistesleben.
Schmidt, A.K. (2019, May 5). *Wo europäische Subventionen afrikanischen Landwirt*innen schaden.* Junge Europäische Föderalisten (JEF) Deutschland e.V. https://www.treffpunkteuropa.de/wo-europaische-subventionen-afrikanischen-landwirt-innen-schaden?lang=fr
Schneider, L. (1990). *Subsidiäre Gesellschaft – Entwickelte Gesellschaft. Implikative und analoge Aspekte eines Sozialprinzips.* Schöningh.
Schneidewind, U. (2011). Auf dem Weg in die resiliente Gesellschaft. *Berliner Republik* 05/2011. http://www.b-republik.de/aktuelle-ausgabe/auf-dem-weg-in-die-resiliente-gesellschaft
Schneidewind, U., & Singer-Brodowski, M. (2013). *Transformative Wissenschaft – Klimawandel im deutschen Wissenschafts- und Hochschulsystem.* Metropolis-Verlag.
Schnur, O. (2013). Resiliente Quartiersentwicklung? Eine Annäherung über das Panarchie-Modell adaptiver Zyklen. *Informationen zur Raumentwicklung, 4,* 337–350.
Schroders. (2022). *European sustainable cities index 2021.* https://www.schroders.com/en/schrodersglobalcities/resources/schroders-european-sustainable-cities-index/
Schultz, S. (2018, June 7). China baut das Weltstromnetz. *SPIEGEL Online.* https://www.spiegel.de/wirtschaft/unternehmen/china-treibt-globales-stromnetz-voran-a-1211629.html
Schultz, S. (2022, March 15). China baut gigantische Solar- und Windparks in der Wüste. *SPIEGEL Online.* https://www.spiegel.de/wirtschaft/soziales/china-baut-gigantische-solar-und-windparks-in-der-wueste-a-7e000676-03e6-4a6b-9577-7e9c1655586b
Schumpeter, J., Röpke, J., & Stiller, O. (Eds.). (2006). *Theorie der wirtschaftlichen Entwicklung. Neuausgabe.* Duncker und Humblot.
Schwane, F. (2009). Weltwirtschaft: Mythos Abkopplung. *FOCUS Money, 34.* http://www.focus.de/finanzen/news/weltwirtschaft-mythos-abkopplung_aid_425272.html

Senge, P. (1996). *Die fünfte Disziplin: Kunst und Praxis der lernenden Organisation*. Klett-Cotta, Stuttgart [original: The Fifth Discipline: The art and practice of the learning organization. Doubleday].

SenWTF und Landesinitiative Projekt Zukunft. (2013). *Innovations- und Kreativlabs in Berlin – eine Bestandsaufnahme. Räume und Events als Schnittstellen von Innovation und Kreativität* (Bearbeiter: S. Schmidt, V. Brinks, S. Brinkhoff). Senatsverwaltung für Wirtschaft, Technologie und Forschung, Landesinitiative Projekt Zukunft. http://www.berlin.de/projektzukunft/uploads/tx_news/130626_Innovations-_und_Kreativlabs_in_Berlin_-_eine_Bestandsaufnahme_02.pdf

Sewell, J. B., & Salter, M. B. (1995, September/December). Panarchy and other norms for global governance: Boutros-Ghali, Rosenau, and beyond/James P. Sewell and Mark B. Salter. *Global Governance, 1*(3), 373–382.

Shiva, V. (2005). *Earth democracy: Justice, sustainability, and peace*. South End Press.

Sibeon, R. (2004). *Rethinking social theory*. Sage.

Simon, F. B. (2014). *Einführung in die systemische Organisationstheorie*. Carl Auer Verlag.

Simonis, U. (2004). Weltumweltpolitik. In W. Woyke (Hrsg), *Handwörterbuch Internationale Politik* (S. 569–582). Bundeszentrale für politische Bildung.

Simpol. *Homepage*. http://www.simpol.org/index.php?id=8

Simpson, D. (2013). *Post-human omnibus. Book 2: Post-human*. Post-Human Media LLC.

Singer, M. A. (2016). *Die Seele will frei sein*. Ullstein Taschenbuchverlag.

SIPRI (2021). *SIPRI Yearbook 2021*. Stockholm International Peace Research Institute/Oxford University Press. https://www.sipri.org/sites/default/files/2021-06/sipri_yb21_summary_en_v2_0.pdf

SIPRI. (2022). *Environment of peace*. Stockholm International Peace Research Institute/Oxford University Press. https://sipri.org/sites/default/files/2022-05/environment_of_peace_security_in_a_new_era_of_risk_0.pdf

Slater, D., & Tonkiss, F. (2001). *Market society: Markets and modern social theory*. Polity.

Smith, D. (2012). *Dan Smith's blog: Peacebuilding IN Europe?* http://dansmithsblog.com/2012/01/29/peacebuilding-in-europe/. Accessed 29 Jan 2012.

Snowden, D. (2000). Cynefin: A sense of time and space, the social ecology of knowledge management. In C. Despres & D. Chauvel (Eds.), *Knowledge horizons: The present and the promise of knowledge management*. Butterworth-Heinemann.

Snowden, D., & Boone, M. (2007). A leader's framework for decision making. *Harvard Business Review, 85*(11), 69–76.

Spath, D. (Ed.). (2013). *Produktionsarbeit der Zukunft – Industrie 4.0*. Fraunhofer Institut für Arbeitswirtschaft und Organisation (IAO). https://microsites.schott.com/d/studentchallenge/c7d319bc-3fd9-40d2-85c2-636906b2c2f0/1.0/produktionsarbeit_der_zukunft_-_industrie_4_0__fraunhofer_studie.pdf

Spiewack, M. (2017, February 9). Und jetzt werden alle kreativ. *Die Zeit, 7*.

Spindler, M. (2006). Interdependenz. In S. Schieber & M. Spindler (Eds.), *Theorien der Internationalen Beziehungen* (pp. 89–116).

Srivastava, S. (2020). Effectiveness of Vipassana meditation on meta cognition and resilience in patients of the major depressive disorder a pre post study. *UCC Care Journal, 31*(33), 277–289.

SSI. (2017). *Sustainable society index 2006-2016*. Technische Hochschule Köln. https://ssi.wi.th-koeln.de/historical_data.html#ssf

Stambolovic, V. (2002). The case of Serbia/Yugoslavia: An analysis through spiral dynamics. *Medicine, Conflict and Survival, 18*(1), 59–70. https://doi.org/10.1080/13623690208409606

Steinberg, F. (1985). *Grundbedürfnisstrategie. Wohnen in der "Dritten Welt"*. Magazin Verlag.

Stockholm Resilience Center. *Homepage*. http://www.stockholmresilience.org/21/research.html

Stockmann, R. (1996). *Die Wirksamkeit der Entwicklungshilfe. Eine Evaluation der Nachhaltigkeit von Programmen und Projekten der Berufsbildung*. Westdeutscher Verlag.

Stokols, D., Hall, K., Taylor, B. K., & Moser, R. P. (2008). Overview of the Field and Introduction to the Supplement. *American Journal of Preventive Medicine, 35*(2), 77–89.

Störig, H. J. (2002). *Kleine Weltgeschichte der Philosophie*. Stuttgart.

Surowiecki, J. (2004). *The wisdom of crowds. Why the many are smarter than the few and how collective wisdom shapes business, economies, societies and nations*. Little, Brown.

Survivalblog. *Homepage*. http://survivalblog.com/

Sustainable Development Solutions Network. (2020). *2019 US cities sustainable development report*. https://s3.amazonaws.com/sustainabledevelopment.report/2019/2019USCitiesReport.pdf

Sustainable Jungle. *Hitting green goals: The 15 Greenest cities in the world*. https://www.sustainablejungle.com/sustainable-living/greenest-cities/

Sustainable Society Index. (2016). *Homepage*. http://www.ssfindex.com/

Taleb, N. (2008). *The black swan: The impact of the highly improbable*. Penguin.

TEU. (2008). Konsolidierte Fassung des Vertrags über die Europäische Union – TITEL I: GEMEINSAME BESTIMMUNGEN – Artikel 3 (ex-Artikel 2 EUV). *Amtsblatt, 115*.

Treaty on European Union (TEU [German: Vertrag über die Europäische Union – EUV])

Thaler, R. H., & Sunstein, C. (2008). *Improving decisions about health, wealth and happiness*. Penguin Books.

The Australian. (2011). *Crushed, but true to law of 'gaman'*. http://www.theaustralian.com.au/archive/in-depth/crushed-but-true-to-law-of-gaman/story-fn84naht-1226022079002

The World Bank. (2018, March 29). *Overview: BRI at a glance*. https://www.worldbank.org/en/topic/regional-integration/brief/belt-and-road-initiative

Tolle, E. (2012). *Jetzt! Die Kraft der Gegenwart*. Bielefeld.

t-online dpa. (2021, August 5). Warum diese Nationen als Corona-Musterländer gelten [Why these nations are considered Corona model countries]. *t-online*. Available at: https://www.t-online.de/nachrichten/panorama/id_90571514/schweiz-und-norwegen-warum-diese-nationen-als-corona-musterlaender-gelten-.html

Tönnies, F. (2005). *Gemeinschaft und Gesellschaft*. Wissenschaftliche Buchgesellschaft.

Transition Network. (2015). *Homepage*. http://www.transitionnetwork.org/initiatives

Transparency International. *Homepage*. http://cpi.transparency.org/cpi2011/results/

Traub, J. (2015). The hillary clinton doctrine. *Foreign Policy*. http://foreignpolicy.com/2015/11/06/hillary-clinton-doctrine-obama-interventionist-tough-minded-president/

Treehugger (n.d.). *Economics*. https://www.treehugger.com/economics-4846045.

Treichel, D., & Mayer, C.-H. (2011). *Lehrbuch Kultur. Lehr- und Lernmaterialien zur Vermittlung kultureller Kompetenzen*. Waxmann.

Trend Micro. (2022, February). *Trend Micro blockiert über 94 Milliarden Cyberbedrohungen im Jahr 2021*. https://www.trendmicro.com/de_de/about/newsroom/press-releases/2022/2022-01-31-trend-micro-blockiert-uber-94-milliarden-cyberbedrohungen-im-jahr-2021.html

Trentmann, N. (2015, December 8). Perfide Überwachung ist in China Wirtschaftsfaktor. *Die WELT*. https://www.welt.de/wirtschaft/article149753135/Perfide-Ueberwachung-ist-in-China-Wirtschaftsfaktor.html

Tshidimba, D., Lateur, F., & Sneyers, N. (2015). *Frugal products*. Roland Berger Strategy Consultants. https://www.rolandberger.com/publications/publication_pdf/roland_berger_tab_frugal_products_20150601.pdf

Tulsi A., & Mahāprajña, A. (2001). *Ācārya, Tulsi; Ācārya, Mahāprajña, Acharangabhasyam*. Translated by English Translation of the Original Text of Ayaro Together with Its Roman Transliteration and Bhasyam (Sanskrit Commentary). Ladnun by Jain Vishwa Bharati.

Tusaie, K., & Dyer, J. (2004). Resilience: A historical review of the construct. *Holistic Nursing Practice, 18*, 3–10.

U.S. Department of Defense. (2002). *DoD News Briefing – Secretary Rumsfeld and Gen. Myers*. Presenter: Secretary of Defense Donald H. Rumsfeld. http://archive.defense.gov/Transcripts/Transcript.aspx?TranscriptID=2636

U.S. Department of Defense. 2015, Juli 29). DoD releases report on security implications of climate change. *DoD News, Defense Media Activity*. https://www.defense.gov/News/Article/Article/612710/

UK Cabinet Office. (2011). *The UK cyber security strategy – Protecting and promoting the UK in a digital world*. https://www.gov.uk/government/uploads/system/uploads/attachment_data/file/60961/uk-cyber-security-strategy-final.pdf

UK Government. (n.d.). *Emergency preparation, response and recovery*. www.gov.uk/government/policies/improving-the-uks-ability-to-absorb-respond-to-and-recover-from-emergencies

UK Ministry of Defence. (2014). *Strategic trends programme global strategic trends – Out to 2045. Global Strategic Trends* (5. Aufl.). Ministry of Defence (MOD) (30.04.2015).

Umweltbundesamt. (2017). Kernbotschaften des Fünften Sachstandsberichts des IPCC. Klimaänderung 2013: Naturwissenschaftliche Grundlagen (Teilbericht 1). Letzte Änderungen: Dezember 2017. www.bmub.bund.de/fileadmin/Daten…/ipcc_sachstandsbericht_5_teil_1_bf.pdf

UN. (2015). *Millenium development goals and beyond 2015*. http://www.un.org/millenniumgoals/bkgd.shtml

UNCTAD. (2018). *Creative economy outlook. Trends in international trends and creative industries. 2002–2015. Country Profiles 2002–2014*. https://unctad.org/en/PublicationsLibrary/ditcted2018d3_en.pdf

UNDP. (n.d.). *Human development reports*. http://hdr.undp.org/en/statistics/

UNESCO. (2002). *Teaching and learning for a sustainable future*. https://unesdoc.unesco.org/ark:/48223/pf0000161849

UN-Habitat. (2017). *Slum Almanac 2015–2016*. https://unhabitat.org/slum-almanac-2015-2016/

UNISDR. *Homepage*. http://www.unisdr.org/campaign/resilientcities/

United Nations. (2010). *We can end poverty 2015.* Fact Sheet. High-level plenary meeting of the general assembly. New York. 20.–22.09.2010. https://www.un.org/en/mdg/summit2010/pdf/MDG_FS_1_EN.pdf

United Nations. (2015). *The millenium development goals report 2015.* Summary. New York. https://www.un.org/millenniumgoals/2015_MDG_Report/pdf/MDG%202015%20Summary%20web_english.pdf

Van Aken, J., & Hammond, E. (2003). Genetic engineering and biological weapons. *EMBO Reports, 4*(1), 57–60. https://doi.org/10.1038/sj.embor.embor860

van Almsick, J. (1981). *Die negative Einkommensteuer: Finanztheoretische Struktur, Arbeitsangebotswirkungen und sozialpolitische Konzeption.* Duncker & Humblot.

Vanhanen, T. (2014). *Global inequality as a consequence of human diversity. A new theory tested by empirical evidence.* Ulster Institute for Social Research.

Vanolo, A. (2013). Smartmentality: The Smart City as Disciplinary Strategy. *Urban Studies, 51*(5), 883–898. https://doi.org/10.1177/0042098013494427

Vattenfall. (n.d.). *Sustainable cities.* https://group.vattenfall.com/de/zukunft/sustainable-cities

Veenhoven, R. (2010). *World database of happiness.* Erasmus University of Rotterdam. http://www1.eur.nl/fsw/happiness/

Vieweg, W. (2015). *Management in Komplexität und Unsicherheit.* Springer.

Vogt, M., & Ostheimer, J. (2006). *Die Suche nach der guten Gesellschaft. Re-Vision – Nachdenken über politische Vordenker. Politische Ökologie. 100.* Oekom Verlag.

Volkan, V. (2006). *Killing in the name of identity: A study of bloody conflicts.* Pitchstone Publishing.

von Clausewitz, C. (2010). *Vom Kriege. Vollständige Ausgabe.* RaBaKa Publishing.

von Hentig, H. (1987). Polyphem oder Argos? Disziplinarität in der nicht-disziplinären Wirklichkeit. In J. Kocka (Ed.), *Interdisziplinarität: Herausforderung – Praxis – Ideologie.* Suhrkamp.

Voros, J. (2006). Nesting social-analytical perspectives: an approach to macro-social analysis. *Journal of Futures Studies, 11*(1), 11–13.

Walach, H. (2017). Secular spirituality – What it is. Why we need it. How to proceed. *Journal for the Study of Spirituality, 7*(1), 7–20.

Walker, B., Abel, N., Andreoni, F., Cape, J., & Murdock, H. (2014). *General resilience: A discussion paper based on insights from a catchment management area workshop in south eastern Australia.* http://www.resalliance.org/files/General_Resilience_paper.pdf. Accessed 23 April 2014.

Wallace, R. S. (2009). The anatomy of A.L.I.C.E. In R. Epstein, G. Roberts, & G. Beber (Hrsg.), *Parsing the turing test* (S. 181–210). Springer. https://doi.org/10.1007/978-1-4020-6710-5_13.

Wallerstein, I. M. (2014). *The uncertainties of knowledge.* Temple University Press.

Ward, P., & Kirschvink, J. (2018). *Eine neue Geschichte des Lebens. Wie Katastrophen den Lauf der Evolution bestimmt haben.* Pantheon, Verlagsgruppe Random House GmbH

Warwitz, S. A. (2016). *Sinnsuche im Wagnis. Leben in wachsenden Ringen. Erklärungsmodelle für grenzüberschreitendes Verhalten.* Schneider.

Washington, H., & Cook, J. (2011). *Climate change denial. Heads in the sand.* Earthscan.

WBGU (Ed.). (2011). *Welt im Wandel. Gesellschaftsvertrag für eine Große Transformation. Hauptgutachten.* WBGU.

Weber, M. (1965). *Die Protestantische Ethik I. Eine Aufsatzsammlung.* Siebenstern.

Weick, K. E., & Sutcliffe, K. M. (2001). *Managing the unexpected: assuring high performance in an age of complexity*. Jossey-Bass.
Weiskrantz, L. (1986). *Blindsight. A case study and implications: Bd. 12, Oxford psychology series*. Clarendon Press.
Weizenbaum, J. (1966). ELIZA – A computer program for the study of natural language communication between man and machine. *Communications of the ACM, 9*(1), 36–45.
Wellensiek, S. K. (2011). *Handbuch Resilienztraining*. Beltz.
Welsch, W. (2000). Transkulturalität. Zwischen Globalisierung und Partikularisierung. *Jahrbuch Deutsch als Fremdsprache, 26*, 327–351.
Welsch, W. (2009). Was ist eigentlich Transkulturalität? In L. Darowska & C. Machold (Hrsg), *Hochschule als transkultureller Raum? Beiträge zu Kultur, Bildung und Differenz*. transcript.
Weltbank. *Homepage*. http://data.worldbank.org/data-catalog
Weltbank. (2018). *Groundswell – Preparing for internal Climate Migration*. World Bank Group.
Wendt, W. R. (2003). Transdisziplinarität und ihre Bedeutung für die Wissenschaft der Sozialen Arbeit. *Studium & Praxis, 4*(2), 93–105.
Werner, G. (1911). *Unternimm die Zukunft*. Wheeler. W. M. http://www.unternimm-die-zukunft.de/
Werner, E. (1977). *The Children of Kauai. A longitudinal study from the prenatal period to age ten*. University of Hawai'i Press.
Weyer, J. (Ed.). (2011). *Soziale Netzwerke. Konzepte und Methoden der sozialwissenschaftlichen Netzwerkforschung*. Oldenburg Verlag.
Wheeler, W. M. (1911). *The ant-colony as an organism* (A lecture prepared for delivery at the Marine Biological Laboratory, Woods Hole, Mass., August 2, 1910). In *Journal of Morphology, 22*(2).
White, A. G. (2007). *A global projection of subjective well-being: A challenge to positive psychology?* University of Leicester. http://data360.org/pdf/20071219073602.A%20Global%20Projection%20of%20Subjective%20Well-being.pdf
Whittle, R. (2010). *Learning the lessons from flood recovery in hull. Final project report for flood, vulnerability and urban resilience. A real-time study for local recovery following the floods of June 2007 in Hull*. Lancaster.
Wielens, H. (2004). *Im Brennpunkt: Geld & Spiritualität*. Via Nova.
Wilkinson, R. G., & Pickett, K. (2010). *The spirit level: Why greater equality makes societies stronger*. Sydney Bloomsbury Press.
Willke, H. (1998). Organisierte Wissensarbeit. *Zeitschrift für Soziologie, 27*(3), 161–177.
Willke, H. (2016). Zur Relevanz der Systemtheorie von Niklas Luhmann. *Agora, 42*(1), 9–14.
Wilson, H. (2013). Banks put to the test over cyber security. *The Telegraph*. https://www.telegraph.co.uk/finance/newsbysector/banksandfinance/10359520/Banks-put-to-the-test-over-cyber-security.html. Accessed 6 Okt 2013.
Wilson, M. (2017). *Smarter cities challenge aims to make lasting urban improvements*. IBM. https://www.ibm.com/blogs/cloud-computing/2017/02/17/smarter-cities-challenge-improvements/. Accessed 17 Feb 2017.

Wink, R. (2014). Regionale wirtschaftliche Resilienz und die Finanzierung von Innovationen. In J. Krüger, H. Parthey, & R. Wink (Hrsg), *Wissenschaft und Innovation* (S. 57–72). Wissenschaftsforschung.
Wink, R. (Ed.). (2016). *Multidisziplinäre Perspektiven in der Resilienzforschung*. Springer.
Wink, R., Kirchner, L., Koch, F., & Speda, D. (2016). *Wirtschaftliche Resilienz in deutschsprachigen Regionen*. Springer.
Winterhoff, M. et al. (2014, December). *Simple, simpler, best. Frugal innovation in the engineered products and high tech industry*. Roland Berger. https://www.rolandberger.com/publications/publication_pdf/roland_berger_tab_frugal_products_e_20150107.pdf
WirvsVirus. (2020). *#WirvsVirus Hackathon*. Homepage. https://wirvsvirushackathon.org/
Wo-Lap, W. L. (2016, April 12). Getting lost in 'one belt, one road'. *Ejinsight*. http://www.ejinsight.com/20160412-getting-lost-one-belt-one-road/
Wolff, E. (2017). *Finanz-Tsunami: Wie das globale Finanzsystem uns alle bedroht*. Büchner-Verlag.
Wood, D. (2006). *The "everyday" resilience of the city. Human security and resilience* (ISP/NSC briefing paper, no. 6(1)).
Woodley, A. W., & Malone, T. (2011). What makes a team smarter? More women. *Harvard Business Review, 89*(6), 32–33.
Woolley, A. W., Chabris, C. F., Pentland, A., Hashmi, N., & Malone, T. W. (2010). Evidence for a collective intelligence factor in the performance of human groups. *Science, 330*(6004), 686–688. https://doi.org/10.1126/science.1193147
World Citizenship Movement. *Homepage*. http://www.garrydavis.org/h_worldcitgov.html
World Economic Forum. (2019a, January). *Global risk report 2019*. Davos. https://www.weforum.org/reports/the-global-risks-report-2019
World Economic Forum. (2019b). *The global competitiveness report 2019*. https://www3.weforum.org/docs/WEF_TheGlobalCompetitivenessReport2019.pdf
World Economic Forum. (2020, January). *Global risk report 2020*. Davos. https://www.weforum.org/reports/the-global-risks-report-2020
World Economic Forum. (2021, January). *Global risk report 2021*. Davos. https://www.weforum.org/reports/the-global-risks-report-2021
World Economic Forum. (2022, January). *Global risk report 2022*. Davos. https://www.weforum.org/reports/global-risks-report-2022/in-full/chapter-1-global-risks-2022-worlds-apart
World Federalist Movement. *Homepage*. http://www.wfm-igp.org/
World Health Organization. (2020, January 30). *Statement on the second meeting of the International Health Regulations (2005) Emergency Committee regarding the outbreak of novel coronavirus (2019-nCoV)*. Geneva. https://www.who.int/news/item/30-01-2020-statement-on-the-second-meeting-of-the-international-health-regulations-(2005)--emergency-committee-regarding-the-outbreak-of-novel-coronavirus-(2019-ncov)
World Institute for Disaster Risk Management. *Homepage*. http://www.drmonline.net/
World Ocean Review. (2015). What is sustainability? *World Ocean Review, 4*(1). https://worldoceanreview.com/en/wor-4/concepts-for-a-better-world/what-is-sustainability/
Yabushita, S., & Hatta, N. (1994). On the possible hazard on the major cities caused by asteroid impact in the Pacific Ocean. *Earth, Moon, and Planets, 65*(1), 7–13. https://doi.org/10.1007/BF00572195

Zakaria, F. (2004). *The future of freedom: Illiberal democracy at home and abroad.* Norton.
Zämma leaba. *Homepage.* http://goetzis.at/gesundheit-soziales/zaemma-leaba
Zander, M. (2011). *Handbuch Resilienzförderung.* Springer.
Zeschky, M., Widenmayer, B., & Gassmann, O. (2011). Frugal innovation in emerging markets. *Research-Technology Management, 54*(4), 38–45. https://doi.org/10.5437/08956308X5404007
Zhang, Z. (2018, October). *The belt and road initiative: China's new geopolitical strategy?* SWP working paper. https://www.swp-berlin.org/fileadmin/contents/products/projekt_papiere/Zhang_BCAS_2018_BRI_China_7.pdf
Zillien, N., & Haufs-Brusberg, M. (2014). *Wissenskluft und Digital Divide.* Nomos.
Zolli, A., & Healy, A. M. (2013). *Resilience. why things bounce back.* Headline Publishing.
Zollinger, C. (2005). *Die Debatte läuft: ganzheitliche Thesen für Gesellschaft, Wirtschaft und Politik.* Via Nova.
Zotz, V. (2000). *Auf den glückseligen Inseln. Buddhismus in der deutschen Kultur.* Theseus

GPSR Compliance

The European Union's (EU) General Product Safety Regulation (GPSR) is a set of rules that requires consumer products to be safe and our obligations to ensure this.

If you have any concerns about our products, you can contact us on

ProductSafety@springernature.com

In case Publisher is established outside the EU, the EU authorized representative is:

Springer Nature Customer Service Center GmbH
Europaplatz 3
69115 Heidelberg, Germany

www.ingramcontent.com/pod-product-compliance
Lightning Source LLC
LaVergne TN
LVHW010335260326
834688LV00036B/719